RIVERS OF THE WEST

A Guide to the Geology and History

Elizabeth L. Orr
William N. Orr

Eugene, Oregon 1985

LC 85-61228
ISBN 9606502-1-0

Additional copies may be ordered from:
P.O. Box 5286
Eugene, Oregon, 97405

Other books by the authors:
Handbook of Oregon Plant and Animal Fossils.
Bibliography of Oregon Paleontology, 1792-1983.

Cover: One of several waterwheels along the lower Owyhee River in Oregon.

Dedicated to the downstairs people.

CONTENTS

ACKNOWLEDGEMENTS

This book was written as the result of an ongoing university summer course on the geology of selected watersheds. The stimulus and encouragement to begin writing was given by Bob Doppelt of Oregon River Experiences in Eugene. Participating on river trips with Bob over several Summers convinced us that such a book could enrich these experiences.

Since the text covers a wide range of subjects, the assistance and suggestions of many individuals were appreciated. Special thanks are due to Bob Carson of Whitman College, Washington, Richard Heinzkill, University of Oregon, Jack Kelso, University of Colorado, Gil Lang, Roseville, California, Steve McWilliams, College of the Desert, California, and John Vitas, Tualatin, Oregon. Jean Orr drew many of the illustrations, and Alison Orr provided helpful suggestions.

INTRODUCTION

This book was written to help make the experience of rafting — or backpacking — along western rivers more enjoyable.

These rivers are famous for their whitewater rapids, long stretches of beautiful scenery, well-exposed geology, and fascinating history. To many, rafting offers the first real look at areas they have passed by many times. When rafting or hiking, we are all curious about the rocks we see. Are they volcanic or have they been deposited by the river? Where did gold in the rivers come from, and why does the canyon have its peculiar shape?

We are also curious about people who lived here in prehistoric times. How did they find food and shelter? What became of them?

Historically, rivers themselves were the main routes of travel and trade. Populations and trading centers were at the confluence of major streams, and tribal Indian boundaries fell along the limits of fluvial watersheds. Equally important, the streams were a major source of food for early Americans. Water in the streams was the magnet that drew game of every description. The river was life itself.

Unfortunately for the Indians, river valleys were also the avenues to material wealth providing clues to hidden riches in the rocks or streambeds. For most of the western rivers, the advance guard of white invasion was the fur trapper, followed by the prospector.

Miners everywhere took over the waterways diverting the water for use in the mining operations or often just to expose and exploit the streambed itself. As the Indians reacted in a predictable manner to the take-over of their environment, the policy was usually to exterminate them. Often little better than misfits in white society, miners made their own rules of conduct on the frontier where compromise and understanding were in short supply.

Rivers have survived these changes and today can give the rafter one of the most relaxed ways to examine the natural environment. Drifting down a stream in a raft or boat, away from the noises of modern metropolitan centers, the visitor can come to appreciate true solitude.

It is difficult to imagine a more environmentally sound means of passing through an area without leaving a trace. Low impact camping in most of the watersheds has given way to no impact camping, where all waste products are removed by the visitor.

When rafting for the first time, it is best to begin with a professional guide. Licensed guiding companies are to be found throughout the western states. Participatory, or "row your own", guided tours are easily the most rewarding. By obtaining a guide, the rafter has been relieved of the problems of equipment (except personal gear), cooking, and, for the most part, unexpected surprises along the waterway. No amount of literature can really prepare an individual for the excitement of the first encounter with whitewater. With a guide, the rafter is purchasing experience — and safety.

Idaho Rivers

Nez Perce Indians

The most numerous and powerful Indian group in the Northwest, the Nez Perce, ranged over a vast area from the Bitterroot Mountains in western Montana to the Blue Mountains in Oregon. Their territory included the Salmon, Clearwater, Snake, and Grande Ronde river valleys where they had lived for more than 10,000 years. Population of Nez Perce has been estimated somewhere between 3,000 to 4,000 in 1780. In 1853 there were 1,700 persons, and the 1910 census reported 1,259.

The Cho-pun-nish, or Pierced Nose Indians, were admired by explorer William Clark in 1805 as "stout likely men, handsom women, and verry dressey in their way." Remaining in Nez Perce villages because of winter snows, Lewis and Clark were impressed with the industry, cheerfulness and dignity of these people. They wore decorative pieces of shell in their nostrils, hence their original French name, Nepercy.

Winter villages were located deep in the river canyons where there was protection from cold winds. Mountain peaks offered relief during hot summer months, and highland plateaus provided plentiful root plants and grazing land for horses. Villages were governed by the oldest qualified male. Groups of villages were organized under an hereditary leader who met with other leaders in a council for major decisions. War chiefs, selected by the council, usually came from one of the larger villages and from the ranks of those who had proven themselves able warriors.

Houses for protection against cold, winter temperatures were semi-subterranean, circular, and covered with reeds and grasses. For comfort and social purposes, more than one family would live in a large house.

Society was divided into three classes, the lowest or slave members, the middle class with the majority of numbers, and the rich upper class. Upper class positions were generally hereditary.

As with most northwest Indians, the Nez Perce practiced no agriculture. Food was supplied from hunting and fishing accompanied by plant gathering. Camas and kaus (kouse) roots were prefered over other plants. In the Spring, with patches of snow still on the ground, women dug the kaus out of the muddy ground with a sharpened stick. Kaus, in the parsley family (Lomatium), resembles a radish and was eaten raw or boiled for mush which was made into cakes.

Young man of the Nez Perce (Drawn by George Catlin).

Camas plants were dug in Summer after their lily-like blue flowers had faded. The root, sometimes eaten raw, was more frequently cooked in a specially prepared pit filled with hot stones. The root was washed and peeled, put into the pit and covered with three inches of grass. Water was poured over the roots then everything was covered with dirt and allowed to cook for up to three days. Camas tastes sweet and sticky.

Camas gathering was the time for a communal get-to-gether with many tribes coming to traditional camas grounds for games and meeting friends. Weippe Prairie, near Lapwai, Idaho, was one of the largest camas digging areas shared by tribes from the Snake, Wallowa, and Salmon rivers. Wild carrot, bitterroot, berries, and other plants were eaten when in season.

Game — deer, elk, bear — for meat was hunted throughout the year. The hunters used a powerful bow perfected by the Nez Perce and coveted by other tribes. The bow, about three feet long, was made from the horn of a mountain sheep and deer sinew. Glue was concocted from salmon and sturgeon. Salmon, lampreys, and many other varieties of fish were caught and dried. Nez Perce annual consumption of fish has been put at 400 pounds per individual.

The Nez Perce were friendly with other tribes

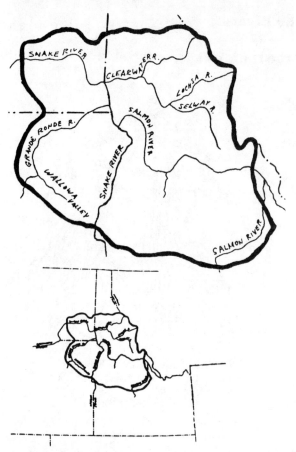

Distribution of Nez Perce Indians in Idaho, Washington, and Oregon

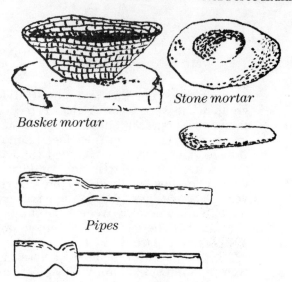

Basket mortar

Stone mortar

Pipes

and engaged them in trade in the cosmopolitan atmosphere at The Dalles, boisterous trading center of the Northwest along the Columbia River. Here they met among the "dogs and children and foul . . . stench of decaying fishheads lining the rocky shores" (Josephy, p.21).

Between 1700 and 1730, the Nez Perce acquired horses from neighbors to the west, purchasing, as legend has it, a white mare and colt which became the basis for their herds. Outstanding horsemen, the Nez Perce developed the largest horse herds in the Northwest. They practiced selective breeding, developing horses for strength and endurance. The horse gave the Nez Perce an enormous advantage in trade, increasing the distances they could travel and the number of items carried thus adding to their wealth. One rich person might own several hundred horses, and the Nez Perce were supposed to have on the average of five to seven per individual.

Shells from The Dalles, salmon oil, fish, bows, and dried plants were transported east and traded for buffalo skins, feathers and other items from the Plains Indians. Nez Perce even took part in buffalo hunts on the Plains, some of

their long profitable journeys lasting up to two years. At the time of Euro-American contact, they had become well-known travellers and traders, their language widely used in trade negotiations.

In September, 1805, William Lewis and Meriwether Clark, on their famous exploring trip, crossed the Bitterroot Mountains from Montana into Idaho, the first Americans to enter the Nez Perce homelands. "The most terrible mountains I ever beheld," wrote one member of the expedition, contained no game to feed the starving men, and winter storms were already bringing snow to the mountain tops. They "proceeded on through a Countrey as ruged as usial," Clark wrote until finally reaching a Nez Perce village, three miles south of the present

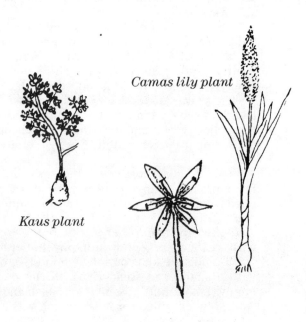

Camas lily plant

Kaus plant

Meriwether Lewis and William Clark, who led the first American exploration to the West Coast in 1805.

town of Weippe, Idaho. Here they were given berries and roots to eat. They remarked on women gathering camas roots in what was a traditional spot.

Fur trappers followed shortly after Lewis and Clark, with Captain Benjamin Bonneville as one of the most colorful to be found along the Salmon. Bonneville's adventures here were related by Washington Irving. "For the greater part of the month of November, Captain Bonneville remained in his temporary post on Salmon River. Besides his own people, motley in character and costume — Creole, Kentuckian, Indian, half-breed, hired trapper, and free trapper — he was surrounded by encampments of Nez Perces and Flatheads, with their droves of horses . . ."

Possibly more respectable than fur trappers, missionaries arrived in 1836. Henry and Eliza Spalding set up a Protestant mission at Lapwai, Idaho, on the Clearwater River, and a second mission was founded near Walla Walla, Washington, about 120 miles away. The missions had operated for only a year when Cayuse Indians massacred almost everyone at Marcus Whitman's Walla Walla mission. Following this, a faction of Nez Perce then attacked the Lapwai settlement after which efforts to establish missions were abandoned.

In 1855 twenty-seven Mormons were sent to convert and civilize the Indians. They built a fort, thirteen cabins, and began gardening along the Salmon, naming their colony Fort Lemhi after a person in the Book of Mormon. All went well, crops prospered, the settlement increased until the winter of 1858 when a band of Shoshone attacked killing two men and capturing a number of livestock. The fort was strengthened and messages sent to Brigham Young at Salt Lake City. Young ordered the closure of Ft. Lemhi, and a relief party was sent to assist the settlers. The area reverted to the Indians until 1866 when a gold mining settlement here grew into the town of Lemhi.

Dissatisfaction with increasing white settlement and the spread of disease were factors contributing to unrest among the Indians. In 1855, Issac Stevens, Governor of Washington Territory and Superintendent of Indian Affairs for Oregon Territory, concluded a treaty with the Nez Perce tribes. The treaty was signed after long negotiations, creating a reservation which included more than half of the aboriginal lands of the tribe ceding the remainder to the U.S. government. Unfortunately the terms of this treaty were neither upheld by the government in Washington, D.C., nor by white emigrants. When gold was discovered on Nez Perce lands in 1860, Steven's treaty obligations came to an end.

During the period of turmoil which followed, the U.S. government failed to protect reservation territory from white intrusions. White squatters were not expelled, and civil pressure was exerted to remove the Indians to smaller reservations so white interlopers could have free claim to their land. This meant more treaties. A new commission in 1863 demanded nine-tenths, or 90,000 square miles, of the Nez Perce lands reserved in the earlier agreement. Northern Indians, under Chief Lawyer, agreed. How-

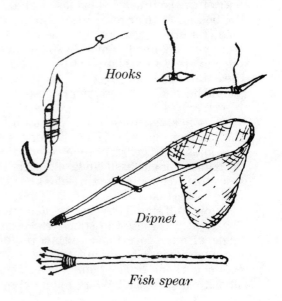

Hooks

Dipnet

Fish spear

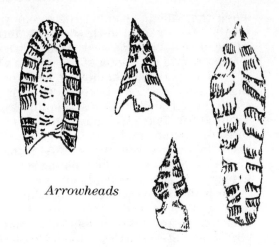

Arrowheads

ever, southern groups under Chief White Bird refused to sign. The effect of this new treaty was to divide the Nez Perce nation permanently into the treaty and non-treaty Indians.

A number of years later, Superintendent Calvin Hale described how these Indians were treated after sixty years of dealing with white people. "How has this faithfulness been requited on our part. Has any suitable recognition been made by the government for the protection which these Nez Perces afforded Governor Stevens and his little band in the winter of 1855 . . . their claims for horses supplied to the Oregon volunteers are yet unsatisfied.

In our treaty stipulations we have done no better. The appropriations for (fencing, houses, and provisions) has, in good part, been squándered or withheld. Their annuities have not been paid fully . . . and much of that received . . worthless trash, bought at exorbitant prices.

Whilst we were thus failing to execute our part of the contract, gold is discovered within the bounds of their reservation. Application is made for privilege to mine on their land. Their consent is obtained to make a steamboat landing and erect a warehouse at the mouth of the Clearwater . . . however . . . no settlements should be made . . . and their lands (protected) from trespass.

No sooner were these privileges granted than the landing and warehouse became a town, now known as Lewiston; their reservation was overrun; their enclosed lands taken from them; stock turned into their grain fields and gardens; their fences taken . . . "

Yellow Wolf stated the views of non-treaty Indians. "It was these Christian Nez Perces who made with the government a thief treaty. Sold the government all this land. Sold what did not belong to them. None of our chiefs signed that land-stealing treaty." White immigration onto Indian lands continued at the same time non-treaty Indians continued to live outside the reservation, subsisting independently in their homelands.

The situation between these two groups worsened until 1877 at which time General Oliver Howard from Ft. Vancouver gave all non-treaty Indians thirty days to move onto the Lapwai Reservation. Leaving the Salmon River, Chief White Bird was camped near the Snake River at the same time Chief Joseph crossed over from Oregon. White Bird's men were angry and took revenge by killing some white men who had mistreated them in the past.

Joseph was appalled, hoping to keep peace, although an Indian council meeting at present day White Bird was told "it is already war." A third leader, Chief Looking Glass, joined the hostiles only after his village, under a flag of truce, was attacked by soldiers, looted, and destroyed.

The war began auspiciously for the Nez Perce with a victory in White Bird Canyon on June 17, where 60 to 70 Indians armed with "bows and

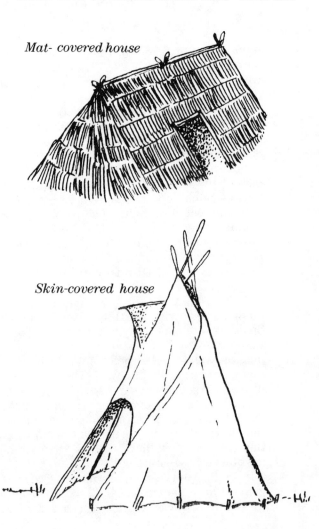

Mat- covered house

Skin-covered house

4

arrows, shotguns, old muzzle-loading fur trade muskets, and a few modern rifles'' defeated 99 troops led by Capt. David Perry. Thirty-four soldiers were killed with three Indians wounded.

A plaque on the battlefield reads ''Before you . . . lies the historic White Bird battle ground . . in which 34 men gave their lives in service for their country June 17, 1877.'' Perhaps a more equitable statement would be that ''while the whites died bravely in the service of their country, the Indians also fought bravely for their country'' (Josephy, p.514).

At the Battle of Clearwater, Idaho, in July approximately 200 Indian men delayed General Howard's 560 troops for two days while families escaped. In August, after indicating their peaceful intentions, and retreating to Montana, the Nez Perce camp at Big Hole was attacked by Montana volunteers and army troops under General Gibbon. Many women and children were killed in the initial confusion although the Nez Perce warriors outflanked and divided the troops, holding them back while the remaining families fled further east and north.

At a council in September the Indians decided to move to Canada and join Sitting Bull's non-treaty group there. Unfortunately their camp just forty miles south of the Canadian border at Bear Paw, Montana, was attacked by soldiers under General Miles. In the confusion and falling snow, heavy casualties resulted. Under a flag of truce, Miles and Joseph conferred. Joseph refused to surrender, and, in violation of the truce, was arrested. Joseph's men then captured Lt. Lovell Jerome who was exchanged for Joseph two days later.

On October 5, General Howard arrived with yet another treaty counselling the Nez Perce to

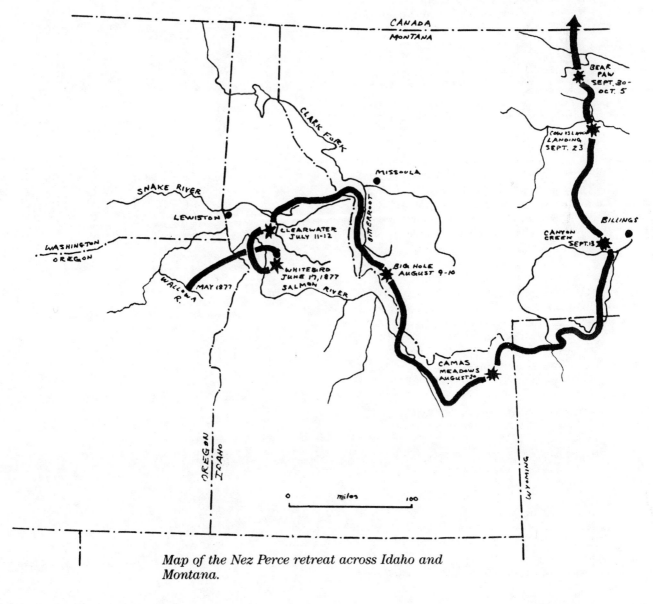

Map of the Nez Perce retreat across Idaho and Montana.

surrender and promising they could remain in the Northwest. Continued shelling by the army cannon, the death of Chief Looking Glass, and cold and hunger convinced Joseph he should surrender with his group of 86 men, 184 women and 147 children. Chief White Bird secretly led a party of 200 over the border from the battlefield into Canada where they were fed and housed by Sitting Bull.

Army promises were not honored, and the Nez Perce were removed to Bismark, South Dakota, then by train to Ft. Levenworth, Kansas, and to the Quapaw Reservation in Kansas. Many had died by the end of the trip. Only in 1885, with 268 people left, were the Nez Perce allowed to return to the Colville Reservation in Washington State where Commissioner for Indian Affairs, Major Anderson, angrily reported "they have been persistent in following their ancient traditions and indulging in their primitive customs . . . they have no religion, believe in no creed, and their morality is at a low ebb . .

Closer restrictions should be thrown around them . . . " Anderson went on to say that Chief Joseph did not "occupy the house erected on his farm . . . He, with his handful of unworthy followers, perfers the traditional tepee, living on the generosity of the government. History has been partial to Joseph in chronicling his atrocious acts. The appalling wrongs done by him are crying from the bloodstained soil of Idaho for restitution." Joseph spent his last days in Washington where died on September, 21, 1905, and was buried in Nespelem.

Lower Salmon River

(Riggins to Hellers Bar, Snake River)
Lewis to Nez Perce County, Idaho
Length of run: 130 miles
Number of days: 7-8

The Salmon River which rises in the peaks of the Sawtooth Mountains reaches the Snake River after flowing through 390 miles of steep canyons. The 6,000 foot Salmon canyon is one of the deepest in North America, exceeded only by the 7,900 foot deep Snake River. The Salmon River drainage remained unaccessible and un-settled until the arrival of miners in 1861 brought an end to this isolation.

Mushrooming mining towns at Pierce, Elk City, and Florence were filled with eager prospectors who came with the news of gold discovery. Other towns as Lucile and Riggins grew up because they were on the route from the northern Salmon River mines to the Boise Basin in the south. Lucile was named by these early arrivals after a child, Lucile Ailshie, who played in the lobby of the city hotel. Riggins where Nez Perce Indians held councils, dances, and feasts, was named for a local businessman, Richard L. Riggins.

This interval of the Salmon is famous for its huge waves, spectacular scenery and abrupt weather changes. In a geological sense the area is a fascinating chapter to understanding Pacific Northwest history.

Early History of the Lower Salmon River

Diplomatic, powerful Nez Perce Indians had lived in the Salmon, Snake and Clearwater valleys for centuries, travelling and trading as far

Fort Hall, center of fur trapping activity, was always well-stocked with beans, bacon, and whiskey for trading with Indians.

Nathaniel Wyeth, American fur trader.

away as the midwestern plains. In 1805 Lewis and Clark wintered over in Nez Perce villages where they were well-treated and impressed with these industrious people. Their explorations brought this region to the attention of fur trappers with the British owned Hudson's Bay Company gaining a monopoly on trade in the Northwest.

American trappers were not deterred by the competition and built temporary posts at Ft. Henry and Ft. Boise on the Snake River. Not all were successful, however. American trapper, Nathanial Wyeth, attempting to break into the fur market here, built a trading post at Ft. Hall, Idaho. Two years later Wyeth sold out to Hudson's Bay Company and returned east after being outbid on fur prices by the British. Ft. Boise, probably consisting of only one cabin, served as a base for trappers in Idaho who removed 75,000 beaver from the streams. Beaver, mink, otter, muskrat and even buffalo disappeared. Nothing was left but the "bones of thousands of beaver and other furry animals" (Beal, p.73).

The responsibility for bringing gold seekers into Idaho and Nez Perce country can be attributed to the determination of E. D. Pierce who had an obsession to open up gold mining in the forbidden Indian territory. Pierce was a veteran of gold fields in both California and the Frazer

River district of British Columbia. For several winters he lived in the village of a friendly chief, Timothy, all the while smuggling mining equipment into the camp. Eventually gold was indeed discovered on the Clearwater River.

Nez Perce bands were violently opposed to any intrusion into their lands, however, Pierce persevered and organized twelve men who furtively entered Idaho from the north on a little used trail. They were led by Chief Timothy's daughter, Jane Silcott. On Orofino Creek which flows into the Clearwater River one of the men, W. F. Bassett, found gold.

First married to a Nez Perce who drowned, Jane Silcott later married John Silcott who had come to Lewiston in 1860 to help in building construction at Ft. Lapwai. He stayed on after finishing this job to operate a ferry near the confluence of the Snake and Clearwater rivers. Jane died from burns at age 53 when her clothes caught on fire from an open fireplace.

After prospecting for a time, the party ran short of supplies and had to return to Walla Walla, Washington. Pierce organized a larger group of 33 men for the return to Canal Gulch where they named the new diggings Orofino, meaning fine gold. When one of the men snowshoed out in March, he was carrying $800 in gold dust. This news was enough to bring several thousand to Canal Gulch where the towns of Orofino and Pierce were established.

Orofino officially became a district in 1861 when mining laws were adopted and claims recorded. The towns of Orofino and Pierce were built that winter when idle miners set to work digging ditches for water and cutting down trees to be used for the next mining season.

These early miners also had problems of an unusual nature. There wasn't enough prospecting ground to go around. In addition, heavy, newly fallen snow on May 3, covered even those

A buffalo, painted by an imaginative Dutch artist in 1630.

claims already begun. Miners had to wait for the snow and high water to disappear in June when serious prospecting could begin.

Mining at Pierce was hard work. Gold in the rocks was dispersed so that rich, concentrated amounts were not to be found. To add to the prospectors' problems, two to three feet of soil had to be removed even to expose the placers. By the middle of June, a miner could make $5 to $20 per day sluicing, with $50 the top amount made on some claims. Even housing costs rocketed during the Summer when a town lot in Orofino cost from $100 to $200 each, and log

Old Fort Walla Walla, along the Columbia River, where miners stocked-up before starting for the Idaho gold fields.

houses started at $500 and reached $1,000.

Prices ran high in the district where supplies had to be brought in from Walla Walla over Indian trails. A yoke of 24 oxen was needed to pull a loaded wagon over the grade to Orofino. In Pierce, flour cost .16 cents a pound while the same amount could be bought in Portland for .04 cents. Beans in Pierce cost .35 cents a pound but were .08 cents in Portland. Nails at .33 cents a pound at Pierce were .28 cents a pound elsewhere.

A franchise to operate a ferry near the mouth of White Bird Creek was granted for eight years

Jane Silcott, daughter of Chief Timothy.

Ferry at White Bird along the Salmon (Idaho Historical Society).

to Samuel Benedict and Grat Bernamayon. They erected a cable ferry and charged .75 cents for a horse and rider to cross, hogs and sheep cost .12 cents, and .25 cents was charged for a man afoot.

On one occasion the operator went on a visit ten miles away leaving the ferry on the opposite side of the river from potential customers. The first to arrive was the mail carrier who plied the week long route from Mt. Idaho, White Bird, Slate Creek, and Florence. On this trip he had $40,000 worth of gold and was being pursued by robbers so was reluctant to return the way he had just come. In desperation he climbed the post to which the cable was attached and worked his way 400 feet across the water hanging by his hands and feet. In this way he was able to elude the thieves. A bridge was built over White Bird Creek in 1888 at a cost of $750 with $250 allotted for grading approaches.

With money to be made supplying the miners, The Oregon Steam Navigation Company, operated by two enterprising businessmen, decided to build a steamboat to run from The Dalles on the Columbia River down the Snake River to Boise, Idaho. The 120 foot long (25 feet wide in the beam), $8,000 stern wheeler named Shoshone, left Boise in April, 1870. The Dalles was reached in five days time, but the ship was so crushed and battered she never made the trip back to Boise. After several years of use on the Columbia, the Shoshone wrecked, and, as the story is told, her cabin floated downstream, was rescued, and used as a chicken house henceforth.

Toward the end of the summer season in 1861, Orofino and Pierce were beginning to lose men to Elk City and then Florence to the south. Although 400 houses and 1,500 men made up Orofino one correspondent remarked "fortunately there are no lawyers yet." October brought on the rush to southern Salmon River mines and the demise of these towns.

In October 4, 1861, miners in bands of twenty or more, in violation of U.S. treaty law and braving fierce resistance from the Indians, examined bars of the Salmon past Riggins to Slate Creek. The Salmon proved disappointingly barren, although gold was found at present day Elk City, then at Florence.

In retrospect, perhaps the greatest accomplishment of miners and settlers in this untouched region of the Salmon River was the removal of those natives who had lived here from times past. In June, 1877, Nez Perce bands under Chiefs White Bird and Looking Glass had been given thirty days to move from their traditional lands to the reservation. Gathering their goods and moving north, they made what has been called the "fatal pause" about six miles west of Grangeville (Josephy, p.499). Joined by Oregon Nez Perce, about 600 Indians were camping on the Camas Prairie.

Angry young Indians discussed their griev-

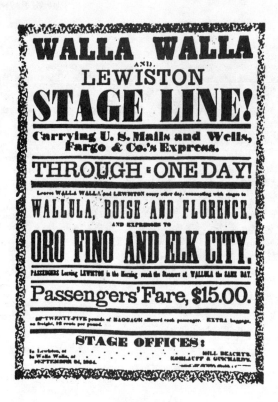

ances, exchanged stories of wrong doing and humiliation. Days passed, and a warlike feeling crept into the camp. These feelings were violently released on June 12, 1877, when a party of young men rode out to revenge themselves. They murdered a man near the mouth of Slate Creek, a rancher near John Day Creek, as well as two other men before returning to camp.

A council was held, and war was begun. In a series of battles from June to September, Nez Perce skillfully outmaneuvered U.S. troops while moving in a northeasterly direction. At a September council the Indians decided to move to Canada where they could join Sitting Bull's non-treaty Indians. Such was not to be the case, however, when the Indian camp was attacked just forty miles south of the border at Bear Paw, Montana. Most of the Indians surrendered or were captured by General Howard, while Chief White Bird led a party of approximately 200 over the border safely.

After being moved to South Dakota, then to Kansas, the remaining 268 people were allowed to return to the reservation at Colville, Washington. (A more thorough account of the Nez Perce is related in the chapter on Nez Perce).

Indian Life Along the River

Although the Nez Perce lived high in the mountains as well as on the plateaus and in river valleys, most the their villages were lo-

cated near the confluence of small streams and rivers. Occasionally a village might be located on an island in the river, but permanent villages were never established high in the cold mountains or windy plateaus.

Some house, village, and burial locations have been found, but, interestingly enough, no large shell mounds have been reported. A few small heaps were probably where a single meal was eaten. Fireplaces, sweathouses and camas steaming ovens were built along the river and are sometimes washed out by action of the water. These sites are easily identified by pieces of charcoal, burnt stone, and burned animal bones.

Villages can be recognized by a series of rings where circular houses had been built. These rings are from eighteen to thirty feet in diameter and up to three feet deep in the center. Long oval rings, sometimes eighteen feet wide and sixty to eighty feet long, indicate a communal house. Tipi rings are rounded river rock placed in circles ten to fifteen feet in diameter. There is virtually no center depression in a tipi circle. Camp sites may contain from 100 to 400 rings indicating a large number of people spent some time here.

Villages were named by their geographical location. Lamtama was on the Salmon River where the White Bird band lived. Wewi-me was at the mouth of the Grande Ronde on the Snake River, Esnime was on Slate Creek, and Tewepu was at the mouth of Orofino Creek.

Several places examined by anthropologists are of interest. West of Grangeville, the Weis Rockshelter, near the confluence of Rocky Creek and Grave Creek, was occupied from 7,500 years ago up to the year 1,400. The rockshelter was near the geographical center of the Nez Perce territory and in the midst of their camas root gathering fields. Unusual amounts of animal bones are from deer, bear, bison, sheep, coyote, rodents, and birds. A few fish bones were scattered among these, but deer bones were most numerous.

Most artifacts from the rockshelter were made of worked, rounded river stone identical to those found on the floor of Rocky Canyon. Some bone and antler items are recorded as well as a few shell objects and obsidian flakes. The source of obsidian was unknown and may have been acquired in trade.

River terraces along Rocky Creek and at its mouth on the Salmon were also inhabited during the same time period. These sites have yielded similar material. Nearby Grave Creek had scattered remains of chipped stone points, worked stone and some shell, bone, and antler tools.

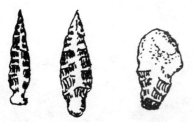

Arrowheads and scraper from Coopers Ferry.

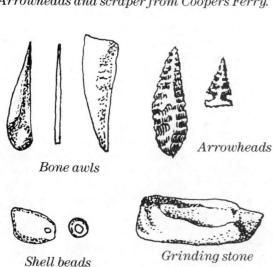

Bone awls

Arrowheads

Shell beads

Grinding stone

Artifacts from Weis Rockshelter.

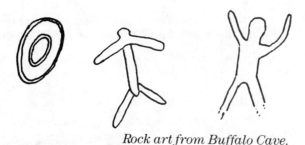

Rock art from Buffalo Cave.

Rock art from Rocky Cave.

Many recorded and doubtless unrecorded rock art sites exist along the Salmon River. Of the pictographs studied, most are paintings of geometric designs, human figures, and animals in colors from red-brown to orange and black. Frequently the designs are in caves. Some have a single element drawn, whereas others have several hundred elements in one place. Exposure to weathering as well as the heavy vegetation cover makes many of the pictographs difficult to see. Pictographs are likely to be found along small canyons and at the mouth of creeks where they enter the river.

One good example can be seen on the Snake River near the mouth of the Salmon where designs painted onto the walls of a cave show human figures, horses, dots, and geometric designs. Artifacts were found on the cave floor which indicated Indians had lived here at least 4,000 years ago. Woven grass mats, ropes, skin moccasins, arrowheads, wooden needles, and curved wooden cooking implements were among the items unearthed. A few other pictographs have been recorded on the cliffs in the same vicinity. These paintings are weather damaged but human figures and circles can still be deciphered.

Geology Along the Lower Salmon River

The runnable section of the lower Salmon from Riggins to Hellers Bar on the Snake River is separated from the remainder of the Salmon by Carey Falls just twenty miles upstream from Riggins. The river from Riggins to Hellers Bar zigzags north, south, then north again where it joins the Snake south of Lewiston, Idaho. In a general way, the Clearwater River watershed to the east and north of the Salmon drainage parallels the trend of the latter draining a much larger area and winding up at the Snake near Lewiston.

The oldest rocks here, called "basement" rocks by geologists, are dated at more than 200 million years (early Mesozoic). Indications are that this basement suite of rocks has been transported from the southwest Pacific area. Rocks include an array of ocean floor sediments and volcanics. Most have been subject to a low degree of heat and pressure (metamorphism) and as a result display distinct signs of alteration.

These rocks probably collided with North America during the middle Mesozoic after being carried across the Pacific basin by a process related to continental drift. Fossils here are altogether different from fossils of the same age found nearby in Nevada and Utah. Rock masses

Evolution of the lower Salmon topography.

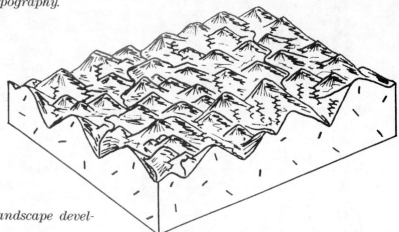

A. Fully mature, erosional landscape developed on older rocks.

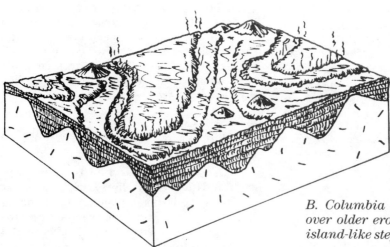

B. Columbia River lavas fill in and smooth over older erosional topography leaving only island-like steptoes.

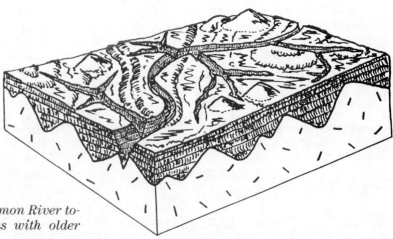

C. Erosion of the lavas by the Salmon River today yields steep-walled canyons with older rocks exposed deep in the valleys.

of this type which have been transported by the movement of earth crustal plates are referred to as "exotic terranes". Typically they are completely different from the local or "country" rocks and are confined to separate areas of exposure. There is growing evidence that the foundation, "basement" bedrock over much of eastern Oregon and western Idaho may have been deposited elsewhere before being transported and annealed to North America.

The metamorphic processes of heat and pressure which distorted and altered this rock have destroyed many of the original structures, but some, including fossils, are occasionally visible. Most of the fossils are of oceanic animals from ancient continental shelf environments. These are largely molluscs, but recently large marine reptiles have been discovered.

The overlying rock series in the geologic layer cake of the Salmon are lava flows which were erupted 15 million years ago (Miocene). These lavas are part of the Columbia River volcanics exposed over much of eastern Washington and Oregon as well as in southern Idaho. During this volcanic episode, very fluid lava poured like syrup over wide areas of the landscape from low volcanic cones and fissures or cracks.

Just prior to these lava eruptions, a spectacular rugged topography had been eroded into the existing older terrain. This erosional surface is estimated to have been as much as 4,500 feet from the canyon bottoms to the crests of the divides. Careful mapping of this ancient surface shows that streams in the Salmon area roughly paralleled the present pattern. The flood-like lavas innundated the valley system covering all but the highest peaks with layer upon layer of hard black columnar basalt. After the eruptions, the Clearwater embayment including the Salmon was a flat expanse of cooling lava with island-like "steptoes" projecting here and there above the lava plain.

Lava eruptions were not continuous. Between major eruptions and flows, vast inland lakes developed and slowly filled with volcanic mud, ash and dust. These lake sediments, referred to as the Latah Formation, are situated at several levels within the lava flows and bear a distinctive preserved flora of fossilized Miocene leaves. Near White Bird, Idaho, exposures of Latah sediments are famous for beautifully preserved deciduous (hardwood) leaves quite different from the local coniferous forests growing there today.

The geologic map of the Clearwater embayment and drainage of the Salmon shows alternating exposures of basement rock along the river. This patchiness is an expression of the roughness of the underlying topography. As the river erodes through the lavas, it first exposes the old basement "highs" that were divides before the lava covered them. Two such divides parallel the present Salmon in its last thirty-nine mile traverse before meeting the Snake River.

Very young, unconsolidated sands and gravels of the river system as well as a layer of very fine dust known as the Palouse Formation overlie these lake sediments and lavas in the Salmon River drainage. The Palouse is a powdery, buff-colored, wind blown, dune deposit (loess). This partially consolidated deposit is very thick in the vicinity of southeast Washington. As it decomposes it forms the foundation for the rich palouse soil which is the basis for eastern Washington's soft wheat industry.

The dust that makes up the Palouse was derived from streams draining glaciers to the north during the Pleistocene ice age up to a million years ago. As these streams piled up the soft glacial "flour" in front of the glaciers, winds picked up the drying dust and carried it south to deposit in a vast area of eastern Oregon and Washington as well as southern Idaho.

The poor drainage of southeast Washington has preserved the soft, easily eroded loess, but in Idaho in the Clearwater embayment area of Idaho almost all of the previous loess cover has been stripped off by rapidly eroding mountain streams. A small remnant of the Palouse may be seen along the river meanders just six miles south of White Bird.

In the interval between Riggins and Hellers Bar, gold has turned up several times, but strikes of major size are not recorded. Much of the active mining took place in the late 1800's and into the very early 1920's. Gold occurs in trace amounts along the entire length of the Salmon with very scattered richer pockets. As a rule, the gold is very fine almost to the point of being flour gold. Mining of gold of this size is usually marginal, and great care must be exercised in the extraction process not to lose the fine metal. Some rich pockets occur, but most of the gravel and sands processed for gold average from .30 cents down to .05 cents per cubic yard processed.

Many types of mining strategies have been attempted along the Salmon here, but the very small, low overhead operations are invariably the ones that show even a modest profit. Although gravels in the river bottom have been extensively explored, higher terrace deposits on the canyon walls that might have been exploited by hydraulic mining techniques have barely been touched.

In addition to fine gold, small amounts of platinum have been recovered along the river.

Curiously, little, if any, of this gold is related to the granites of the Idaho Batholith. Instead mineral rich fluids from younger volcanic intrusives are believed to be responsible for the metals.

Geology Along the Route

The interval of the Salmon from Riggins to the Snake does not directly cut across the Idaho Batholith, however, exposures of granites are only a few miles away to the east. Subsequently, much of the river gravel is quartz-rich, grey to white crystalline granites.

Many north-south faults occur in the canyons north of Riggins. The river here has followed these breaks in the rock in a straight line for miles before meandering away. The faults are visible in the canyon walls as smooth, striated surfaces that reflect the sun like polished metal. Along the old faulted surfaces, mineralization is evident where stains discolor the rocks.

Just north of White Bird, the river cuts into the top of an old divide below the skyline lavas. Two hundred million year old (Triassic) rocks in this twelve mile stretch bear a distinctive, dark greenish hue due to characteristic heat treated minerals (metamorphic).

Metamorphic rocks are typically very hard, and one effect of this is the narrow canyons through this interval. With the constricted channel, the stream velocity increases substan-

tially. River erosion of these rocks causes curious shapes. One of the more striking of these is a fluting or sand-blasting effect the river gives the wall rocks. Sand borne in the water cuts long, finger-size grooves or flutes parallel to the direction of stream flow. The rocks are almost incredibly smooth to the touch where the water and sand have polished the surface. Near the waterline the rocks receive a final patina of "desert varnish" or a jet black color due to exposure alternating from wet to very dry and hot.

A stretch of lavas after the basement rocks represents a spot where the river is crossing, at nearly right angles, an old stream channel in the lower erosional surface. Twenty miles of basement rocks follows with spectacular scenery and steep, sheer cliff walls. A short interval of basalt before entering the Snake River is an example of an old stream channel now filled with a pod of lava.

In the short distance rafted along lower Hells Canyon a good representation of basement rocks can be seen. Just a few miles south of Hellers Bar on the Snake, limestones appear on both sides of the river. These limestones are cooked in places to form the metamorphic rock, marble. The limestone is folded up and is lying nearly on its edge. This orientation gives the eroded rocks a very sharp series of ridges. Locally in topographic maps the word "lime" appears often as in Limekiln, Lime Hill, and Limestone Point.

Pierce after the 1861 gold rush (Idaho Historical Society).

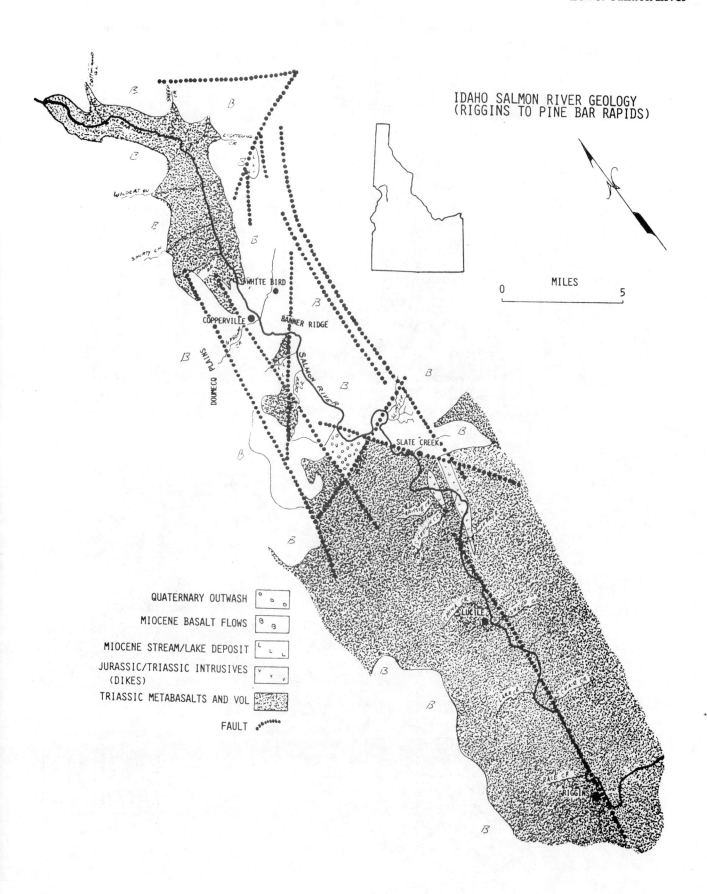

IDAHO SALMON RIVER GEOLOGY
(RIGGINS TO PINE BAR RAPIDS)

MILES

0   5

QUATERNARY OUTWASH

MIOCENE BASALT FLOWS

MIOCENE STREAM/LAKE DEPOSIT

JURASSIC/TRIASSIC INTRUSIVES
(DIKES)

TRIASSIC METABASALTS AND VOL

FAULT

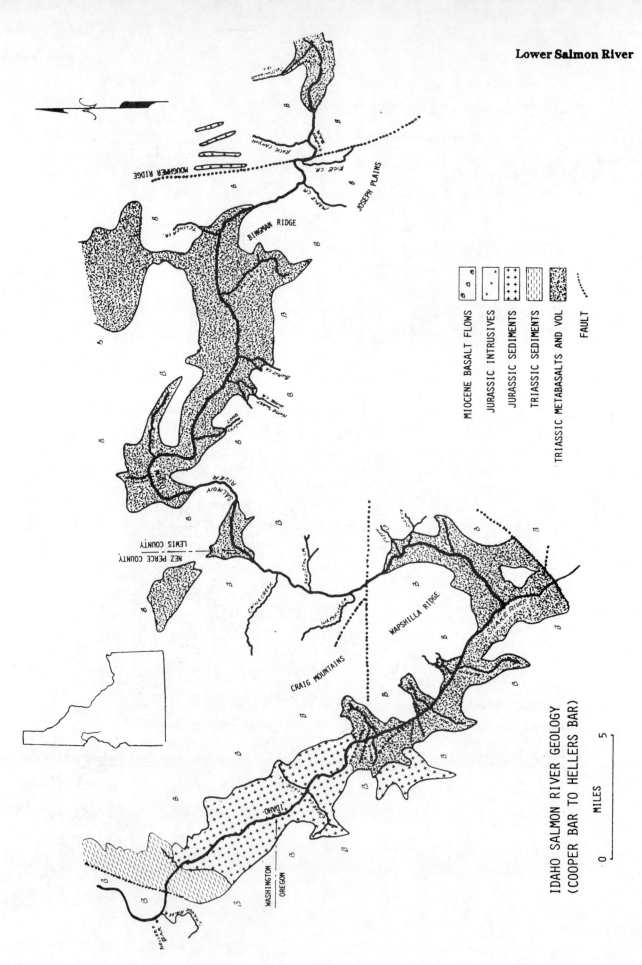

IDAHO SALMON RIVER GEOLOGY
(COOPER BAR TO HELLERS BAR)

MIOCENE BASALT FLOWS
JURASSIC INTRUSIVES
JURASSIC SEDIMENTS
TRIASSIC SEDIMENTS
TRIASSIC METABASALTS AND VOL
FAULT

0   MILES   5

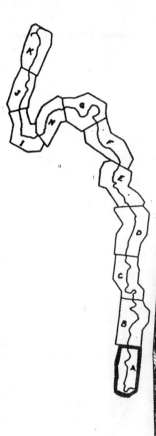

Lower Salmon River

A

0   mile   1

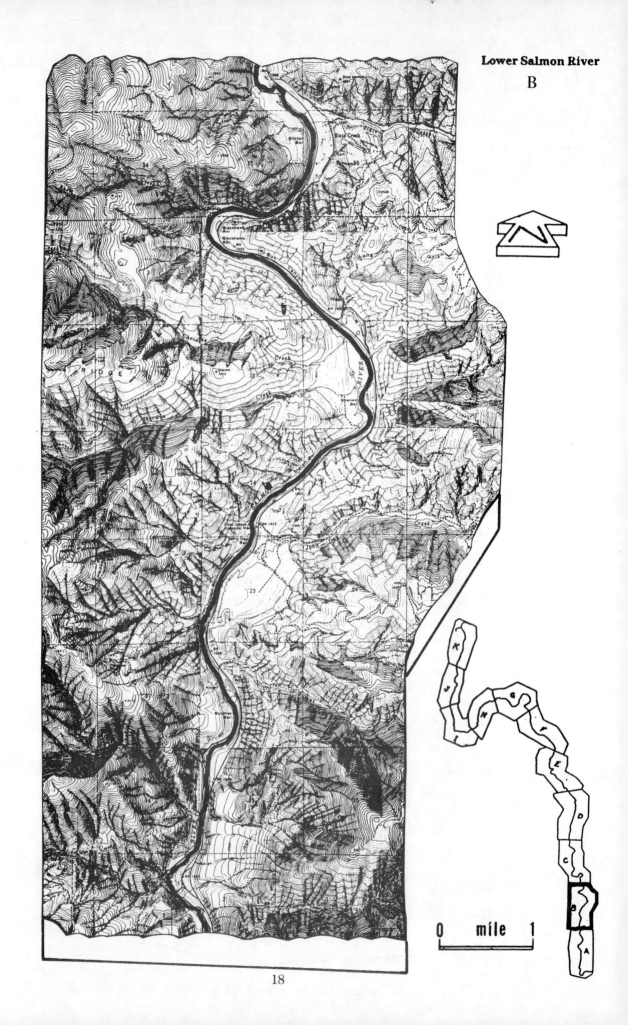

0   mile   1

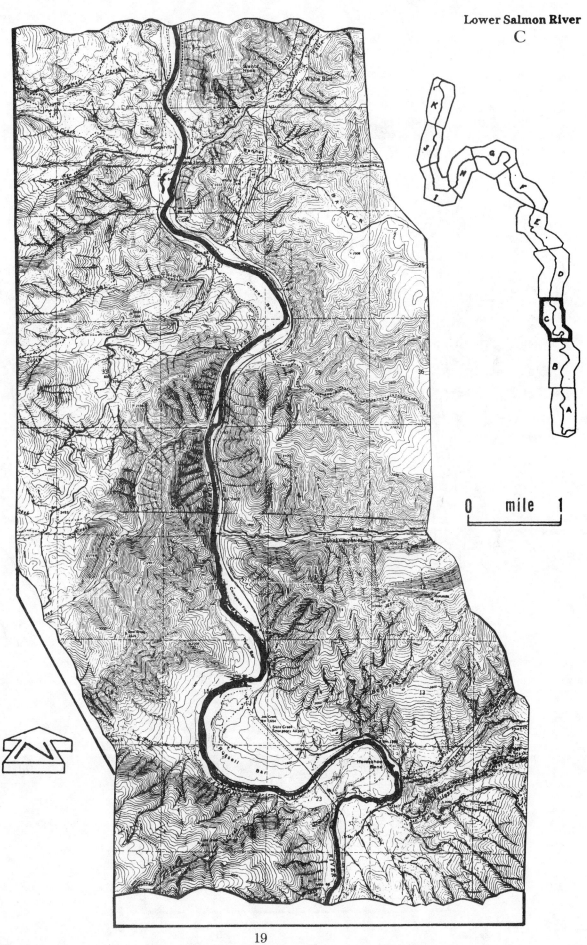

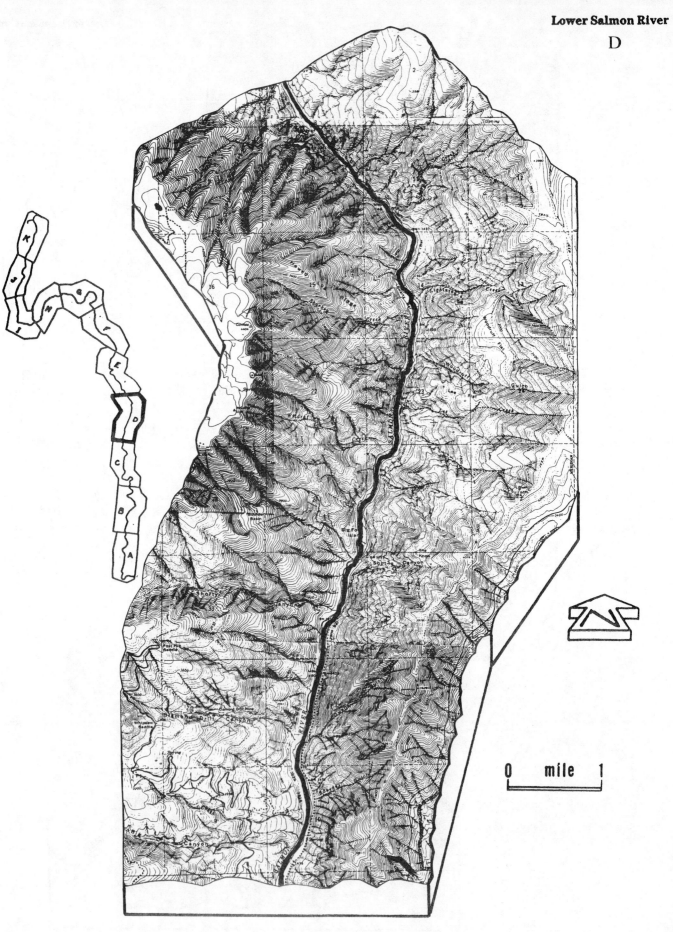

20

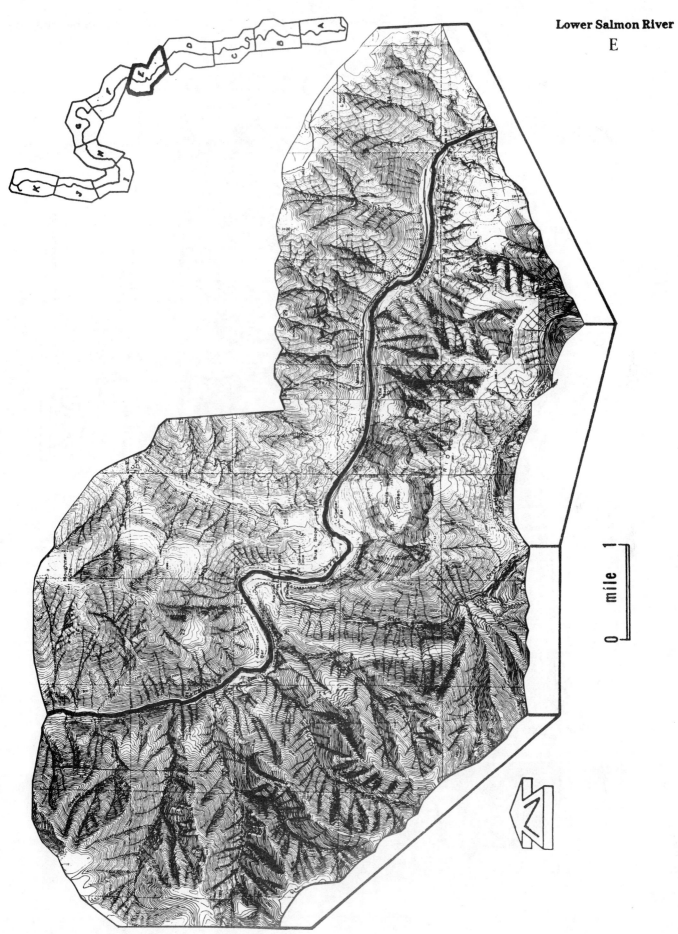

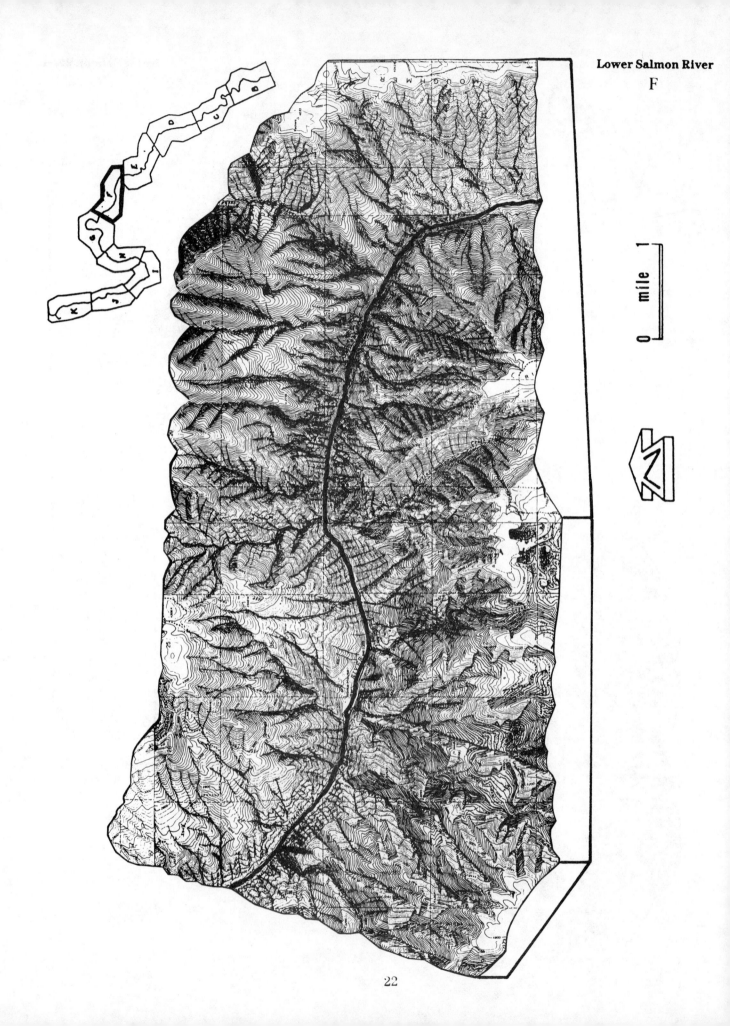

0   mile   1

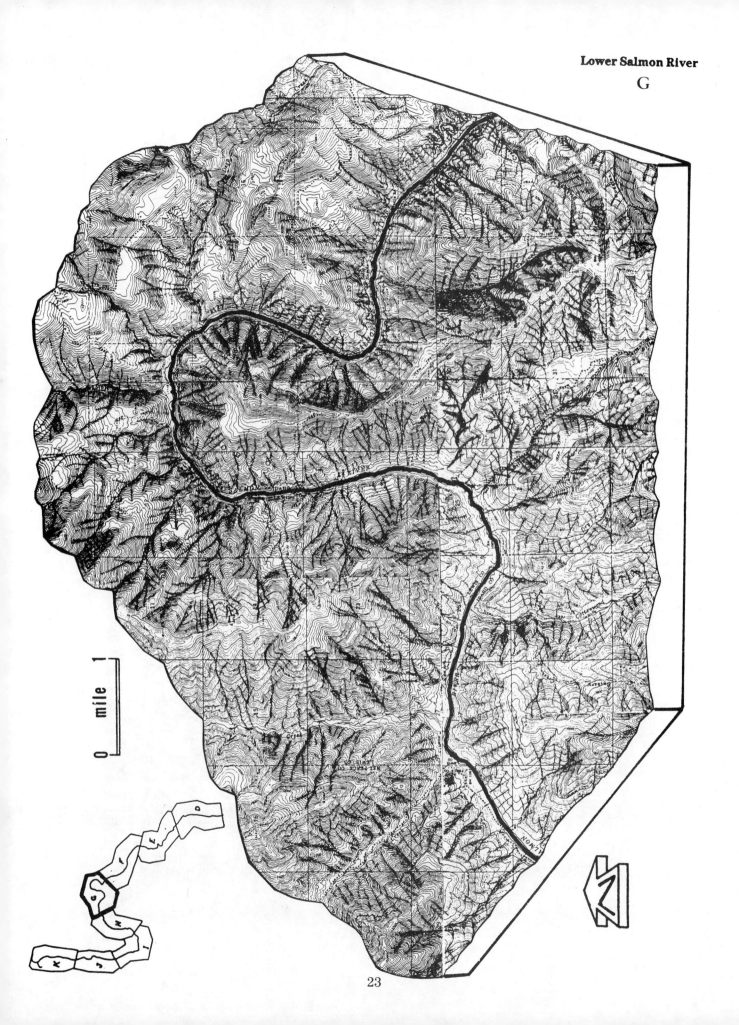

mile

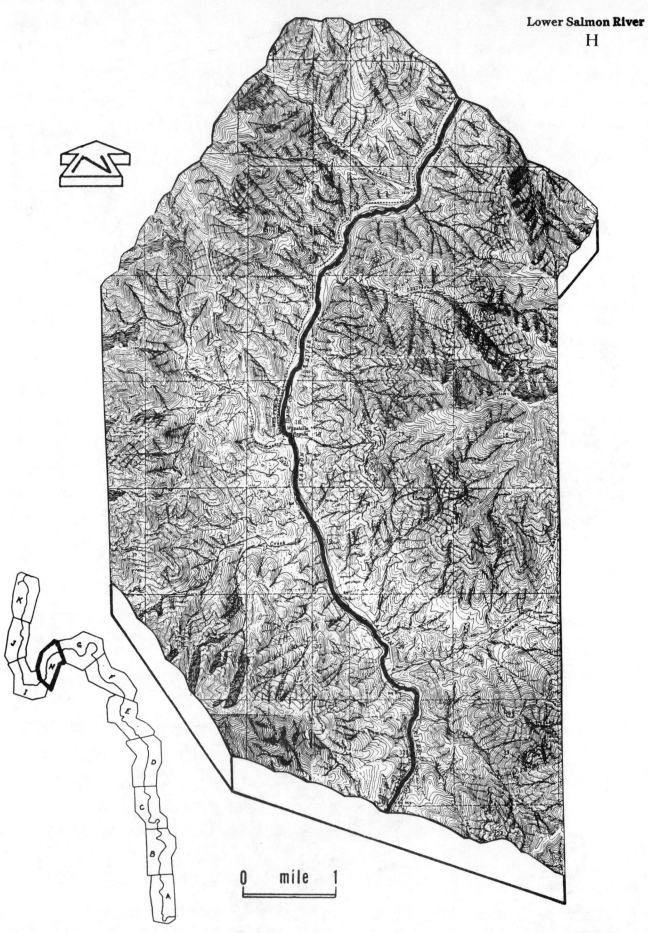

0 mile 1

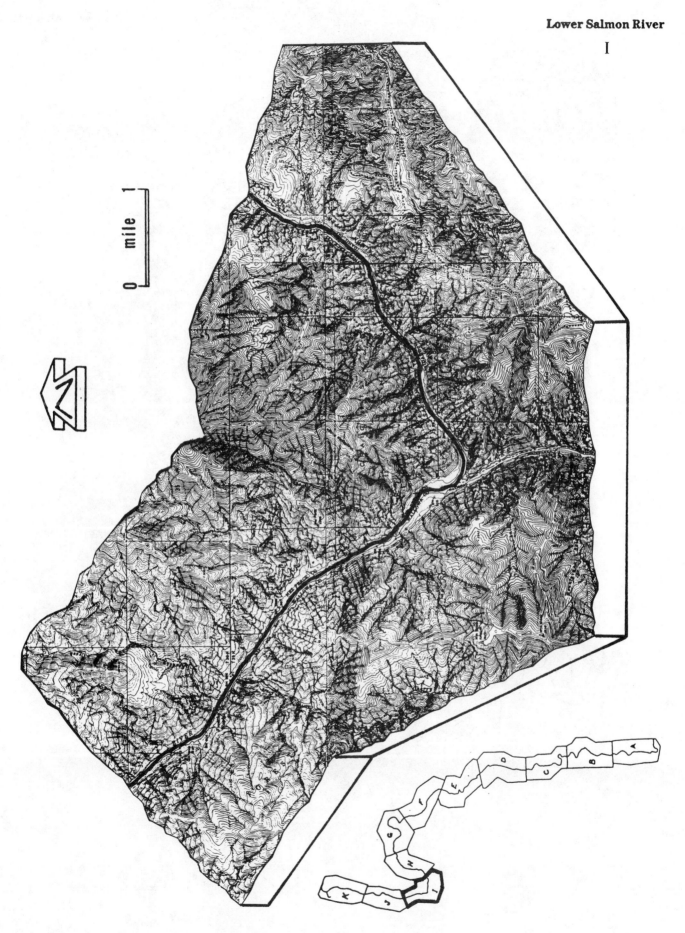

0   mile   1

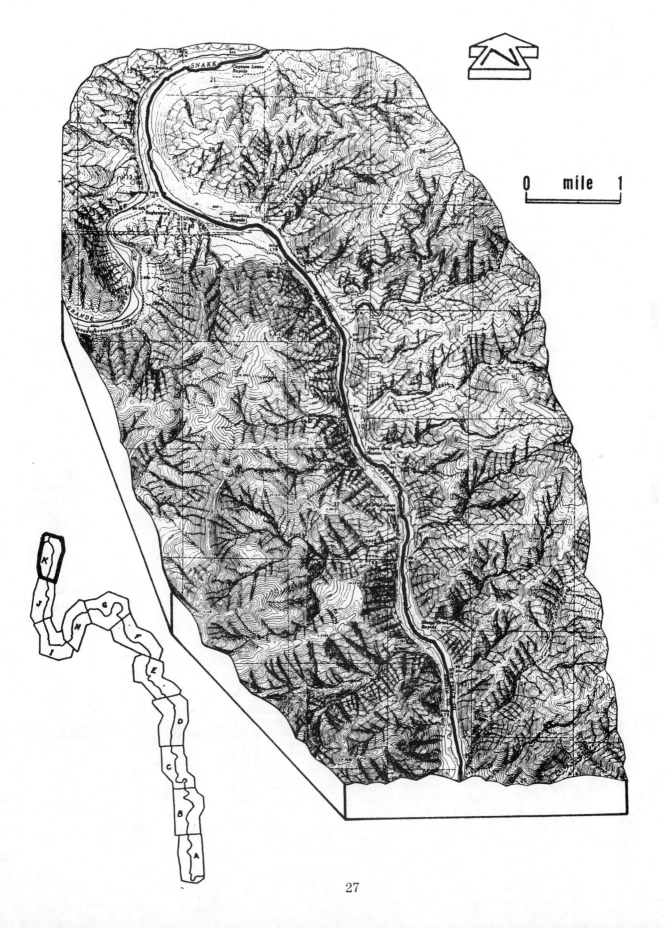

Middle Fork, Salmon River

(Dagger Falls to Cache Bar)
Custer, Valley, and Idaho Counties, Idaho
Length of run: 100 miles
Number of days: 5-6

Cold, clear water of the Middle Fork of the Salmon, draining from the mountains of central Idaho, drops over 3,000 feet in a northeastward, meandering direction before mixing with waters of the main Salmon. The river canyon exposes an immense granite mass which was once an underground chamber of fluid magma. This mass of granite altered and baked most of the central Idaho rocks as it slowly cooled 70 million years in the geologic past. In contrast, ancient hunters appeared in this region as recently as 11,000 years ago.

The Middle Fork was preserved in its natural state by an act of Congress in 1968 which declared the river to be Wild and Scenic. Additionally the river is surrounded by one and one-half million acres of the Idaho Primitive Area.

Young Indian man.

Early History of the Middle Fork of the Salmon River

Sheepeater Indians, *Tukudeka*, meaning "mountain sheepeaters," inhabited the rugged canyon terrain of the Middle Fork of the Salmon River. Their name was derived from their reliance on Rocky Mountain sheep for food and clothing. The *Tukudeka* belonged to the mountain Shoshone whose area probably extended from Idaho into southern Montana.

Population information on the *Tukudeka* is more sketchy than for most. The 1886 census of Indians at Lemhi Agency in Idaho reports 422 individuals. This number included Shoshone, Bannocks, as well as Sheepeaters. The total Shoshone population in mountainous regions of Idaho and Montana was thought to be 1,200 in prehistoric times, but by the reservation period that number had been reduced by half.

Because of their physical isolation, the Sheepeaters had retained an older, slow singsong speech and practiced intermarriage between first cousins, both traits not approved of by the surrounding Shoshone. Perhaps this contributed to their reputation of being morally and intellectually inferior, although in actuality, the term "sheepeaters" implied

a certain respect for these people who were big game hunters.

Historically this group has been misunderstood and misrepresented. At different times they were described as "outlaws expelled from various tribes," "wily and treacherous, though cowardly," "cunning and treacherous renegades," "not far removed from the lower animals," and "mere scabs, so contemptible and mean that the other tribes would not allow them to remain in their camps."

The *Tukudeka* lived in winter villages along rivers and streams. These villages were occupied by families and organized loosely under a headman. Families traditionally used the same villages from year to year. During summer months, small family groups moved about in search of food an activity which involved most of their time.

Powerful bows and arrows, the tips treated with plant poisons, were used in game hunting. Dogs accompanied the hunters. Communal hunting took place only for birds as sage hens or water fowl, while antelope, deer or sheep were killed by the individual hunter. Game hunting took place all year round,

covery of gold between 1866 and 1873 brought hordes of miners to this area.

At first ebullient miners confined their activities to the western part of the state. However, gold discoveries in 1866 in the Salmon River drainage at Leesburg and near present day Salmon on Panther Creek were in close proximity to *Tukudeka* winter homes. With this contact the Indians retreated further into the canyons. In addition, a small number of ranchers and farmers had settled along the river. The Sheepeaters developed a reputation for being a "constant menace to prospectors passing through or camping in their territory; they ran off stock or stole anything they could lay hands on. They were not beyond murdering . . ." (Arnold, p.223).

In 1869 gold placer finds on Loon Creek, a tributary of the Middle Fork sent prospectors deep into Sheepeater lands. Led by Nathan Smith, 70 men travelled overland from Leesburg and Idaho City to organize claims in the new district. By August, 2,000 to 3,000 men lived in tents, but the town of Ora Grande was being built in anticipation of winter. Perhaps 200 miners planned to winter over. Ora Grande grew to be a fair-sized establishment of 1,500 whites and Chinese. There were five stores, five saloons, and fifteen private houses.

The only one to make a fortune here was Gentle Annie, a 200 pound restranteur who encountered a stage on a narrow trail as she was headed toward the mines with her pots and pans. When the stage knocked her off her mule, Annie sued and collected several

Rocky Mountain sheep, used for clothing and food.

Mountain goat was also hunted.

whereas, salmon fishing and plant gathering were seasonal.

Fur was tanned by a process of rubbing animal brains into the hides. *Tukudeka* were considered superior furriers, and their finely worked skins were made into clothing, each skin used for a particular article. For example, badger skin was made into moccasins. Men's caps and leggings were made from coyote skin. Mountain sheep skin was used for women's dresses, two skins making one dress. Their fur clothing was highly prized by neighboring tribes who acquired it in trade.

Rough topography and deep canyons of the Middle Fork kept it geographically isolated from influences of other Indians as well as from contact with whites. Probably members of the Alexander Ross fur trading expedition explored the Salmon River drainage in 1824, but contact was limited until the dis-

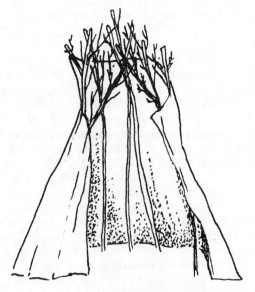

Winter shelter

Map showing how gold claims were laid out.

thousand dollars for her broken arm. Once again the rumor of gold was more exciting than the real find.

Placers at Loon Creek had been worked out by 1871, and the Ora Grande was abandoned to the Chinese. A few hundred worked here until attacked by Indians. Two Chinese who hid in a rootcellar among potatoes stored there survived. Indians burned the town which was never rebuilt.

The slaying of the Chinese as well as several white ranchers along the Salmon River in what was called the Loon Creek Massacre, precipitated the need for a solution to the "Indian problem." It was never proven the Indians were involved in the killings, and statements of witnesses were later retracted. However, the white population wanted action, and troops were sent in from forts at Boise, Grangeville and Pendleton, Oregon.

The ensuing conflict, known as the Sheepeater War of 1879, was described as an Army campaign "in which they (whites) won no classic military victory over their adversaries but merely conducted an expedition that cost the government much in money and prestige. Eventually the brutal terrain forced both hunter and hunted to capitulate. The campaign was hardly a high spot in United States or Idaho history" (Pavesic, p.13).

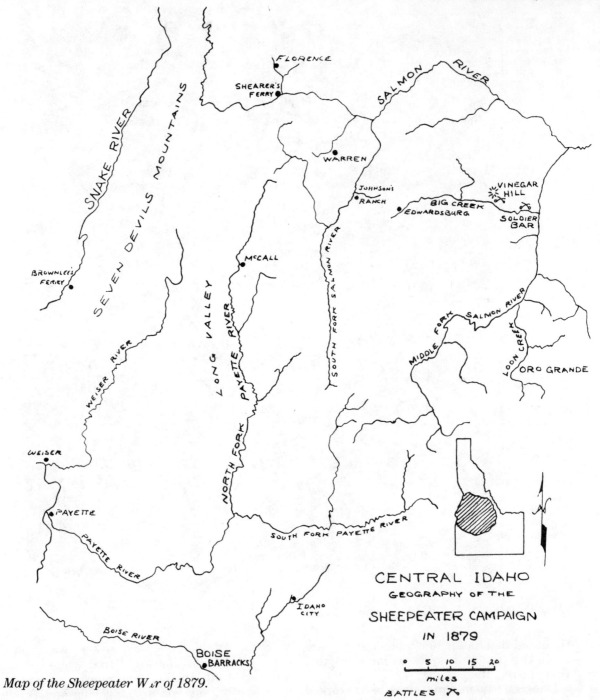

Map of the Sheepeater War of 1879.

General Howard directed the war to capture the "bloodthirsty and thieving" Sheepeaters who had reportedly been joined by renegade Bannocks and other refugees from Indian wars in Idaho and Oregon. The General and his men set off "like hunters in search of wildest game," but the Indians were able to elude them and vanish completely into the wilderness.

June 4, found Captain R. S. Bernard on the North Fork and Lt. H. Catley and his men on the Middle Fork near Big Creek. Both men were delayed by deep snows and rushing waters of a late Spring. Supply mules had been lost. No signs of Indians were found until July when Catley came across remnants of an Indian camp along Big Creek. The soldiers followed down the steep canyon only to be ambushed. Two soldiers were wounded. Unable to see the Indians hidden along the steep walls, the soldiers reversed position with much difficulty and retreated.

Captain Bernard.

Lieutenant Farrow who finally defeated the Sheepeater Indians.

The next day, camped on a hill near the confluence of the Middle Fork and Big Creek, Catley was again attacked by Indians who set fire to the brush and trees. Surviving the flames and heat, the soldiers only had vinegar from kegs to drink, hence the present day name of Vinegar Hill. The troops eventually retreated all the way back to Warren, Idaho. Catley was court martialed for his role, but President Hayes dismissed the charges.

Capt. Bernard chased the elusive Indians into September until a lack of supplies and exhausted horses forced his men back to Ft. Boise. The first military casualty occurred when Bernard's men were fired upon at Soldier Bar on the Middle Fork. Private Harry Eagan was wounded and died while being operated on. Years later a memorial stone, transported 75 miles, was erected on the spot commemorating his passing.

The field was left to Lt. E. Farrow, his Umatilla scouts, and soldiers who had marched back toward the Middle Fork in spite of beginning September snowstorms. By persistance he used a military strategy designed to destroy the *Tukudeka* food supply and keep them "on the jump . . . to demoralize them."

By October 1, most of the Indians had surrendered unconditionally to Farrow. They were taken to Pendleton, Oregon, where they spent the winter before being sent permanently to the Ft. Hall Reservation. There, as one author commented, "they seemed satisfied to take up the arts of peace." A total of fifty-seven Sheepeaters were captured. Of these only fifteen were warriors.

Indian Life Along the River

The Middle Fork of the Salmon had been the home of ancient hunters for at least 1,000 years. An archaeological survey records 93 Indian cultural locations. These included house pits, caves, rockshelters, tipi rings, and sites of unknown usage. House pits are numerous. These pits are from ten to twenty feet in diameter and of varying depths. Most are in good condition and represent an original excavation from one to four feet below the ground where a house was built with a pole frame and mats. More caves and rockshelters were recorded than anything else. Many had been used for dwellings and were fairly undisturbed.

House pit or depression

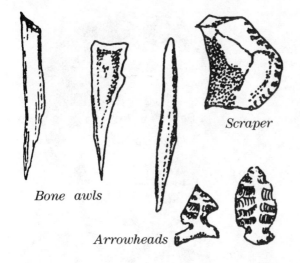

Scraper

Bone awls

Arrowheads

The tipi ring, a circular series of river stones on a flat surface, is also evidence of an early Indian dwelling used after 1700. These places are superficial and easily destroyed. One near Loon Creek, consisting of thirty tipi rings, had been destroyed by 1971.

Large boulder circles, small rock circles, hunting blinds or pits and undetermined rock structures tell something about Indian culture, although usage of some of these rock arrangments hasn't been determined.

Rock art along the Middle Fork is in a poor state of preservation. Weathering has made many of the pictographs difficult to see. Others are covered by vegetation. All rock art here is pictographs, and both abstract designs as well as designs of people and animals are represented. Colors of the designs range from red-brown to orange.

Geology of the Middle Fork of the Salmon River

The Middle Fork of the Salmon River is run from Dagger Falls where Boundary Creek enters the river to Cache Bar on the main Salmon. This run is 100 miles in length and drops roughly 3,000 feet in elevation over that distance. The most spectacular aspects of the Middle Fork are the deep canyons averaging 4,000 feet from the divide to the river below. Slicing as it does through central Idaho on a meandering northeast trend to meet the main Salmon, the river cuts a variety of volcanic rocks as well others that display evidence of considerable heat and pressure.

Geologically the river route provides a good cross-section across the middle of Idaho beginning with 80 million year old granites of the

Idaho Batholith, cutting through lavas of the Challis Formation formed 50 million years ago (Eocene), before winding up in some of Idaho's oldest rocks over 1 billion years of age (preCambrian). Several hot springs occur along this route, but, in spite of their persistence, they are rated low as potential power sources.

Volcanic rocks of the Challis Formation exposed along the Middle Fork include a variety of thick quartz-rich extrusive and intrusive volcanic rocks. In addition to the volcanics, intermittent low spots in the older erosional surface have accumulated sediments which in turn entombed fossil plants and a few mammal bones. The plant fossils include conifers which grew at an altitude of 4,000 feet or more.

Although central Idaho has been, and still is, the site of intensive gold prospecting, the returns to date have been disappointingly limited. The largest gold strikes have been in areas to the east of the Middle Fork drainage in the Yellowjacket Gold District and to the west in the Thunder Mountain District. The Thunder Mountain District, named for the way thunder bounces back and forth between its steep, narrow canyons during electrical storms, was the site of one of the last authentic gold rushes in the West in 1902. The Yellowjacket District extends for 60 miles west of Salmon City. Placer

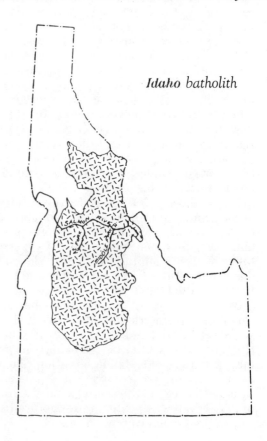

Idaho batholith

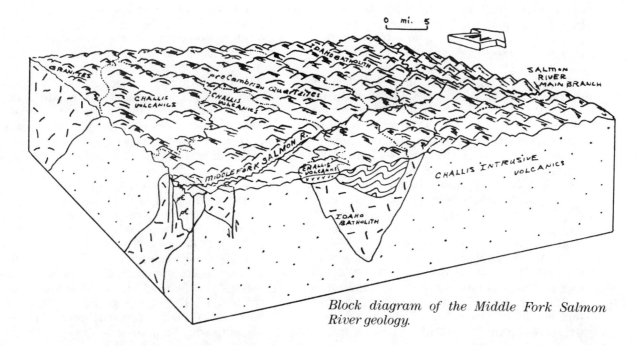

Block diagram of the Middle Fork Salmon River geology.

mines discovered in 1869 were named because a nest of yellowjackets chased off the prospectors.

Precious metals recovered from the Middle Fork and adjacent watersheds are dominantly gold and tungsten, but small amounts of copper, lead, and silver ores occur as well. Along the river, gold has been discovered and mined in both placer (gravel) and lode (vein) deposits, but no large scale discoveries are recorded. Several stories are told of lost mines or of miners carrying large amounts of gold being drowned in the Salmon. Along the route, old mining operations are obvious where debris marks placer sites and dumps. Miners' cabins and rusting equipment are scattered along the banks.

In this area of Idaho the geologic "grain" of the rocks is northwest by southeast. Because the river traverses from southwest to northeast, it slices directly across the major rock types. The overall trend of exposures is from younger, 80 million year old, Cretaceous rocks to older 1 billion year old, preCambrian rocks.

Geology Along the Route

Beginning at Boundary Creek at the put-in, the first twenty to thirty miles are in the Idaho Batholith. Although details of the origin of this massive intrusive granite mass are not yet thoroughly understood, the unit has been regionally mapped. The batholith covers an extensive area of central and western Idaho well up into the panhandle. The middle Salmon section begins close to the geographic center of the batholith, and the granitic rocks here cooled 70 to 80 million years ago.

Quartz veins cut into the unit everywhere, but gold production from these veins is surprisingly limited. One interesting aspect of the Idaho Batholith is the presence of intermittent thermal springs along most of the major deep streams cutting the structure.

Light-colored granites of the Idaho Batholith are readily identified by the large grey and white crystals of quartz and feldspar. Thirty miles downriver from Dagger Falls, the change from granite to the 50 million year old Challis volcanic rocks is profound. This suite includes varieties of rocks cooled above ground (extrusive) as well as those cooled below the surface (intrusive). In this interval and for the next twenty-five miles, the intrusive rocks dominate.

Half of the way through the Challis interval, the river briefly cuts two small sections of the Idaho Batholith and a very old exposure of preCambrian rocks. These stretches are easily recognized by the rapid changes from gray granite of the batholith to highly fractured, older metamorphic rocks of the preCambrian. In addition to the color and texture, these rocks also have distinctive fracture patterns.

After yet another brief exposure of the Idaho Batholith, the river flows for three to four miles through a spectacular section of the 1 billion year old preCambrian quartzites. Quartzite is derived from sandstone which has been deformed by heat and pressure (metamorphosed).

Because the mineral quartz is highly resistant to weathering, the rock quartzite is exceedingly tough and durable. This quartzite occurs as a northwest to southeast arcuate exposure covering eighty miles.

A high spot or "stock" in the underlying batholith can be seen briefly for several miles before the river moves back into more preCambrian rock. This exposure is the massive Yellowjacket Formation which covers hundreds of square miles close to the Montana border. The Yellowjacket is composed of metamorphic rocks including coarse-grained, banded gneiss and shiny, mica-bearing schist. These rocks contain the distinctive minerals kyanite and sillimanite which are unmistakable evidence of exceedingly high temperature and pressure.

The Middle Fork traverses three rock groups, the Idaho Batholith granites, Challis volcanics, and preCambrian metamorphics. Within each rock group the fault and fracture pattern differs markedly. The trend of faults and joints (fractures) in the Idaho Batholith is dominantly north/south. Over the intervals where these granites are exposed, many long straight stretches occur where the river's trend is due north. Major northeast traverses by the river occur in the volcanic and metamorphic sections where faults and fractures follow separate trends to the northeast and northwest. Although the overall slope toward the northeast controls the river's ultimate pathway, a strong northwest trend causes many short excursions in that direction giving the canyons a zigzag pattern.

Some of these structural trends produce huge lineations (lines) which are literally too large to be noticed on the ground and which only appear by connecting points over long distances on large scale maps. These lines rarely appear on a small scale maps but stand out particularly well in satellite photographs. The lessons and applications learned by studying lineations in the Idaho Batholith and similar rocks on earth have been used in efforts to interpret the geology of the distant planets of Mars and Venus.

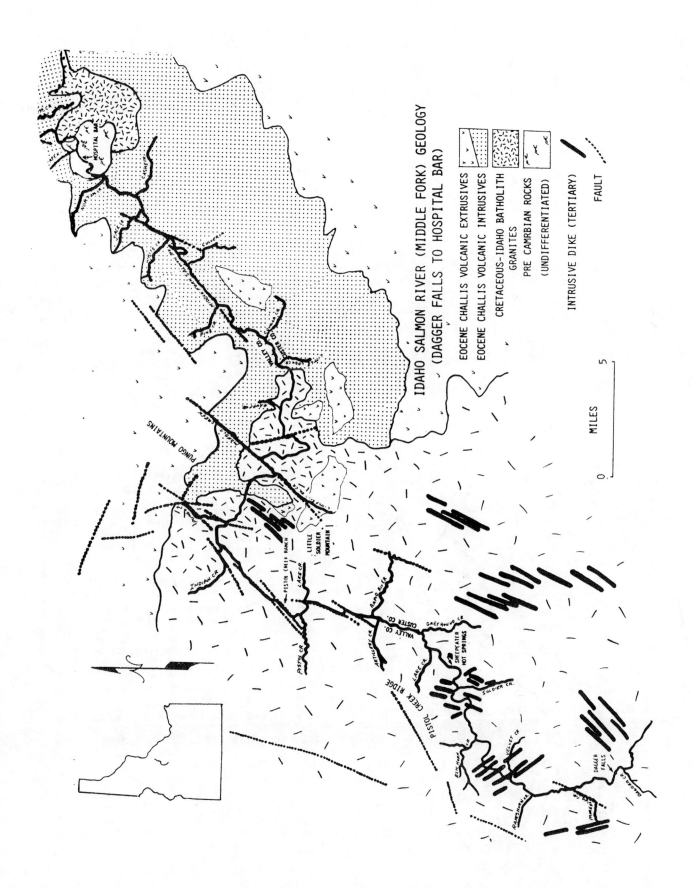

IDAHO SALMON RIVER (MIDDLE FORK) GEOLOGY
(DAGGER FALLS TO HOSPITAL BAR)

EOCENE CHALLIS VOLCANIC EXTRUSIVES
EOCENE CHALLIS VOLCANIC INTRUSIVES
CRETACEOUS-IDAHO BATHOLITH GRANITES
PRE CAMBRIAN ROCKS (UNDIFFERENTIATED)
INTRUSIVE DIKE (TERTIARY)
FAULT

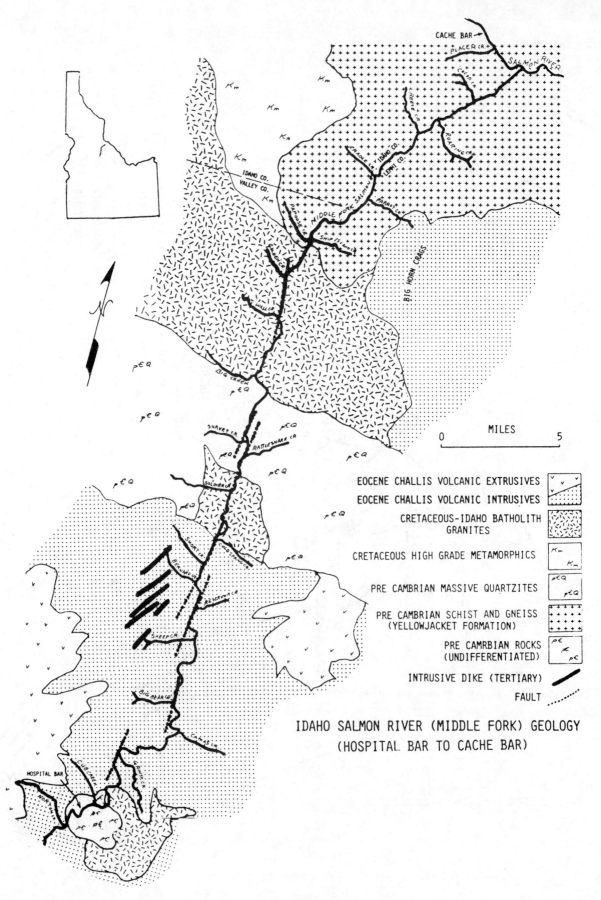

MILES
0   5

EOCENE CHALLIS VOLCANIC EXTRUSIVES
EOCENE CHALLIS VOLCANIC INTRUSIVES
CRETACEOUS-IDAHO BATHOLITH GRANITES
CRETACEOUS HIGH GRADE METAMORPHICS
PRE CAMBRIAN MASSIVE QUARTZITES
PRE CAMBRIAN SCHIST AND GNEISS (YELLOWJACKET FORMATION)
PRE CAMRBIAN ROCKS (UNDIFFERENTIATED)
INTRUSIVE DIKE (TERTIARY)
FAULT

IDAHO SALMON RIVER (MIDDLE FORK) GEOLOGY
(HOSPITAL BAR TO CACHE BAR)

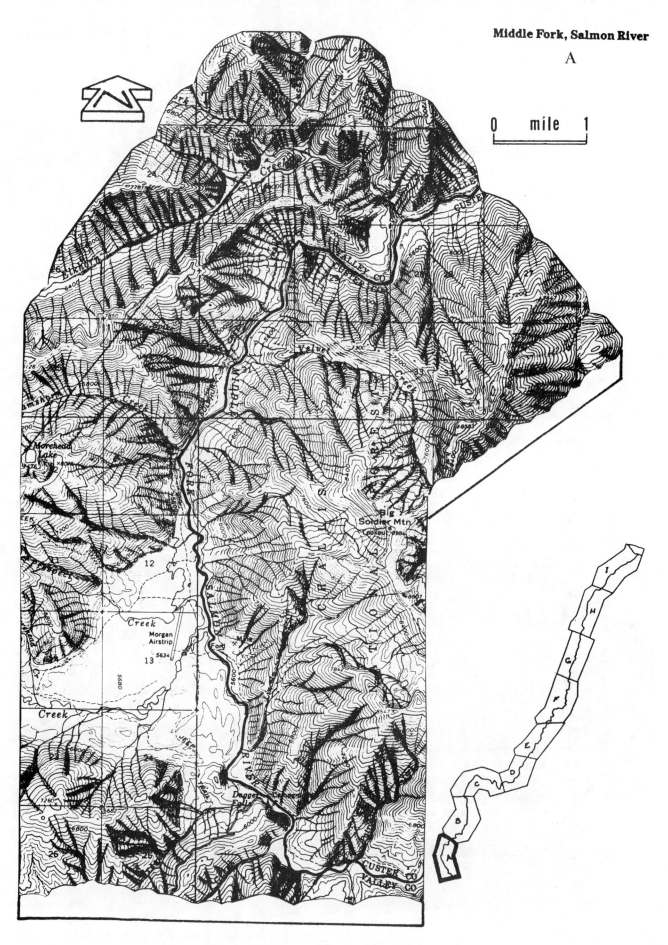

0   mile   1

0   mile   1

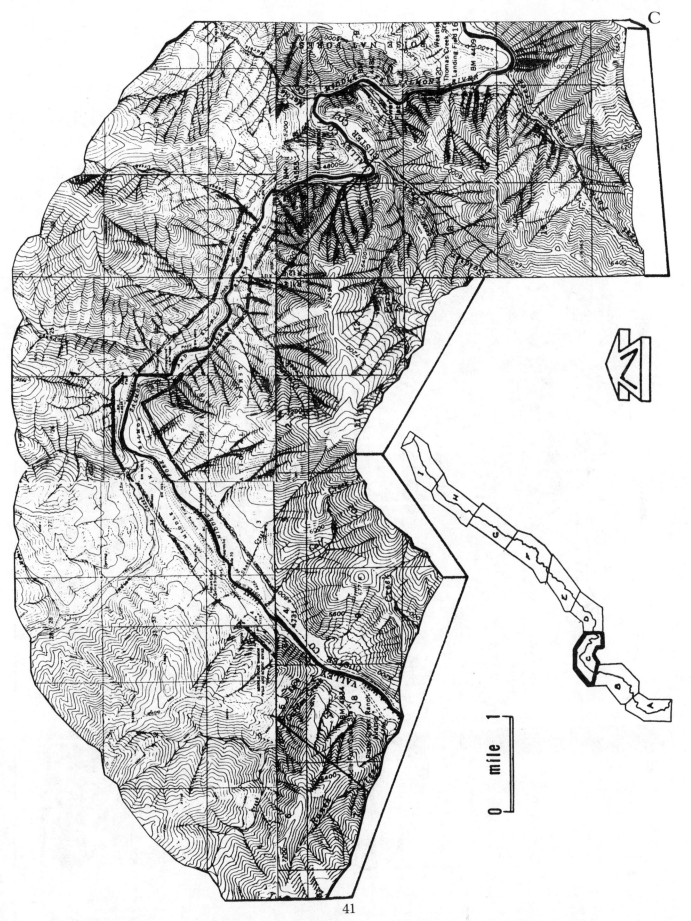

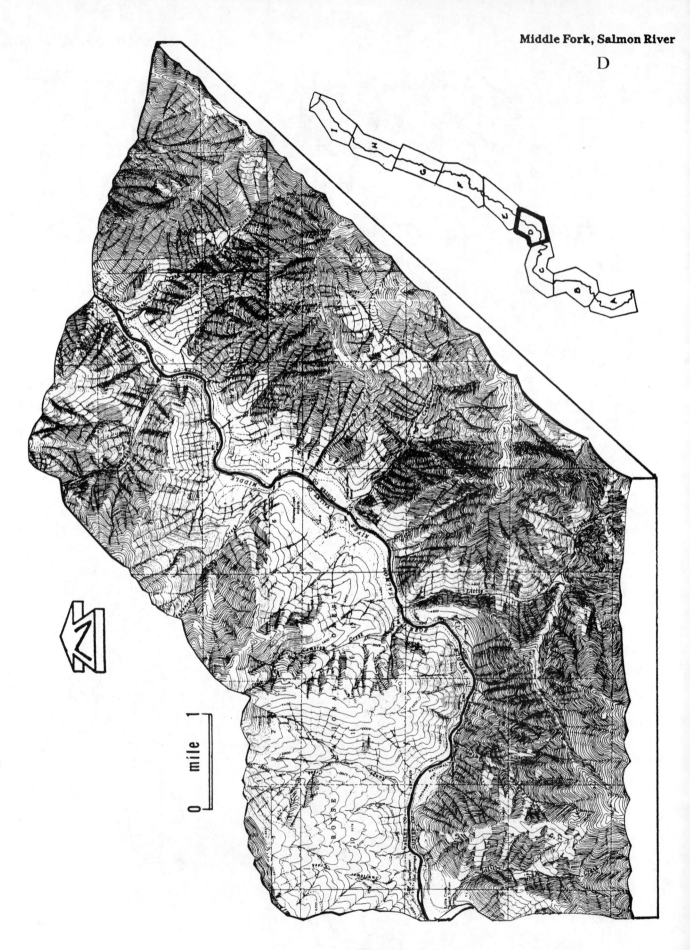

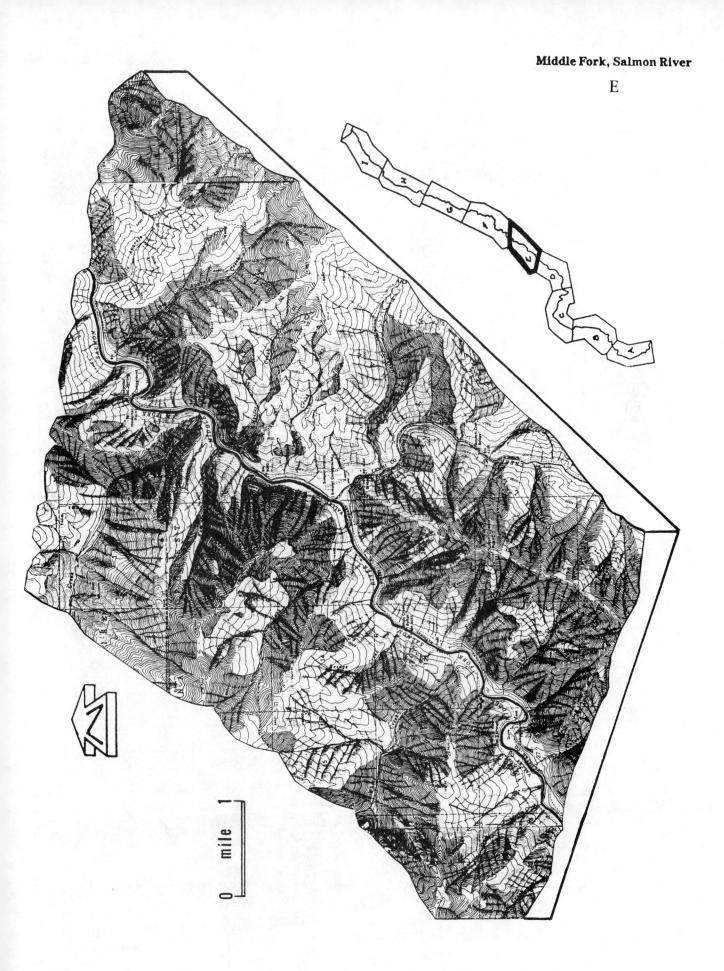

F

0 mile 1

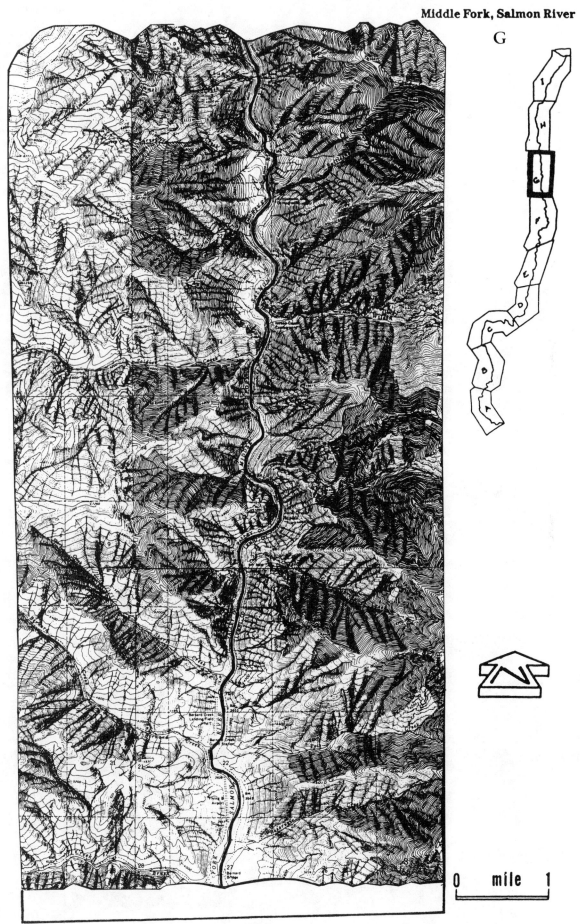

0 mile 1

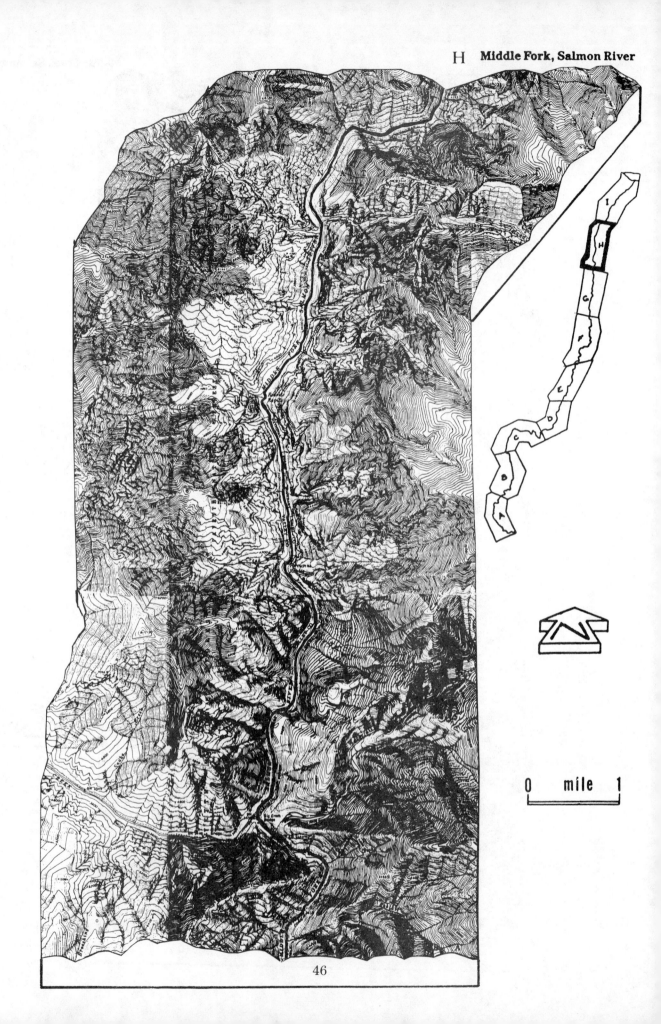

0   mile   1

Middle Fork, Salmon River

I

0 mile 1

47

Upper Salmon River

(North Fork to Vinegar Creek)
Lemhi to Idaho Counties, Idaho
Length of run: 79 miles
Number of days: 3-5

The deep canyon of the Salmon River is exceeded only by Hell's Canyon of the Snake River. The Salmon remained unmapped until 1912, and much of the canyon was still unknown as late as 1949. The chasm of the Salmon, cutting across Idaho, is the largest tributary of the Snake River, draining 14,000 square miles of Idaho. From its origin in the mountains, it drops from an elevation of 8,000 feet to 903 feet at its mouth, and in some places the precipitous canyon walls are 5,000 to 6,000 feet high. Much of the river runs through the Salmon Breaks Primitive Area.

Tricky rapids and whitewater made this River of No Return a one-way trip navigable downriver only. Boats were abandoned at the end of a trip. Many men lost their lives attempting the journey, and even more boats were broken up on the rocks. It wasn't until Harry Guleke conceived the idea of building his scow-like boat in the early 1900's, that the Salmon was conquered. Guleke's boat was made of green lumber so it would have enough flex to withstand shocks from the rapids and falls. It was thirty-two feet long, eight feet wide, and four feet deep.

Mention should be made of the 1935 National Geographic Expedition which went down the length of the Salmon River beginning at Shoup and ending at Lewiston on the Snake. The expedition members included a forester, geologist, State Representative as well as the Geographic crew. The boat they used was one of Guleke's. During this 253 mile long trip the canyon was carefully surveyed and photographed.

To this day, however, a ride down the Salmon is still known as America's Wildest Boat Ride.

Early History of the Upper Salmon River

With the exception of a handful of fur trappers and an occasional explorer, the awesome and remote lands of southwestern and central Idaho were known only to the Nez Perce Indians until after the discovery of gold there in 1860. William Clark and Meriwether Lewis, while traversing the continent on their famous expedition for the government, entered the Salmon River Valley near the present town of Salmon. Searching the valley, they decided it would be impossible to proceed downriver especially after their Indian guide, Toby, convinced them "miles of whitewater" and snow covered mountains would prove impassable. William Clark named the river, Lewis River, after his companion.

Salmon, from the Latin, salmo, means "leaper" because these fish would leap cataracts and rapids. Fur trappers were generous in their use of "salmon" to name falls, rivers, creeks, and towns. The present Salmon River was known from the earliest times for its outstanding salmon. Soldiers, camping nearby and eating these fish, described them as of a "fine rose-colour, with a small mixture of yellow, and

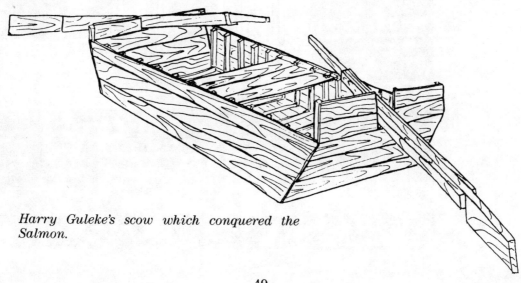

Harry Guleke's scow which conquered the Salmon.

so fat that they were cooked very well without the addition of any oil or grease."

Idaho was settled much later than Oregon and California to the west. Perhaps this was because Hudson's Bay Company dominated the fur trade here much more aggressively than in Oregon. In Idaho, the HBC letters on the Company flag were interpreted by fur trappers to mean Here Before Christ, meaning the British based operation was taking over the most valuable trapping districts.

The climate as well wasn't considered hospitable. One emigrant said little was known of the climate and "that little was unfavorable." Fur trappers were forced to spend long winters in one well-stocked cabin, all other locations being inhospitable.

The first American owned fur trading post in Idaho was built by Andrew Henry in southeastern Idaho north of the present town of Rexburg. Fort Henry, as this collection of two to three cabins was called, housed a small band who wintered here in 1811. Although not troubled by Indians, they found no game. After a severe winter, followed by heavy spring rains, they were forced to eat their horses to keep alive. Dispirited by this time, the party broke up, taking

off in different directions. The abandoned fort was burned to the ground and never rebuilt.

Donald McKenzie led the first trapping party into Idaho in 1812. The 300 pound McKenzie was clever and diplomatic in his dealing with the Nez Perce. He built a post on the Clearwater River and operated here over a period of several years untroubled and highly successful economically. McKenzie removed over 70,000 beaver from streams and rivers before trapping operations ceased in the 1840's.

Fur trappers must have been aware of gold in Idaho streams and rivers, but they were singularly uninterested in exploiting it for wealth. With the discovery and interest in gold elsewhere in the West, it was inevitable that gold seekers would find their way to Idaho. Only fierce resistance from the Nez Perce delayed the Idaho rush until 1860.

E. D. Pierce and a man known only as Martin had already done some panning of the Clearwater River and were trying to interest others in starting up a gold mining operation. While miners were willing in spirit, most were intimidated by the mighty Nez Perce. Settlers as well were reluctant to bring on any disturbance which might lead to war. Of the merchants in Fort

Fort Nez Perce of the North West Company, built by Donald McKenzie.

Donald McKenzie, who built Ft. Nez Perce and established trading routes into Idaho.

Walla Walla, Washington, who were asked to contribute toward the gold prospecting trip, only one was willing to donate at all and that was 1,000 pounds of flour.

By advertising, Pierce was able to gather together a small party of 12 men. Leaving Fort Walla Walla in August, 1860, they sneaked across the border from the north. Following an obscure trail and working the creeks for six weeks, they found a promising placer deposit in Canal Gulch on Orofino Creek. The men recovered $100 during this trip. News of the discovery reached California, and by the following Spring the Orofino District complete with mining rules had been set up.

The ultimate rush was only delayed by heavy snows, but by April several thousand miners worked in western Idaho in the Clearwater watershed. The first party of miners to venture south to the Salmon River in 1861 met only hostile Indians. A second group of 23 men found gold on most of the bars in the Salmon. Prospectors moved south to Florence and Warren along the Salmon River, then east to the Lemhi District, and finally southwest to Ora Grande on Loon Creek a tributary of the Middle Fork of the Salmon.

Rumors of lucrative mines at Florence along Slate Creek slowly reached miners all over the Northwest. Claims were supposed to yield $20 per pan, even $1,000 dollars worth of gold in one hour, and $1,600 in a day. While many of these stories were exaggerated, Jacob Weiser's claim on Baboon Gulch produced $3,360 or 60 pounds of gold on November 21 only and $20,000 in eight days. Four men with two rockers worked in the gulch. Weiser sold his claim three weeks later for $10,000. Baboon Gulch was so named because its discoverer was reputed to resemble his distant ancestors. In actuality, Weiser's finds were exceptional. Of the 30 or 40 active claims around Florence, most averaged $25 to $50 a day.

In 1862, bars all along the Salmon River were re-examined and at one point 400 miners were working along 120 miles of the river, even though production wasn't spectacular. One miner's estimated budget was $1,000 for outfitting, $100 for travelling and ferrying, and $100 and one month time to find a claim. Hiring men to help work the claim and living expenses added up to $26,980 total, with a remaining amount of $3,660 as earnings for thirteen months.

The opening of Milner Trail by Moses Milner from Camas Prairie down to Florence had the advantage of being direct and at the same elevation, making freighting faster and reducing food prices. A toll of $1.00 was charged for each horse and rider. A second trail followed along Slate Creek. Both trails were rough and one rider described them by saying "I have packed

Rush to the Idaho mines.

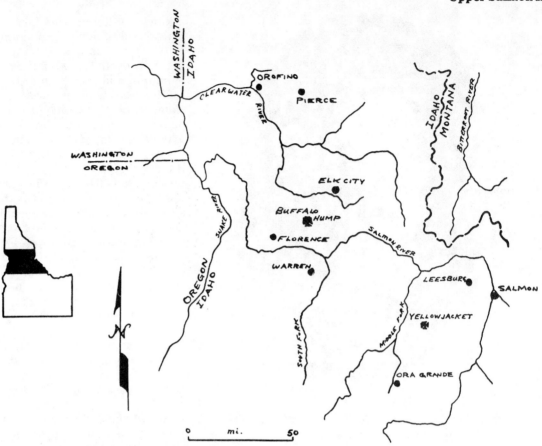

Early mining towns of the Salmon River drainage area.

over both of them, and each time I went over one, I wished I had taken the other."

Flour brought over the new trail dropped in price at Florence from $2 per pound to .50 cents per pound. Sugar went from a high of $2 to a low of .40 cents per pound. Meals at a restaurant cost $52 per week in contrast to $9 per week in a non-mining town.

Murder, robbery, hunger and diseases had to be dealt with. Dr. G. A. Noble gave $5 to a stranger "wasted to a mere skeleton by want and destitution," who was eating a few frozen beans thrown out into the snow. The man later took a job hauling 100 pound sacks of bacon on his back fifteen miles over the mountains into town. One writer said "society here is in a woeful conditioin. Scarce a week passes without a shooting or stabbing affray." Hanging was punishment for a killing.

In the early days, ferries carried miners and settlers across the Salmon. The experience could be far from pleasant as A. F. Brown related "another bad scare I had was the first time I crossed the Salmon River with the pack train between Florence and Warrens. The ferry

boat was small . . . and I went over with the first load. They had to row upstream (because) of the current . . . when the boat struck the current the water rushed over the boat and I thought sure it would sink. I stood in water eight inches deep . . . .but to my surprise, the boat did not sink, but went through the strong current in a hurry and they made the landing."

Work had to be suspended during the bitter winter of 1862 when the Columbia River froze. Miners at Florence tried to keep their equipment thawed out unsuccesssfully with large bonfires. By the March thaw, most miners were sick with scurvy and hopelessly in debt from purchases over the winter. Late snow into May, heavy melting causing floods in the Summer, and water shortages during the Fall discouraged many miners who left Florence for claims elsewhere.

Enterprising prospectors pushed 50 miles south searching for new gold sources. James Warren, a "shiftless individual, a petty gambler, miner and prospector," and a party of 20 men found fine quality gold along Warrens Creek in the Summer of 1862. Two rival towns were started, one called Washington and the other Richmond after the southern Civil War capital. Finds were of uniform richness so the strikes lasted longer than those at Florence, even though these camps never reached the fabulous excitement generated at Florence. Never more than 1,000 people lived here, and most were Chinese. A later quartz strike at Warrens from 1866 to 1868 kept the district in production for several years more.

By 1862 the Florence mines were worked out. "I dare say there are five thousand idle men here, and hundreds of new arrivals daily. About one out of three of these are leaving, swearing that a man could not make his grub" (Wells, p.28). At the peak, 3,000 men were working at claims and production reached $50,000 per day. Today nothing remains of old Florence except building foundations which trace the outlines of the town. Gold discoveries at Buffalo Hump near where the Salmon River turns north proved to be more rumor than fact, and the

amount of gold taken out was negligible.

In 1866 and 1867 further discoveries were made in the eastern Lemhi district. A party of five men from Montana led by F. B. Sharkey found rich gold bearing gravels on Napias Creek in the Leesburg Basin. The Indians called this creek "Nappias" meaning gold.

The Montana Post of August recorded this event. "We were quite unprepared for the stampede that set in yesterday for the Lemhi Valley . . . the tide still rushes on, and unless there is a counter current in a few days our streets will be almost deserted." Ten days later the Post reported the stampede was at an end, and most miners had returned to Montana. A still later story was more optimistic. "Trade is brisk . . . we believe there will be a good camp there next season."

Leesburg, Grantsville, Summit City and Salmon City were founded during this rush. Leesburg was made up of ex-confederate Army soldiers, and Grantsville was named for Ulysses

Mining works

53

S. Grant by Union soldiers. Leesburg was a settlement of twenty-six houses, six stores, and two butcher shops with at least 500 miners. Col. George L. Shoup operated a mining supply store in nearby Challis which was said to do the largest volume of business anywhere in Idaho.

Heavy snows of 1867 and high waters the following Spring discouraged the first wave of eager miners, and many left not to return. However, the population of Leesburg settled down to a stable number of nearly 20 men who were able to make a modest living.

Only snowshoe communication was possible in wintertime, and by February, when supplies were running out, a twelve man shovelling team was organized to dig a trail to Salmon City. With new snowfalls, the project took all of February and a week of March before the trail was clear enough for a supply train and eight head of cattle to come through the twenty foot high drifts. Toll rates were charged at a cost of 75 cents per animal one way.

Boulders in the creeks and river were an ongoing problem, and the year-long construction project of a Napias Creek flume came to naught when big boulders washed down and destroyed the work. Even this disaster didn't bring an end to the Lemhi mines which came with the 1869 discovery of gold on Loon Creek, a tributary of the Middle Fork. This was one of the last wildfire strikes. Hundreds of prospectors searched the creek fruitlessly, but few fortunes were made. The richest Salmon River placers were exhausted by 1874 even though mining continued on for 50 more years.

Indian resistance to these gold seekers and permanent settlers led to the Nez Perce conflict of 1877. The Indians had been ordered onto the Idaho Reservation. Gathering their goods together, families moved northward to camp at Camas Prairie. While there, angry young men of the tribe killed a number of whites in retaliation for wrongdoings in the past.

U.S. Army troops under General Howard were sent in against the Nez Perce led by Chiefs White Bird, Looking Glass, and Joseph. Skillfully able to outmaneuver the Army, the tribe was moving toward sanctuary in Canada when attacked near Bear Paw, Montana. Demoralized by the long trek, the attack and falling snow, most of the Indians surrendered. A small group led by Chief White Bird managed to escape into Canada. (For a complete account, see the chapter on the Nez Perce).

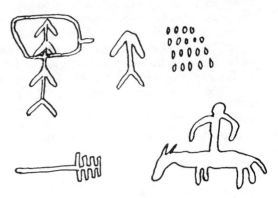

Rock art from Horse Creek.

Indian Life Along the River

The Nez Perce lived in the cool mountains and plateaus during summer months and in sheltered valleys during the Winter. Permanent villages were built near streams and rivers.

Two rockshelters along the Salmon River, seven miles downstream from Shoup, Idaho, were excavated by anthropologists who found evidence of early human occupation going back 8,225 years. In the caves a large number of bird and mammal bones were scattered among freshwater mussel shells. The berry bushes along the river today as well as salmon would have probably been the food supply used by the cave dwellers. Artifacts here included notched points, choppers, worked pebbles, and bone implements.

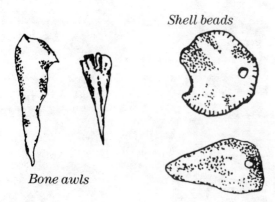

Shell beads

Bone awls

Geology of the Upper Salmon River

The Salmon River system stretches clear across Idaho separating the panhandle from the rest of the state. The river drops 7,000 feet in only 425 miles as it traverses the state, with an average drop of about twenty feet per mile. Several of the falls along the river are unrunnable.

Much of the river follows a zigzag course in response to a pattern of faults and fractures. The topography can be described as mature or at maximum relief with small meanders at the bottom of steep canyons. There are little or no flat uplands which have not been dissected by erosion. Along the meanders of the river, bars of sand and gravel occur at every turn. These sediments move downstream a few feet each year to make their way to the Snake River.

In the middle 1800's when gold was being mined along the Salmon, most supplies to the miners were moved down the river by barges. These early attempts at navigating the Salmon met with varying degrees of success. In one memorable episode a paddlewheel steamer put in at Salmon City barely lasting through the first small rapid before being smashed on the rocks.

The westward run of the Salmon begins at North Fork in rocks which are billions of years old (preCambrian). As it flows east to west, the river cuts through granite rocks of the Idaho Batholith. Batholith, literally meaning "deep rock", describes a large mass of crystalline, intrusive rocks which cooled far below the ground surface. As the batholith magmas melted their way upward into the existing rocks (in this case, the preCambrian rocks), they left isolated patches of them exposed on top of the batholith. These remants of the original rocks are seen today as roof pendants, or pieces hanging down into the younger batholith rocks.

As the batholith cooled down from a molten state enormous quantities of heat were released into the adjacent rocks. This heat cooked and altered the surrounding rocks by a process called metamorphism. The amount of alteration depends on the amount of heat and time of exposure. Many of the older preCambrian rocks along the main Salmon were raised to a near molten state.

Traversing canyons cut through the granites of the batholith it is well to keep in mind that, although it is now exposed to view, 70 million years ago the rock cooled perhaps as much as a mile or two under existing rocks. The batholith's granites were then cooled and hardened under high pressures.

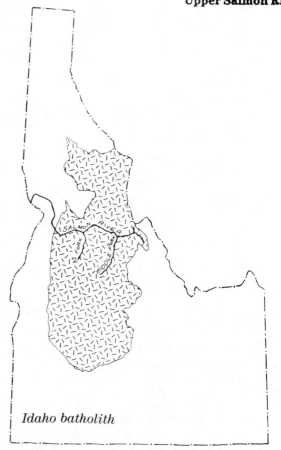

Idaho batholith

As erosion slowly stripped off the overlying rocks, the granite expanded or rebounded. The pressure has been released as a network of faults and joints through the rock. Two distinct sets of rock fractures and faults dominate this section of the river trending northeast by southwest and northwest by southeast. Combined, these fractures yield a crisscross pattern of fractures to the local rocks. As it travels over and erodes these hard resistant rocks, the river is quick to exploit any small crack or crevice to further the erosion process. The zigzag trend of the river over the section from Cunningham Bar to Vinegar Creek follows this structural trend of faults and joints.

In the interval between Myers Creek and the South Fork, the main Salmon follows a prominent fault for more than twenty miles meandering back and forth across the old fault trace. This fault is enormous and may be traced beyond the canyon of the Salmon over a distance of nearly 100 miles in an arcuate pathway from southwest to northeast.

Somewhat younger geologic features along the river are terrace deposits of sand and gravel that were literally left "high and dry" in the canyon walls as the river started downcutting elsewhere. These terrace deposits probably

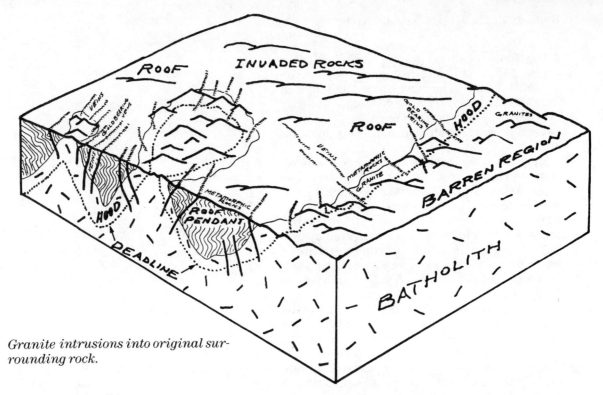

Granite intrusions into original surrounding rock.

have gold in them but were left untouched because of the difficulty of getting water up to them to wash out the gravels. This same problem was solved in the California Sierras at great effort and destruction of the environment by using elaborate flume and tunnel systems to divert water. In some of the California operations, water was transported dozens miles to the gold washing sites.

Idaho was largely spared from this kind of effort for two reasons. The Salmon drainage is very deeply incised or entrenched due to rejuvenation and renewed intensive downcutting. The river and its tributaries are at the bottom of steep-walled canyons hemmed in by hard rock. Only extensive tunneling would sufficiently divert streams for hydraulic gold mining. In addition, assays of most Idaho ores have been far too low to justify this type of operation.

As gold mining proceeded in this area during the middle and late 1800's, the occurrence of gold in veins began to assume a determinable pattern. Just north of Shoup, for example, the edge of the batholith has been mapped, and the change from granite to metamorphic (heat treated) rocks is subdivided into the "deadline", "hood," and "roof". It was found that lode gold in quartz veins almost never occurs below the deadline out in the granite batholith rocks. Instead, the gold is concentrated within a mile or so of the granite-metamorphic rock contact or in the "roof" area. Even as crude as

these early observations were, they provided vauable guidelines to exploration activity and were infinitely better than simply blindly prospecting for gold anywhere in the millions of acres of Idaho.

Many of the early gold prospects were grossly overpromoted by stock companies but never really checked out. It has often been noted that more individuals got rich off Idaho gold by selling stocks than by actually mining. Early valuable discoveries of bismuth and selenium, and other metals, were largely ignored in favor of what little gold could be squeezed from the area.

The principal mining area from North Fork to Cunningham Bar is the now largely abandoned district near Shoup. Gold was discovered here in 1883 and mined until about 1910. Late in Shoup mining history, copper was also produced. In all, roughly three quarters of a million dollars in gold came from the vicinity.

Gold returns in the area west of Cunningham Bar have been remarkably poor, but not because no one has looked. About one-third of Idaho's gold has come from an area cutting across Salmon River drainage from southwest to northeast just at the western margin of the runnable stretch. This gold trend includes the communities and districts of Pierce, Orofino, Elk City, Ora Grande, Ten Mile, Dixie, Florence, Buffalo Hump, Warren and Burgdorf.

In 1898, the Buffalo Hump, then spelled Buf-

falow, gold stampede began with the discovery of high grade ore at the base of a prominent middle Idaho physiographic feature known by the same name. Within six months 5,000 miners were prospecting here. Because very little gold was "free" it had to be processed on site requiring heavy, expensive crushing and refining equipment. With no railroads and only crude roadbeds, most supplies came by mule. As logical realities began to overtake the miners' high hopes, claims were sold, resold, and sold again at ever diminishing prices. By 1903 most of the activity in the area had ceased.

The Salmon has been thoroughly prospected by placer mines and evidence of this is visible at nearly every meander and bar. Old placer workings and pits, miners cabins, ruins, roads, and occasional equipment dot the river from end to end. Much of the placer activity focused upon locating "blacksands" and attempting to exploit them for the small quantities of gold they contained. Blacksands are composed of darker, heavier minerals magnetite and ilmenite. The original source of these minerals may be the granites of the Idaho Batholith. Within the watershed of the Salmon, silver, copper, lead, zinc and tin have been located in addition to gold, but never in very large scale, economic deposits.

Although gold was first discovered here in 1860, the greatest placer mining activity took

Buffalo Hump, scene of early gold mining activity.

place after 1930 at the height of the great depression. In spite of intensive activity, it has been estimated that only $50,000 in gold (in 1930 dollars) was extracted over this entire route. If that amount seems small for all the energy expended, consider that the U.S. Geological Survey's best estimate of the remaining gold in this area of the Salmon is placed between $25,000 and $50,000 (in 1972 dollars). Historically access has been a major problem to modern mining. It seems doubtful that the Salmon area will ever see large scale dredging operations such as occurred in California along Sierra streams in the middle 1800's.

Block diagram of the upper Salmon River geology.

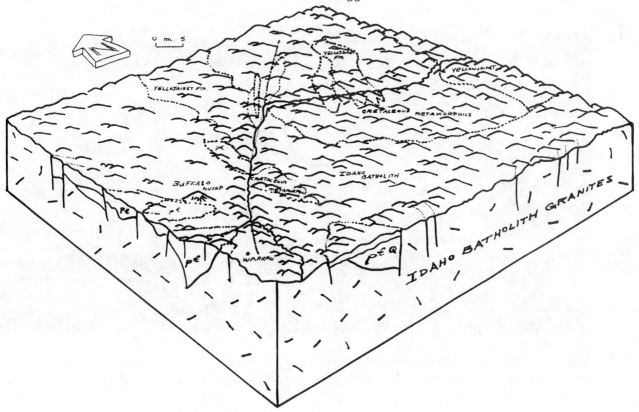

Geology Along the Route

Beginning at North Fork, the rock is largely early preCambrian quartzites. About five miles into the river, rocks of the Idaho Batholith are apparent as gneiss and granites. Gneiss is the product of intense heat and pressure applied to the degree that the rock began to flow like plastic but did not melt completely. Gneiss is readily identified by the characteristic separation and banding of light and dark material in the rock. The wavy and contorted bands of the gneiss are evidence of the semi-molten phase. These rocks are highly fractured and intruded by dike swarms. Faulting and fractures are also obvious in the canyon walls along the route.

Eighteen miles downriver from the put-in, the old mining community of Shoup straddles the "hood" area between the Idaho Batholith and preCambrian roof pendants. Dikes here contain gold, and placer mines along the river have been worked over extensively.

From Shoup to Cunningham Bar, the river remains in gneiss and related preCambrian rocks. In this twenty-five mile stretch, the river's course is controlled by joints and faults. This is particularly true just southwest of Shoup. Straight river stretches and sheer, flat canyon walls provide indisputible evidence of the structural control.

Rafting west from Cunningham Bar, the first rocks seen are known as the Yellowjacket Formation, a mass of very old preCambrian rocks. The Yellowjacket is primarily sedimentary rocks that have been thoroughly metamorphosed (heat treated). This series continues along the river into coarse-grained, banded gneiss which dominates over the next twenty to thirty miles.

Only two to three miles above Barth Hot Springs, the river cuts through the preCambrian rocks and into the underlying batholith. These grey, crystalline, jointed granites stand out in sharp contrast to the metamorphic rocks elsewhere along the route.

Running back into gneisses, the river travels only a distance of nine miles before making an almost 90 degree turn southwest to follow a major fault. At this point near the confluence of Myers Creek and the Salmon a very large and spectacular fine-grained, light colored rhyolite dike cuts diagonally southwest from Boston Mountain for a distance of almost ten miles to where Mallard Creek enters the Salmon.

Running along the faulted zone through gneisses, the river follows a southwest meandering course for nearly twenty miles to the South Fork where it again swings 90 degrees to a northwestward course. In this interval it crosses the Idaho Batholith granites where the best exposures yet can be seen. Near Vinegar Creek, preCambrian roof pendants finish the route.

Texture of the rock gneiss. Intense heat has partially remelted the rock. Minerals have segregated themselves into semiplastic bands which have been bent and folded.

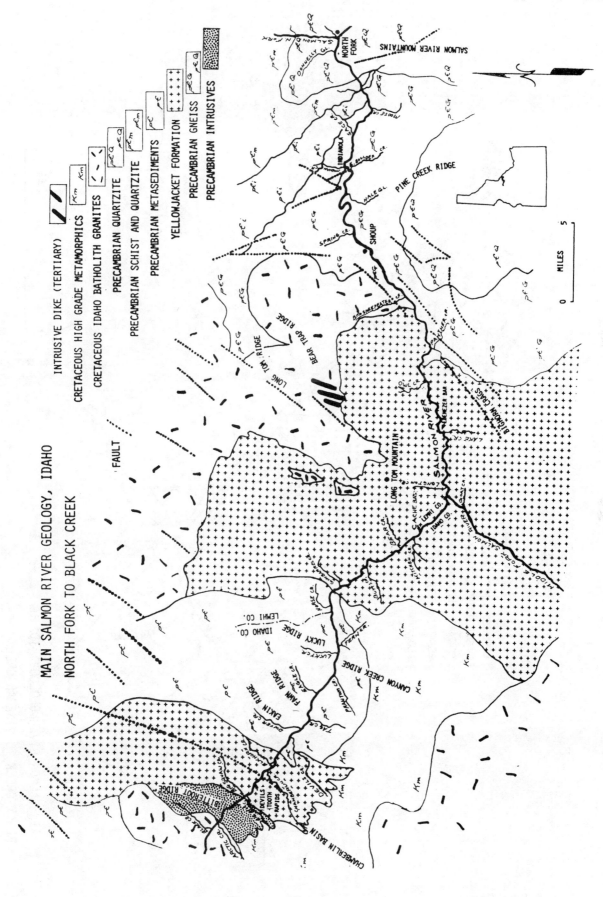

MAIN SALMON RIVER GEOLOGY, IDAHO
NORTH FORK TO BLACK CREEK

INTRUSIVE DIKE (TERTIARY)
CRETACEOUS HIGH GRADE METAMORPHICS
CRETACEOUS IDAHO BATHOLITH GRANITES
PRECAMBRIAN QUARTZITE
PRECAMBRIAN SCHIST AND QUARTZITE
PRECAMBRIAN METASEDIMENTS
YELLOWJACKET FORMATION
PRECAMBRIAN GNEISS
PRECAMBRIAN INTRUSIVES

FAULT

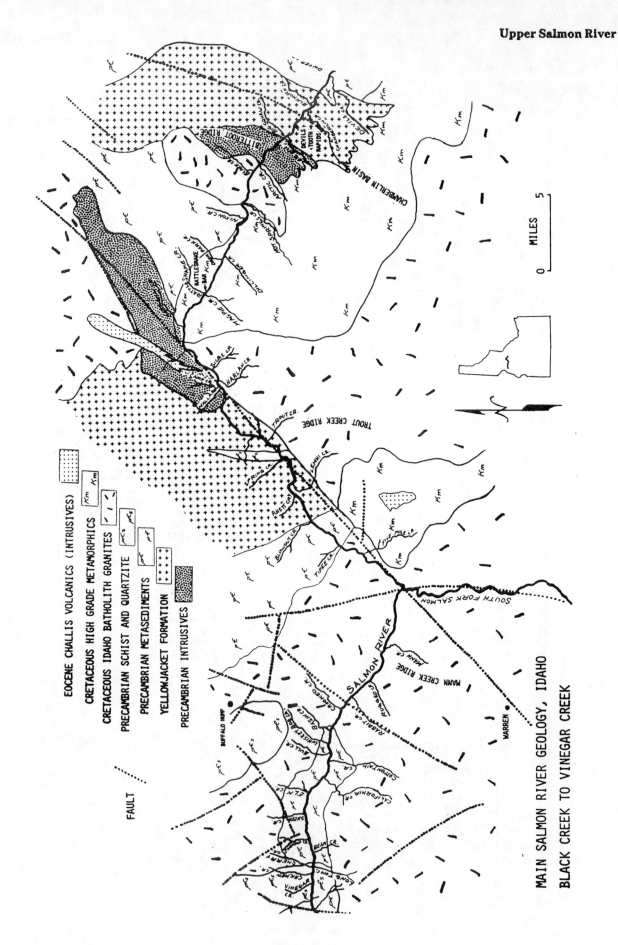

EOCENE CHALLIS VOLCANICS (INTRUSIVES)

CRETACEOUS HIGH GRADE METAMORPHICS

CRETACEOUS IDAHO BATHOLITH GRANITES

PRECAMBRIAN SCHIST AND QUARTZITE

PRECAMBRIAN METASEDIMENTS

YELLOWJACKET FORMATION

PRECAMBRIAN INTRUSIVES

FAULT

MAIN SALMON RIVER GEOLOGY, IDAHO
BLACK CREEK TO VINEGAR CREEK

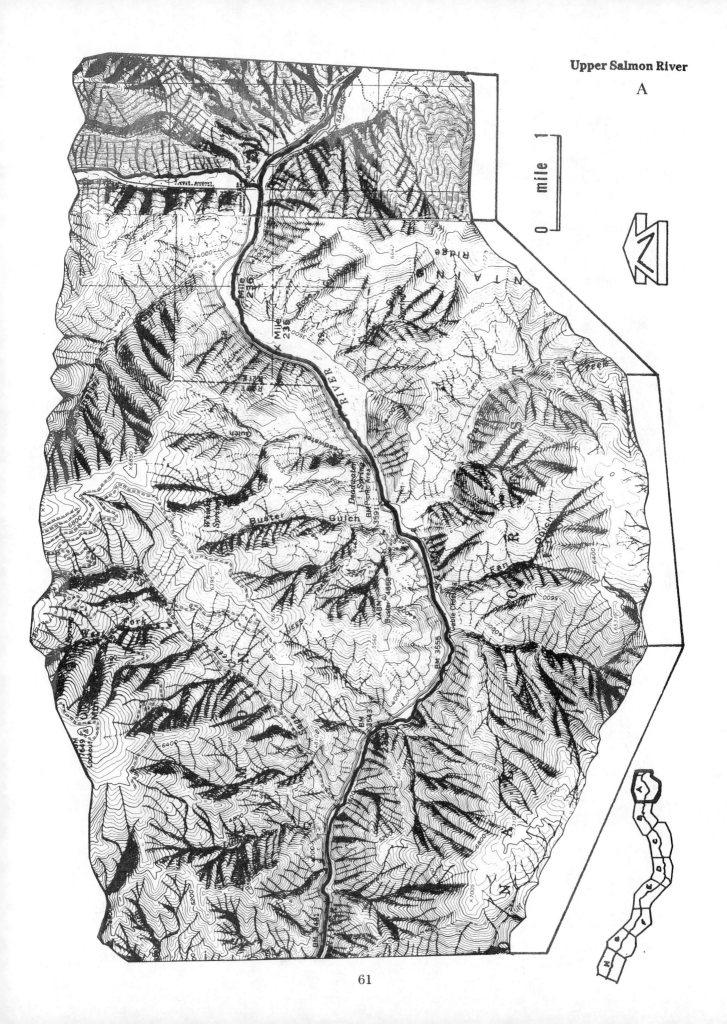

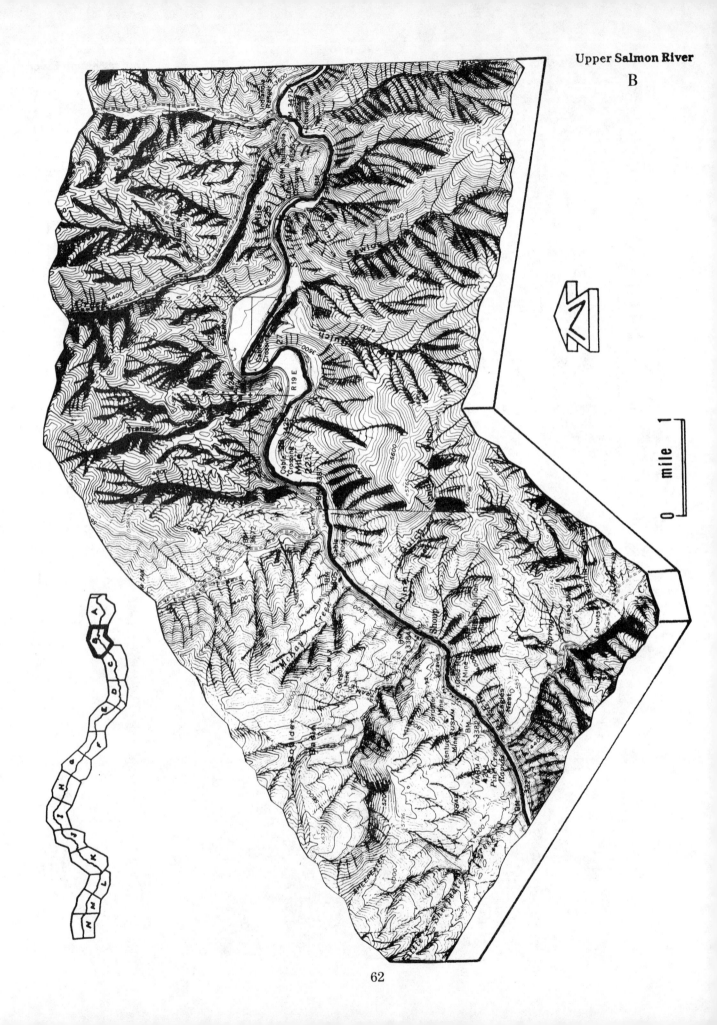

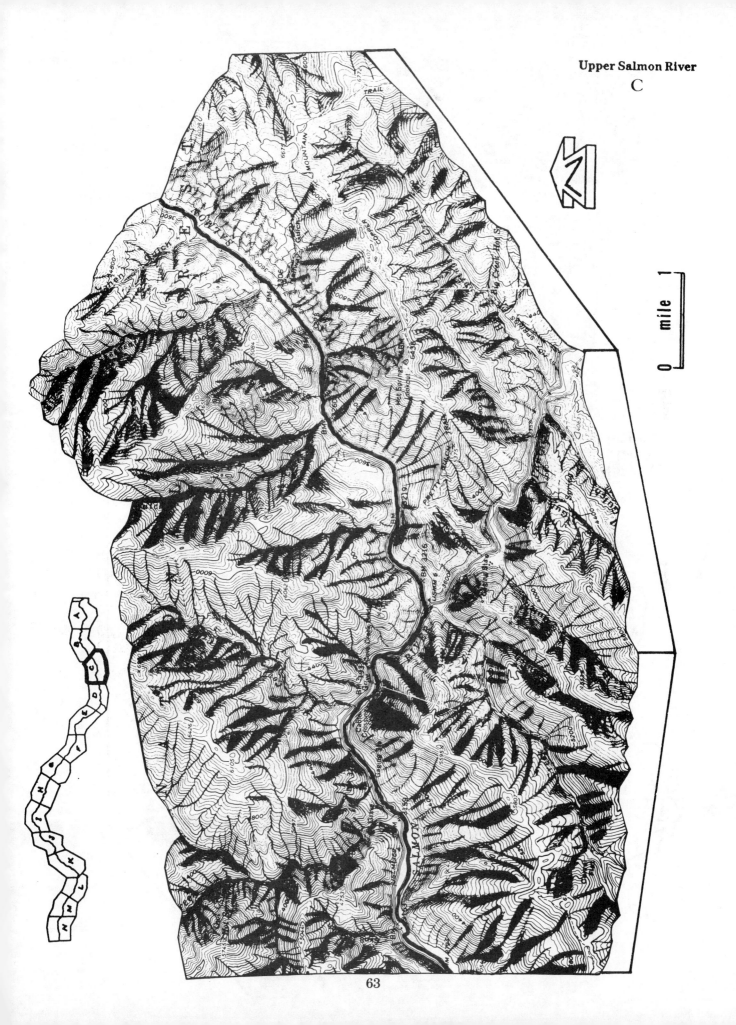

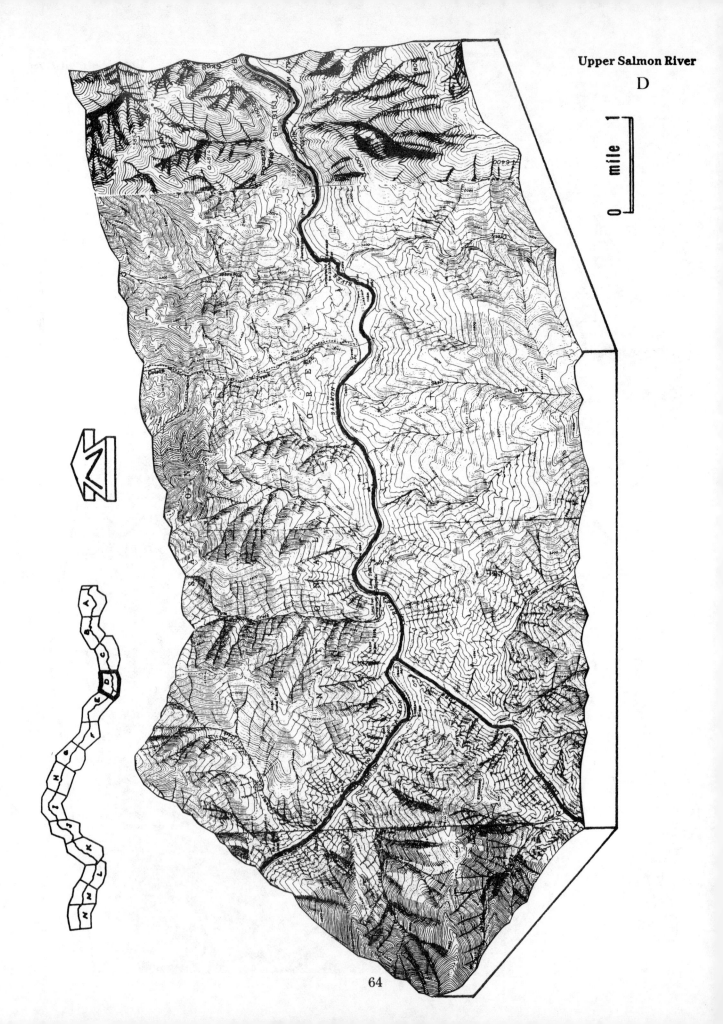

0   mile   1

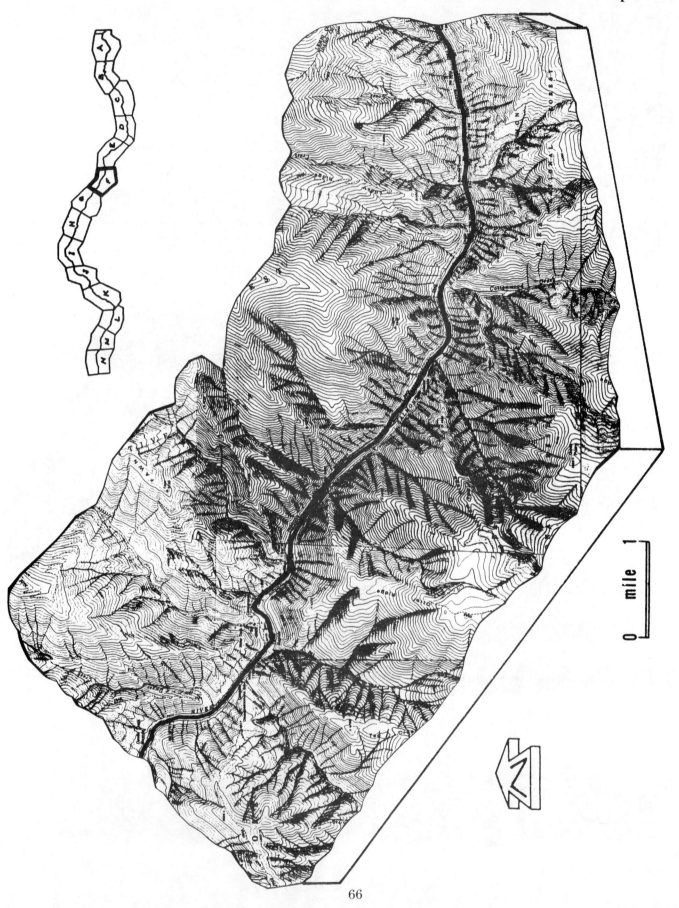

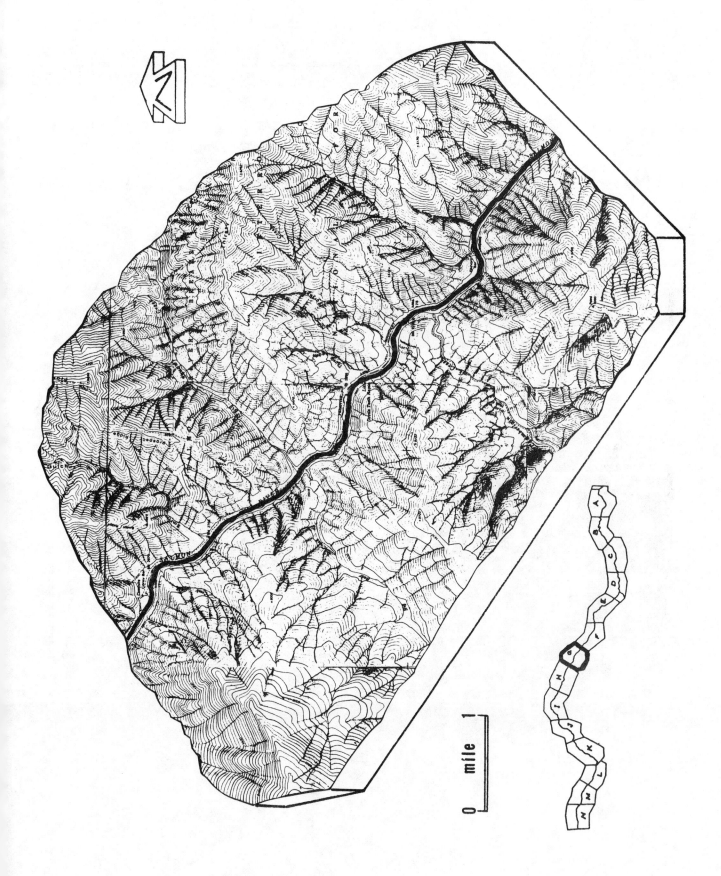

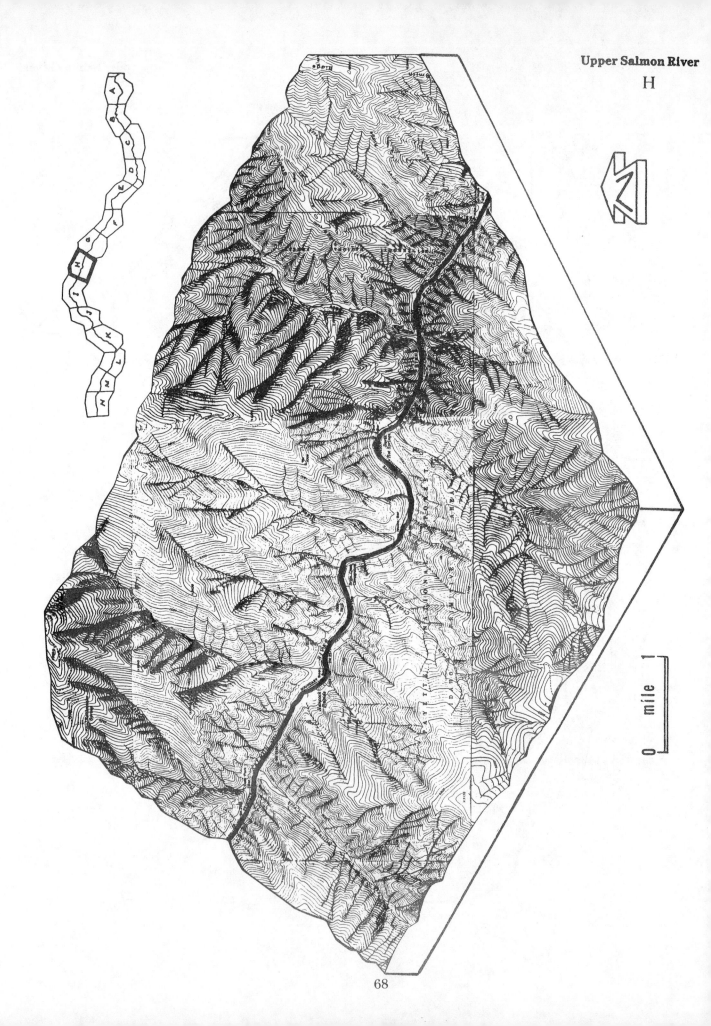

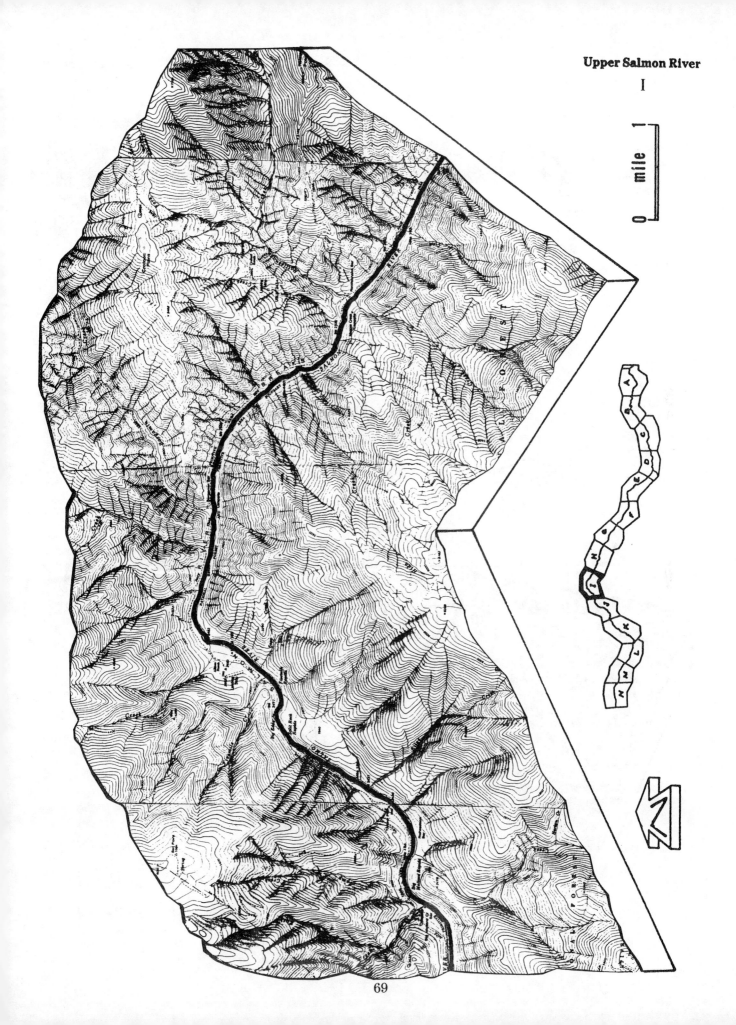

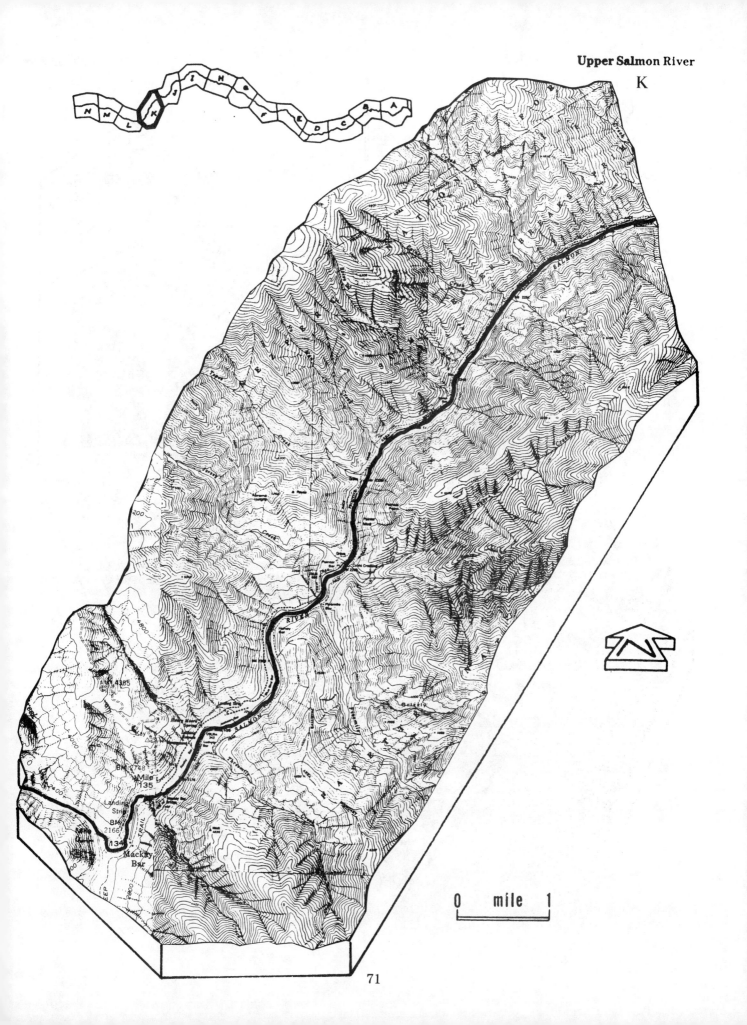

0 mile 1

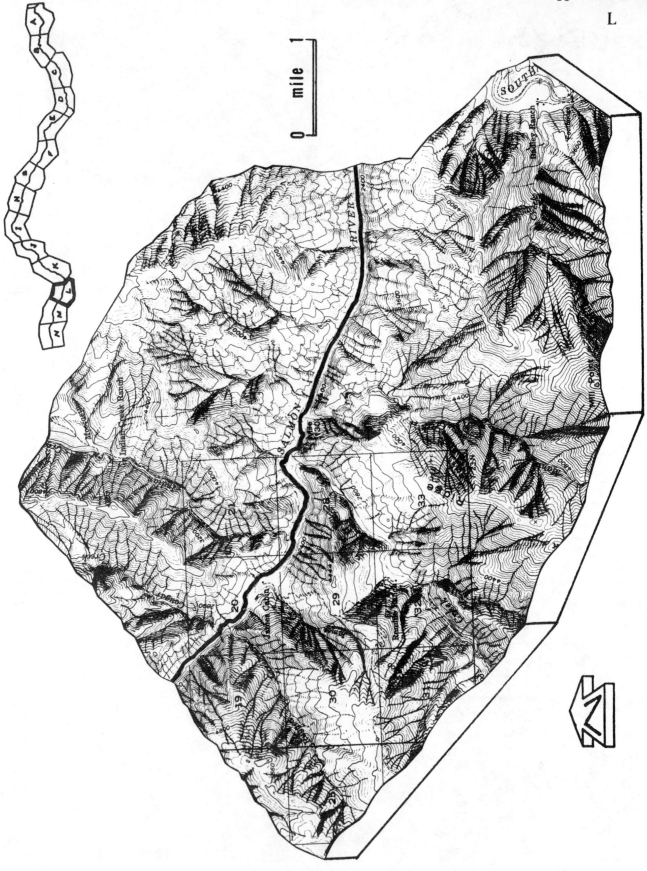

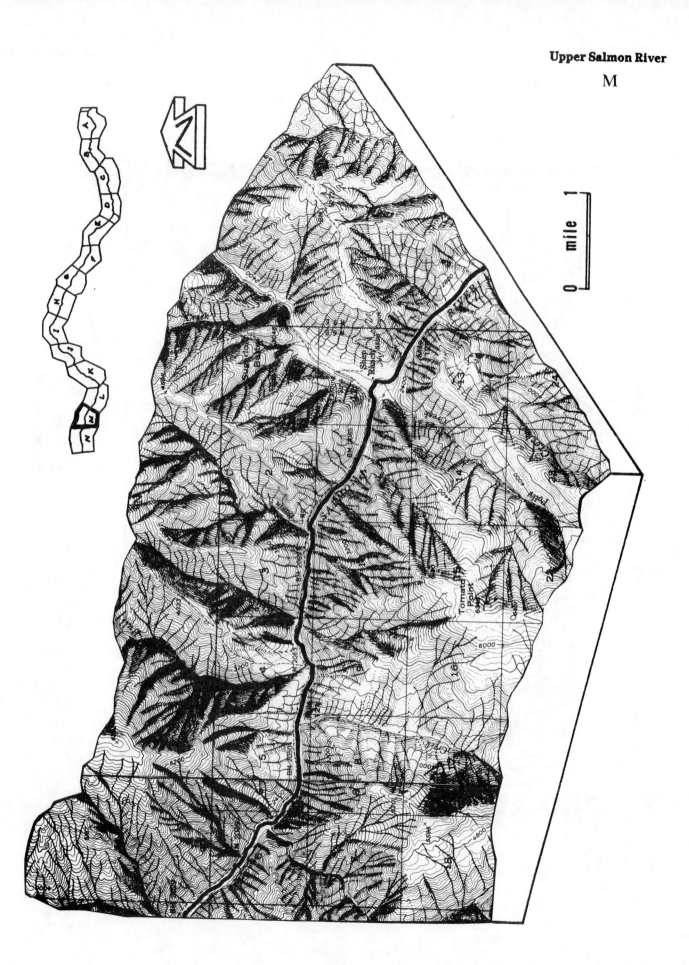

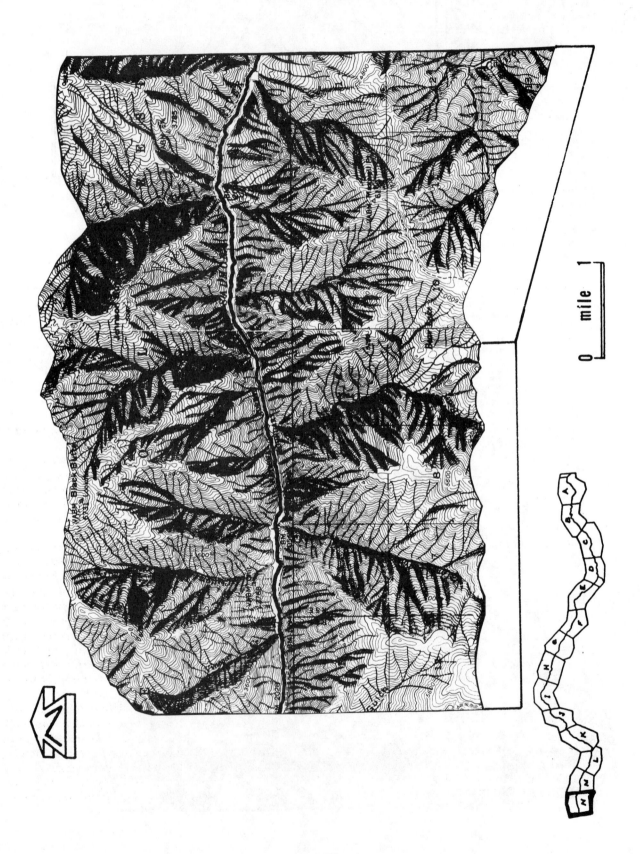

Oregon Rivers

Deschutes River

(Warm Springs to Columbia River)
Jefferson, Wasco Counties, Oregon
Length of run: 105 miles
Number of days: 4-8

One of America's most unique rivers, the Deschutes begins in the eastern Cascade Range of southcentral Oregon and reaches the Columbia River 250 miles later. The even, yearlong stream flow, so unusual for a river, is due to snow melt which is trapped in the sponge-like lava to be slowly released year-round.

Narrow canyons, rapid waters and numerous waterfalls abound along the Deschutes, called the Riviere des Chutes, river of the falls by French explorers. Charles Preuss, a cartographer with John Fremont's expedition, commented "I believe there is hardly a country where one sees as many (waterfalls) as here. As soon as we pitch our camp at a river, we hear the din of some larger or smaller waterfall."

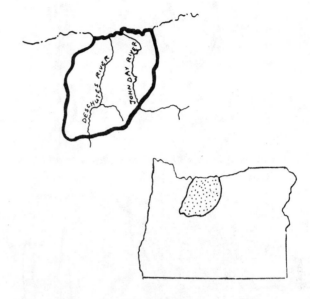

Distribution of Tenino Indians along the John Day and Deschutes rivers of Oregon.

Early History of the Deschutes River

In times past, both the Deschutes and John Day valleys were the home of the Tenino Indians. They lived along the south bank of the Columbia River from Celilo to Arlington, Oregon, and south along the Deschutes and John Day rivers. The Tenino, numbering about 1,200 were well-adapted to river life. Here they fished and dried their catch in temporary racks attached to the houses. Salmon was trapped as well as trout, eel, and whitefish. A variety of fishing devices included weirs, dams, funnel and hoop traps, open-top baskets, nets with wooden floats and stone sinkers, hooks, harpoons, and poison.

Hunting didn't occupy such an important place in their activity, and even though they held group hunts for deer, elk, bear, and other game, more time was spent gathering plant foods. Camas and kouse roots were dug during the Spring and nuts and berries into the Fall. After food had been gathered and dried, summer houses were moved to protected areas of the river valleys where permanent winter villages were reused each year. A household was organized around the family consisting of a man, his wife and children as well as another man, either a brother or son, his wife and children. Once a house became crowded, married children moved out.

Funeral ceremonies were observed by both feasting and weeping. Bodies were painted, wrapped in woven matting, occasionally with valuables, and placed in gravel or river banks. Burial some distance from the village was preferred.

The Tenino were singularly fortunate in living near The Dalles on the Columbia River. The scene of constant activity, The Dalles was at the crossroads of trade in early times, and the Tenino were able to control trading traffic here for centuries. Frequently items were brought over an extensive series of trails from as far away as northern California and the Pacific Ocean. Well used Indian trails running north connected the John Day Valley directly to the Columbia River.

"From the east came buffalo robes, horses, and meat; the Klamath to the south brought elk skins and beads, the Wenatchee and Klikitat of Washington brought goat hair robes, slaves, meat, nuts, and berries, while the coastal Indians came upstream to trade oysters, wappato, and trade goods obtained from the whites" (Toepel, Willingham, Minor, 1980).

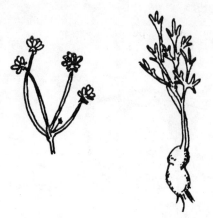

Kouse or cous, biscuit root, harvested by the Indians.

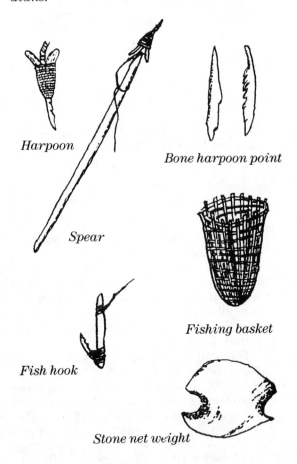

Harpoon

Bone harpoon point

Spear

Fishing basket

Fish hook

Stone net weight

Tenino were naturally friendly and intermarried with their neighbors. The exception to this seemed to have been the Northern Paiute who lived to the south and east of the John Day River basin. Perpetual feuding occurred between these two tribes, although the Northern Paiutes had the reputation for not getting along well with anyone. These feelings may have stemmed from Tenino raids on the Paiutes for slaves to be traded at The Dalles. Historically problems arose when the Tenino pushed the Northern Paiute out of a part of their territory west of the Deschutes sometime around 1805. This was about the same time Lewis and Clark were passing down the Columbia on their way to the coast.

Lewis and Clark didn't stop long to explore. Their journals record the tremendous impression the "Great Falls" at Celilo made on them. The Falls dropped 47 feet from "a high black rock, which, rising perpendicularly from the right shore, seems to run wholly across the river . . . to meet the high hills of the left shore, leaving a channel only 45 yards wide through which the whole body of the Columbia must press its way."

Fur trappers from Hudson's Bay Company at Vancouver, formerly owned by Jacob Astor, explored and trapped along river valleys in eastern Oregon reaching the Snake River by the 1830's. Bunch grass, growing abundantly along the Deschutes, was noted frequently, and one trader, Nathaniel Wyeth, commented that the area could produce "hides, tallow, beef, and wool." The quixotic explorer, John Fremont, working for the U.S. government, praised the grass and soils on his journey down the Deschutes to the Klamath in 1843. Eventually reaching California, Fremont was accompanied by Chief Billy Chinook of the Tenino who went on to Washington, D.C. Most travellers, anxious to reach the Willamette Valley, didn't stop longer than was necessary in the arid country east of the Cascades.

For the most part, permanent American changes were in the form of roadways and military forts. A military post was established at The Dalles to provide protection for emigrants passing through on the Oregon Trail. Wagons taking the Barlow cutoff, a short cut, were ferried across the Deschutes just above Sherars

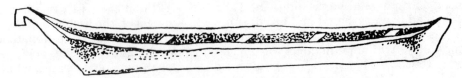

Chinook-style canoe, about 30 to 40 feet long.

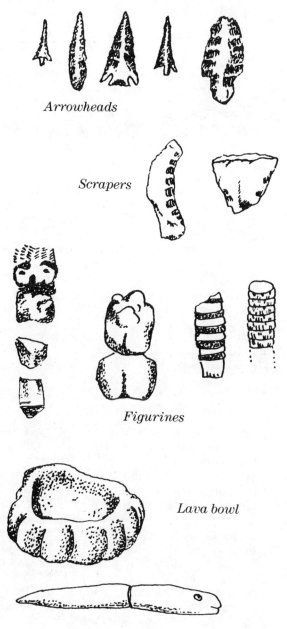

Arrowheads

Scrapers

Figurines

Lava bowl

Carved eel, possibly a fishnet weight.

Settlers were coming to take advantage of Indian lands opened up to them by the federal government. Land was virtually given away away at rates from 16 acres for $1.25 per acre to 80 acres for $2.50 per acre. One hundred and sixty acres were given free to anyone who would homestead for five years.

With land in the Willamette Valley becoming more crowded, sheepherders and cattlemen took advantage of open range east of the Cascades. The first large herds of livestock arrived in 1857. Settlers built cabins and wintered large herds on the plentiful grass of the plateau before sending their animals to the spring market. In 1862 Felix Scott and his brother, Marion, drove 900 cattle over the McKenzie Pass from the Willamette Valley and spent the winter in a cave on Hay Creek near Madras.

In 1861, the discovery of gold at Auburn in Baker County as well as on the North Fork of the John Day brought even greater changes to eastern Oregon. Gold prompted the development and improvement of roads, railroads, and towns. The army spent considerable time looking for better military routes from the Great Salt Lake to the Columbia River. Col. H.D. Wallen's mapping expedition of 1859 identified the east bank of the Deschutes as the best north/south route. This route crossed the lower Deschutes and turned east to the John Day River before going into Idaho. Lt. John Bonnycastle reported "I regard (this) as a very good one for wagons . . . I brought the entire train (17 ox wagons and 6 six-mule wagons) in 12 travelling days."

The most controversial east/west trending road through Oregon was the grandiose sounding Willamette Valley and Cascade Mountain Military Wagon Road, known as the WV&CM Road. The road went from Albany and Lebanon in the Willamette Valley, over the Santiam Pass, across the Deschutes River and eastward. In 1866 Congress granted two military road companies thousands of acres of land in alternate odd numbered sections on each side of the road for a distance of three miles. Settlers were irate at losing land they had just taken from the Indians. Under this contract, roads were never built. Some remained mere trails, while bridges weren't constructed over rivers. Investigation of the fraud was carried to a hearing by the Supreme Court where it was dismissed on a technicality.

In 1910-1911 the Oregon Trunk Railroad owned by James Hill laid a line up the west bank of the Deschutes at the same time that Deschutes Railroad owner E. H. Harriman built up the east side in a race to extend track up the Deschutes canyon. Hill and Harriman were engaged in a

Bridge. The bridge was built in 1860 by John Todd who later sold it to Joseph Sherar. Sherar improved and operated the bridge for 40 years. Because this bridge was kept in good repair, it influenced overland road routes. A ferry was also operated over the Deschutes by Howard Maupin at the present day site of Maupin.

After crossing through eastern Oregon, immigrants were happy to see the water of the Deschutes even though they complained it was "too cold to do us any good, so we carried some up in our hats and cups, built a fire and warmed it before we were satisfied."

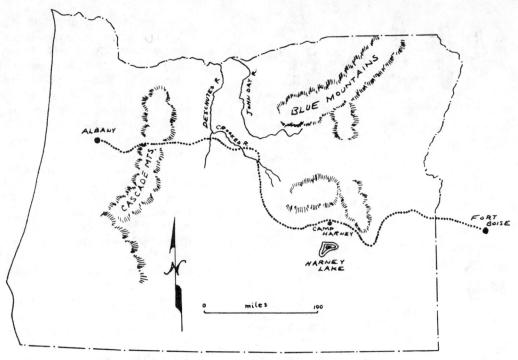

Willamette Valley and Cascade Mountain Military Road.

fierce ongoing competition to outdo each other. Gangs of workers blasted rock and dug tunnels along the riverbank each side striving to win. Occasional shots were even fired back and forth. At Madras only one line could pass through the canyon, so the federal government stepped in to enact the Canyon Act which forced both sides to combine and build only one line. Hundreds of workers and horses helped in the construction, and the railroad reached Bend, Oregon, in 1911.

As early as 1852, Oregon Superintendent of Indian Affairs, Joel Palmer, had the idea of establishing a territory east of the Cascades where Indians from all over Oregon could share land isolated from whites. His plans were frustrated until 1855 when a reservation was set up south of The Dalles along the Deschutes. This was the Warm Springs Reservation, and in the agreement Indians of the Columbia Plateau ceded their lands for those on the reservation and the right to fish in the accustomed places.

Although Palmer was enthusiastic about the reservation, the Indians who had to live there were apprehensive. Farming was a different way of life for them. Land was poor and supplies were short. Tilled land went unused, and Indians left the reservation for long periods of time.

In 1859 the Northern Paiute raided the reservation for horses and supplies. The unarmed reservation residents asked for government protection. In response, Dr. Fitch, in charge of the agency, accompanied by 40 armed Warm Springs Indians, ambushed a Paiute camp near the John Day killing ten. Paiutes continued to harass the reservation until a major expedition was mounted against them in 1864. Volunteers under Capt. John M. Drake were accompanied by Tenino guides, Stockietly (Stoc Whitley) and Simustus, whose wife and daughter had been abducted by the Paiutes. Stockietly was killed in a skirmish, and the Paiutes, led by Chief Paulina, escaped. It wasn't until 1868 that the Paiutes were finally subdued.

The Tenino moved back to the reservation where they settled permanently, learning to cultivate, build houses and attend schools. Older customs were dropped, and their name was changed to Malila, which means "warm springs."

Indian Life Along the River

A short trip south to Round Butte near the confluence of the Deschutes and Metolius rivers is well worth the time. Round Butte used to be the site of a large Indian village, and 48 separate house were found here. These contained extensive collections of obsidian artifacts and tools. Freshwater clam shells of *Margaritifera*

Pearl mussel (Margaritifera).

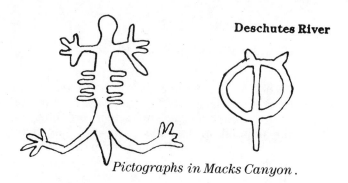

Pictographs in Macks Canyon.

were abundant. The shell was crafted for jewelry, spoons, and knives. The unusually wide variety of artifacts has led to speculation that the west bank of the Deschutes was occupied by a group with a different culture than those on the east bank. A small number of dwelling places were found in hidden caves or lava tubes dating as far back as 7,900 years.

A second stop should be made at the junction of Macks Canyon and the Deschutes in Sherman County where 29 housepit depressions are to be found on a terrace above the river. Numerous artifacts — arrowheads, knives, scrapers, mortars, bone antlers, and beads — recovered by archaeologists led to the conclusion Indians lived here from 7,000 years ago to the first appearance of American settlers. A wide variety of animal bones show the village was a winter site. Red painted pictographs of ''lizards'' and animals with horns have been drawn on a cliff overlooking the creek.

Rock art in Jones Canyon.

More than 100 other archaeological sites along the Deschutes in Jefferson county have been examined. Most have flakes, shells, and some pictographs. Many houses were placed along the Columbia, especially near Celilo Falls. On the upper Deschutes, there were villages near the hot springs on the Warm Springs Reservation and near Sherars Bridge which was considered one of the best all-year fishing spots.

Petroglyphs can be found near Sherars Bridge. These designs are cut or pecked into volcanic rock and represent human figures, spoked wheels, a face, sheep horns, and a figure carrying a rifle. Unfortunately many of these drawings have been damaged by vandals.

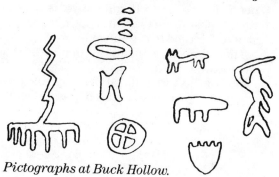

Pictographs at Buck Hollow.

Three other pictograph locations along the Deschutes River display curious designs. At Buck Hollow Bridge, just north of Sherars Bridge, 25 groups of red and white pictographs have a variety of elements representing human figures, circles, plant forms, dots, bow and arrows, a quadruped, and geometric patterns. Human figures, ''lizards'', a bird, lines, and dots have been painted with red pigment on the cliff above Rattlesnake Canyon north of Maupin. In a nearby cave on the south side of Jones Canyon Creek, red and black pictographs of what might be bird figures, wheel spokes, and human figures have been preserved in good condition although the bank in front of the cave was washed out by a flood in 1964.

Pictographs in Rattlesnake Canyon.

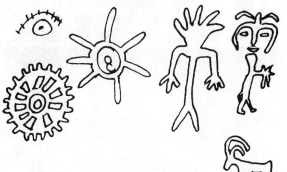

Rock art at Sherars Bridge.

Geology of the Deschutes River

The Deschutes River is run in two sections from Warm Springs to Maupin and from Maupin to the Columbia River. Owing to its proximity to the population center of Portland, this river sees the heaviest traffic of any northwest river. Geology along the runnable stretches of the river is variable and well-exposed to the rafter. Most profound are the variety of topographic features cut into the numerous rock formations by the river.

In the vicinity of Warm Springs, rock layers are nearly flat, and the bed of the river flows in the John Day Formation. Nice exposures of this formation in the canyon wall are visible from the river. It appears as a distinctive, crumbly-layered, buff to tan rock exposed in bluffs. Although the John Day Formation elsewhere in Oregon is known to range from Oligocene to Miocene in age (35 to 18 million years old), these rock exposures in the Warm Springs area are restricted in time to the Miocene.

The John Day is the most famous fossil bearing rocks series in the Pacific Northwest. Sediments of this formation are largely volcanic ash swept up by streams and redeposited as flood-plains and riverbeds along with rich accumulations of fossil mammal bones and leaves. Just above the light, buff colored John Day Formation near Warm Springs is a prominant mesa of thick, dark lava. These are known as the Yakima basalts belonging to the Columbia River lava series. At the top of this sequence are sediments of the Pliocene Dalles Formation which are in turn topped by Pliocene lavas.

Only a few miles downriver, the Columbia River lavas capping the buttes and mesas are left behind as the river enters a wide valley of mostly John Day Formation exposures. On the hills east and west, small outcrops of the unit appear as buff colored scars against the rolling hills and gentle valleys typical of the John Day erosional topography. Only 15 miles from Warm Springs, the river cuts an extensive landslide area known as the White Horse Rapids. The hummocky irregular topography is the hallmark of landslide terrain.

With the right amount of moisture and slopes, the John Day Formation develops landslides commonly along its length. Landslides are typically the sites of large rapids where debris being fed into the river channel clogs the stream creating higher stream velocities and rapids.

Just past the mile long sequence of White

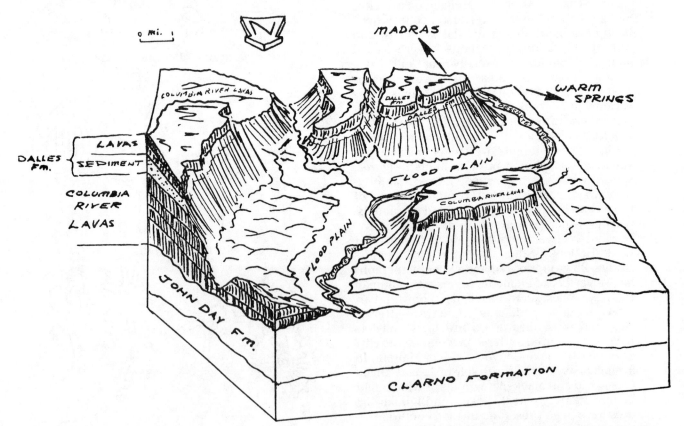

Stratigraphy along the Deschutes River in the vicinity of Warm Springs.

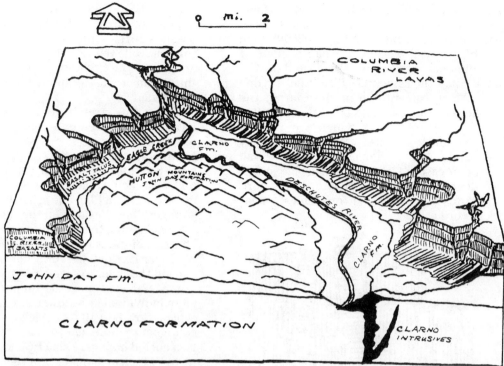

Stratigraphy of layered rocks along the Deschutes River in the vicinity of the Mutton Mountains.

Horse Rapids, the river meanders sharply to the east and cuts into the core of a volcanic rock which has intruded or perforated the Clarno Formation from underneath. The Clarno includes the oldest rocks exposed along the Deschutes, and the contact between the John Day and Clarno Formations can be seen here. Due west of this area, the prominent Mutton Mountains are an uplifted section of the earth's crust and are composed of sediments and volcanic rocks of the John Day.

A prominent rock within the Clarno Formation is rhyolite ash flows. Even on the skyline, the reddish and pink ash flows are easily identified because of their layered structure. Rhyolites, like basalts, include lavas but their similarity ends beyond their common volcanic origin. Whereas basalt is a very high temperature runny flow rock, rhyolite is still molten at substantially lower temperatures. Flows of rhyolite typically thicken and thin over very short distances as the viscous lava piles up upon itself in slow moving flows.

Ash flows of incandescent volcanic debris are also characteristic of this stretch of the river. The flows result from a particularly violent eruption of ash and gas. The resultant cloud is a fiery mass suddenly belched into the sky. Because the mass is heavier than air, it sinks and

flows along the ground for several miles before settling. The John Day and Clarno age landscapes both experienced these types of destructive ash flows.

The Clarno Formation is readily distinguished from the John Day. While the latter is mostly buff to tan sediments, the former includes abundant volcanic rocks displaying a beautiful array of yellow, red, maroon, and very characteristic olive greens. In addition, the John Day erodes more easily and develops wide open valleys compared to the narrow canyons cut into the harder Clarno Formation.

Fiery cloud eruption (nuee ardente) of gas charged, incandescent ash typical of John Day and Clarno volcanoes.

81

Six to eight miles north of White Horse Rapids, the river exposes the highly colored Clarno. The overlying John Day is visible high above on the valley walls as small pockets of buff sediments between the Clarno and the Columbia River lavas which form the skyline ridges. Along this stretch of the Clarno, the river slices through a series of small sheet-like, intrusive, volcanic rocks which are vertical dikes. The latter are easily recognized as they often stand out in erosional relief like the wall of a ruined castle. Columnar shrinkage joints, or basalt columns, in these dikes are lying horizontal, thus the dike often resembles a neat stack of cord wood.

Vertical dikes through layered basalt lavas. Columns in the dikes are oriented horizontally.

Mine dumps visible high up on canyon walls in this interval are perlite ash prospects. This peculiar volcanic rock is named for its grey, pearl-like luster and is commercially used as a decorative stone. Near the end of this stretch, a small rapids with a large standing wave known as Buckskin Mary is created where a resistant dike is intersected by the river.

After only a brief exposure of a mile or so of John Day Formation followed by another mile of landslide debris, the river enters an extensive section of 15 million year old Columbia River lavas which dominate the topography to Maupin. The Columbia River lavas are relatively even, black, fine-grained flows from 3 to 30 feet thick. Basalt columns dominate the scenery here. The columns are formed from vertical cracks which develop as the hot lava cools and contracts. The columns vary in size with the type of lava and the rate of cooling. Ordinarily columns are oriented vertically like stacked posts, but irregular rosettes or fan-shaped arrangements of the columns are indications of where the lava flowed over a rough surface.

Wapinitia Rapids here as well as Boxcar Rapids are in the Columbia River basalt. This basalt is difficult for the river to cut and erode, thus canyons tend to be narrow, steep-walled, and deep. The slightest debris or rock fall into the chasm creates a rapid where the river flow is impeded.

Within this stretch of the river, successive lava flows are often punctuated by baked, reddish soil horizons seen between the flows. At Maupin the river bed is in Columbia River lavas, but The Dalles Formation can be seen in the high east facing bluffs. This Formation is made up of water lain Pliocene ash covered in turn by younger Pliocene lavas on the skyline.

From Maupin to the Columbia River, the Deschutes maintains its course in Miocene Columbia River lavas which form not only the riverbed but the skyline as well. On this stretch there are several good spots to examine the anatomy of an individual lava flow. Close inspection of a typical flow 8 to 10 feet thick shows a massive columnar section near the base of the flow. The lava cooled here very slowly, and often crystals up to 1/4 inch in diameter are visible in the fresh rock. These crystals of dark, translucent plagioclase feldspars alternate with black crystals of pyroxenes and amphiboles that resemble charcoal.

Two-thirds of the way up from the base of the flow, small bubbles appear in the basalt. The hydrostatic pressure of the hot, liquid lava was low enough to allow some of the gas to come out of solution at this point. These bubbles travelled upward in the fluid lava sometimes leaving tube-like tracks. Higher in the flow, the bubbles are larger and more numerous as more gas came out of solution with decreased pressure. This bubbly, or vesicular, basalt is light in weight because of the abundance of pore spaces. Near the top of the flow, the lava may be almost frothy with bubbles (scoria). This portion of the flow often shows evidence of having intermittently formed a cool skin that broke up as the lava continued to flow.

Although the bubbles in lavas are formed by dissolved gas coming out of solution, the holes themselves may fill with mineral material still in solution after most of the lava has cooled. The minerals, quartz and calcite, as well as members of a mineral family called zeolites are the commonest filling agents in these bubbles. The mineral filling thus developed is a miniature "thunder egg" or geode. The minerals may line or completely fill the gas cavity with crystals. Basalt with mineral filled bubbles is called "amygdaloidal basalt". Often the amygdales are more resistant to weathering than the host rock, and it is possible to collect pebbles of gas bubbles cast in agate by breaking away the

Columns of overlying flow

Broken up rubble atop flow capped by a soil layer

Gas bubble layer at top of flow

Pipes (bubble tracks)

Ten foot thick, lava flow :

Lava with less gas bubbles

Massive lava with well-developed columns near base

Rubble zone with soil of underlying flow

Vesicular layer (bubbles) of underlying flow

crumbly matrix.

Mineral components of basalt lava include olivine, amphiboles, and pyroxene which form at very high temperatures of 1100°C to 1800°C. These minerals easily weather, although the dry climate of eastern Oregon slows down the decomposition substantially. Fresh-looking lava flows 10 to 20 million years of age often appear to have cooled only recently.

Just north of Sherars Bridge, the Deschutes canyon drops into a slot created by inward tilting of local rocks. This trough-like structure, called a syncline, traps the river's course in a northeastward curving path for twelve or more miles. In this interval, the river is confined to very limited meandering back and forth. After breaking out of that trough, the river proceeds only 10 more miles before coming under the influence of yet another structure (fold) which causes the river to swing back west at this point.

Near the mouth of the Deschutes, within four miles of the Columbia River, remnants of an intercanyon lava flow may be seen clinging to the canyon wall. An intercanyon flow is the result of molten lava flowing into an existing river canyon. After the new lava cools, effectively

casting the older river valley in stone, the river will often be dammed up until erosion cuts through the flows blocking the valley. Erosion will ordinarily deal with the blockage quickly, and one typical result is an exceedingly steep-walled canyon such as that in parts of the Crooked River to the south. Intercanyon flows are easily identified because the older flow layers and columns are interrupted and covered by the younger flows. The source of these younger lavas lies only a mile east of the river and is recognizable as the volcano towering over the older Columbia River lavas.

A final interesting feature which occurs within the Deschutes canyon near the Columbia is the presence of granitic glacial boulders. These large stones, referred to as erratics, are completely out of place in their present setting. The origin of these erratics might well be the Rocky Mountains, and their presence here is due to one of the multiple ice age Spokane floods. These floods occurred when large lakes in the Missoula Valley of Montana, dammed by ice and debris, suddenly broke and released millions of gallons of water. The cascade of water, mud, ice, and rock flowed southeast across eastern Washington to the Columbia River

Pathway of the Spokane-Missoula flood.

gorge. After scouring out the channeled scablands of eastern Washington, the flood proceeded on to deposit its debris all the way to the mouth of the Columbia River as well as in every canyon along the way.

Adjacent to the Columbia River çanyon, scattered remnants of The Dalles Formation crest the Columbia River lavas. The Dalles Formation here consists of sediments of volcanic ash which was redeposited in streams and lakes. This chain of The Dallas Formation rocks follows another trough-like syncline along the Columbia River.

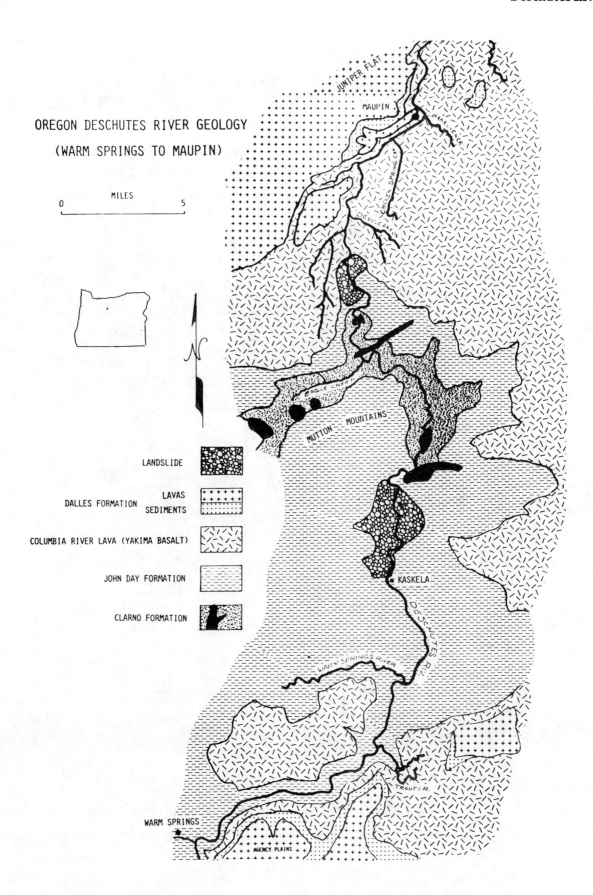

OREGON DESCHUTES RIVER GEOLOGY
(WARM SPRINGS TO MAUPIN)

MILES
0   5

LANDSLIDE

DALLES FORMATION   LAVAS
                        SEDIMENTS

COLUMBIA RIVER LAVA (YAKIMA BASALT)

JOHN DAY FORMATION

CLARNO FORMATION

JUNIPER FLAT

MAUPIN

MUTTON MOUNTAINS

KASKELA

WARM SPRINGS RIVER

DESCHUTES RIVER

TROUT CR.

WARM SPRINGS

AGENCY PLAINS

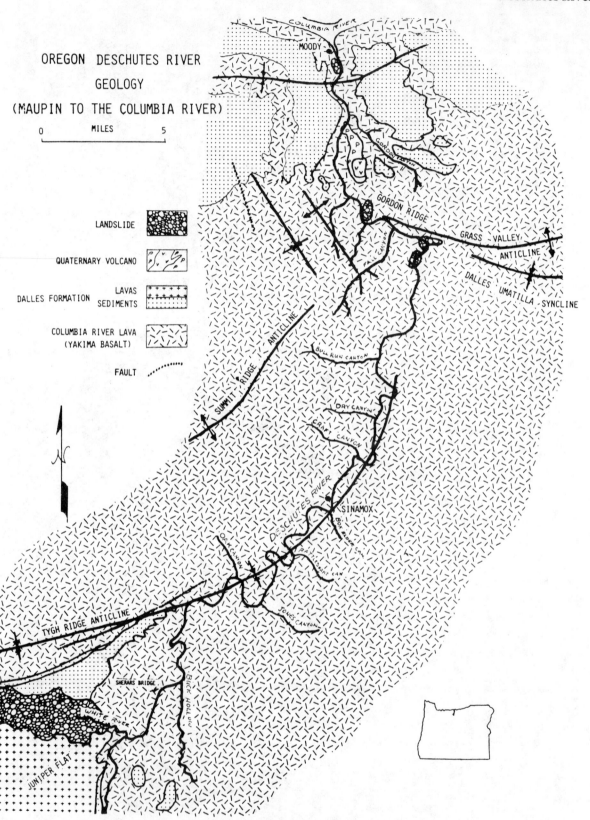

OREGON DESCHUTES RIVER
GEOLOGY
(MAUPIN TO THE COLUMBIA RIVER)

0   MILES   5

LANDSLIDE

QUATERNARY VOLCANO

DALLES FORMATION   LAVAS
                       SEDIMENTS

COLUMBIA RIVER LAVA
(YAKIMA BASALT)

FAULT

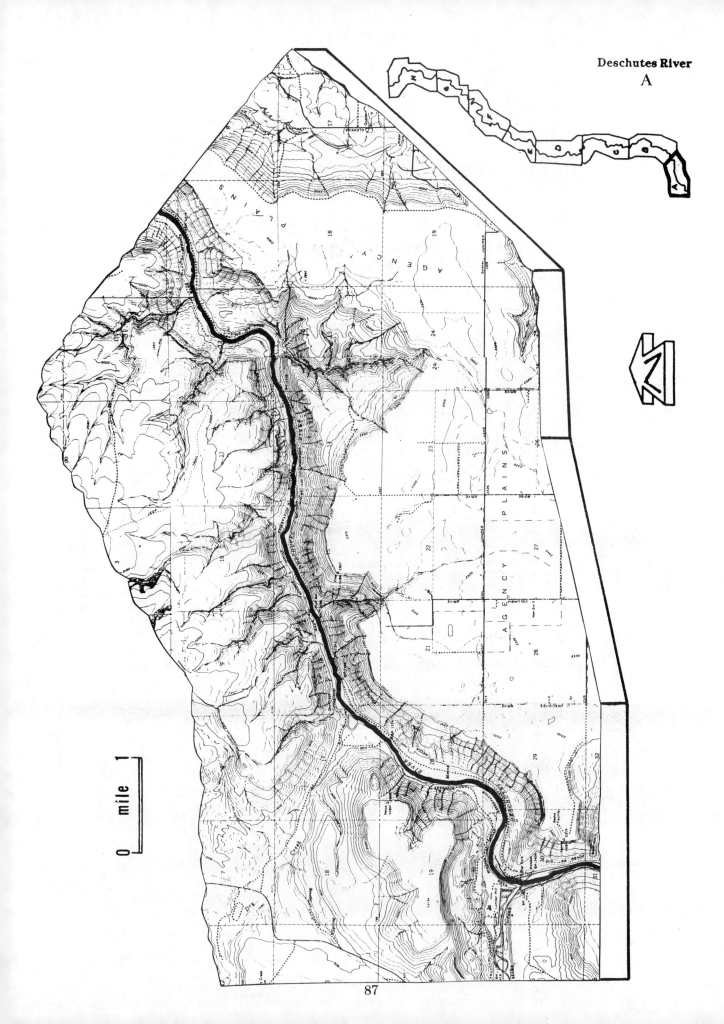

mile

0   mile   1

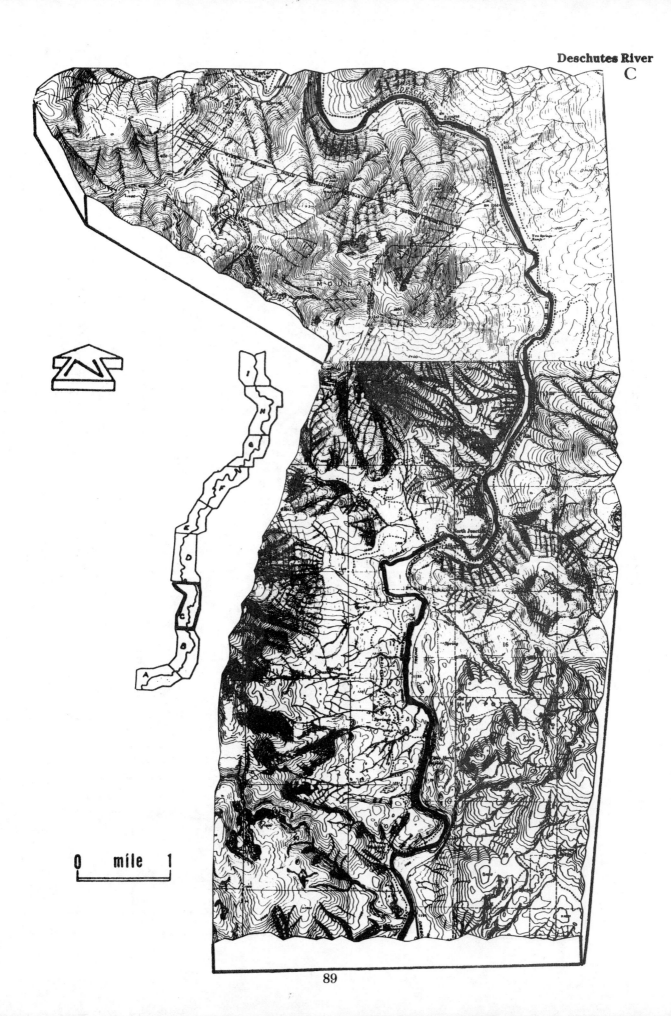

0 mile 1

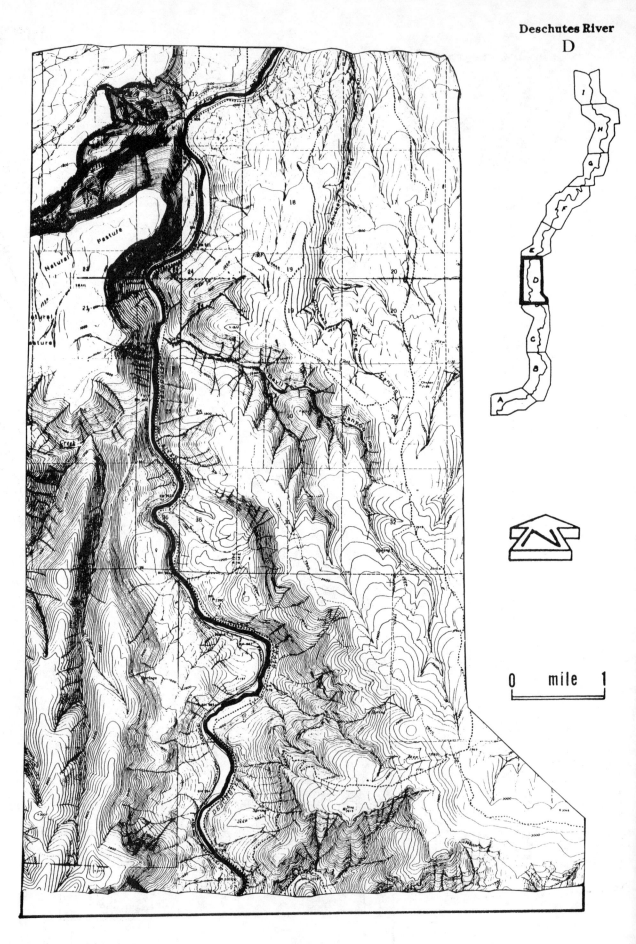

mile

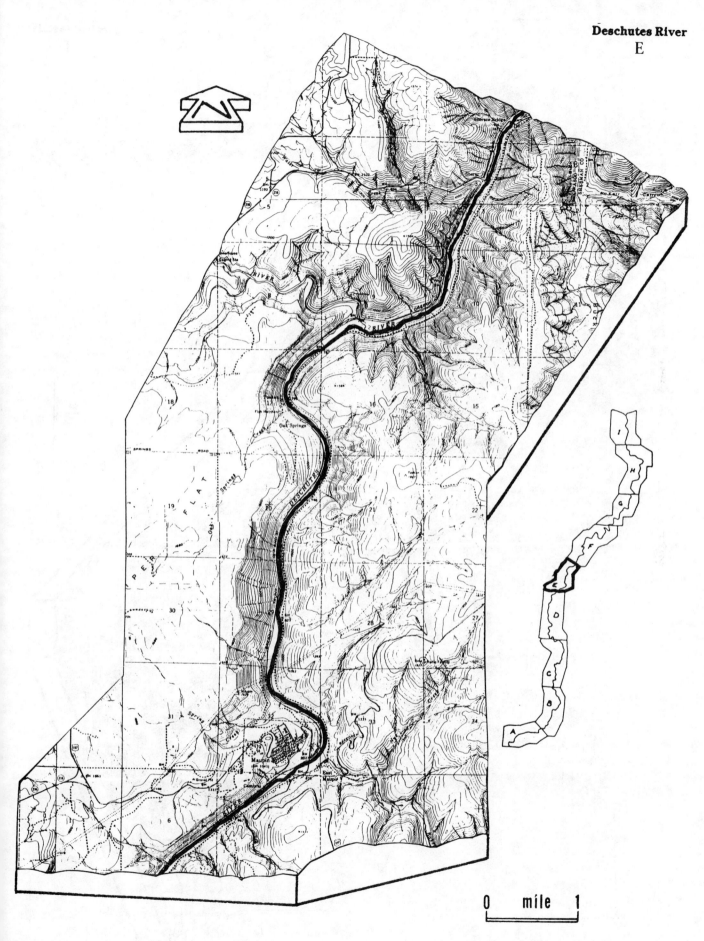

0   mile   1

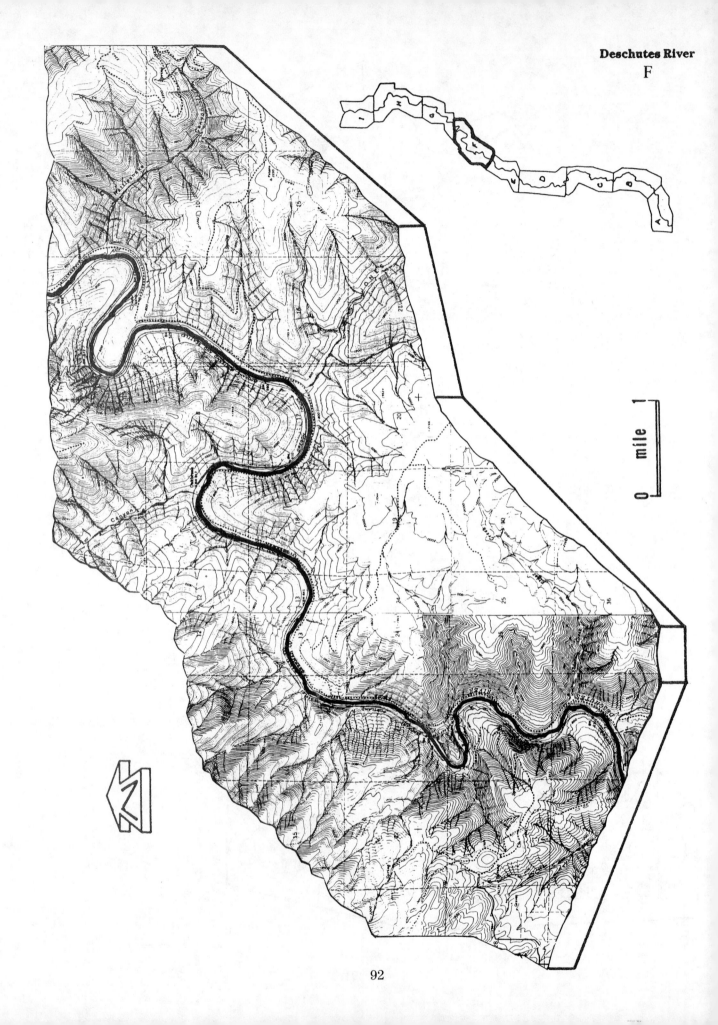

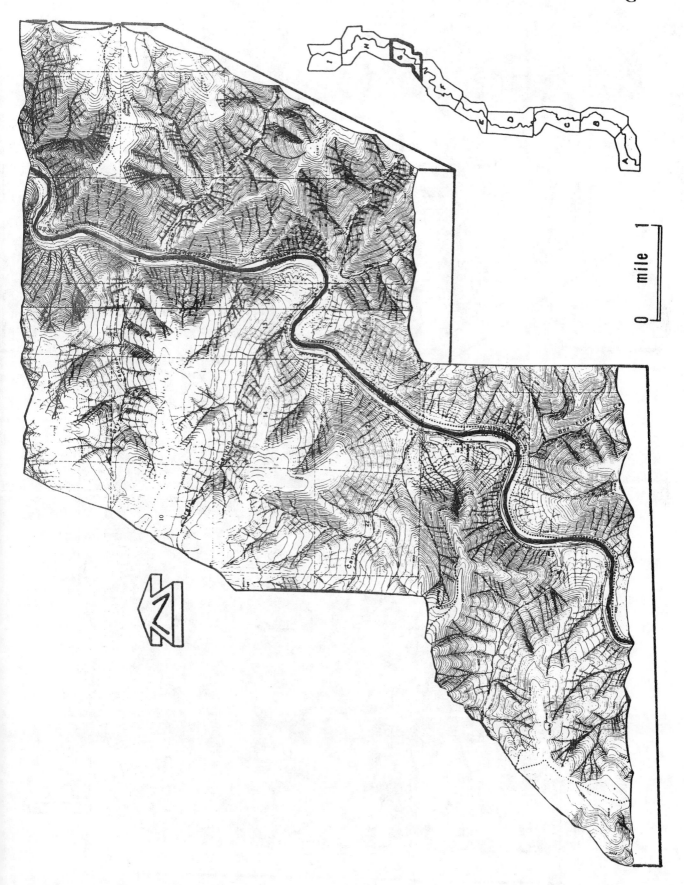

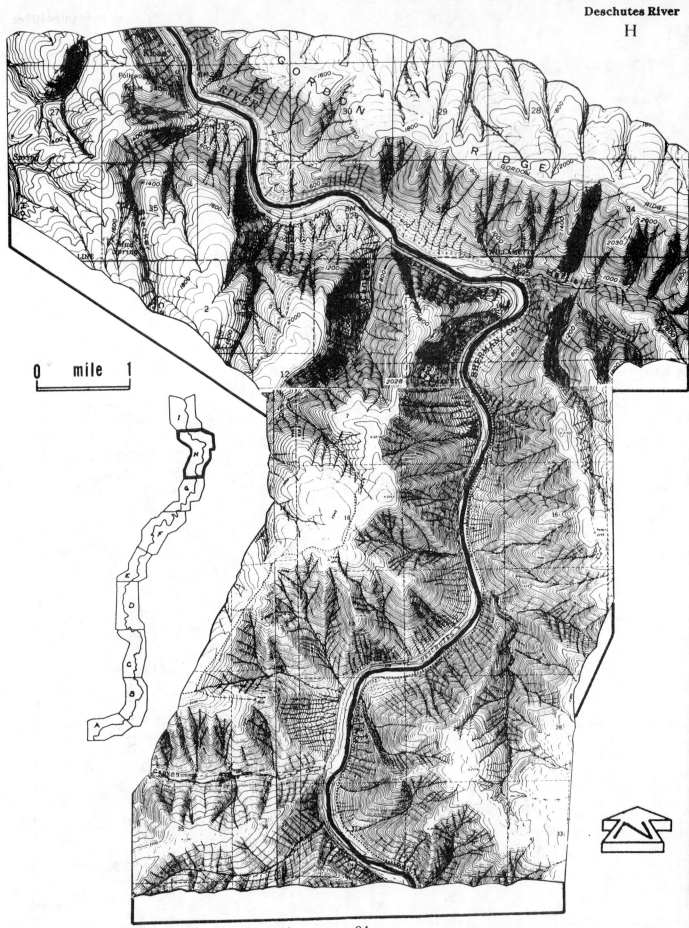

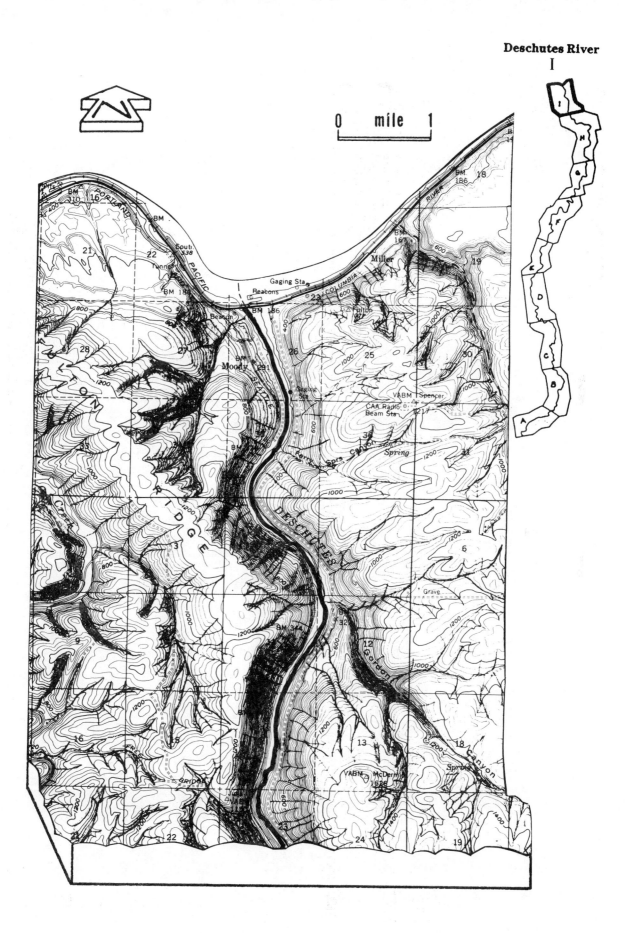

0   mile   1

Grande Ronde River

(Minam to Hellers Bar, Snake River)
Wallowa County, Oregon to Asotin County, Washington
Length of run: 90 miles
Number of days: 3-5

Beginning near Anthony Lakes, the Grande Ronde flows northeastward for 185 miles, crossing into Washington State for a short distance before joining the Snake River where it forms the boundary between Idaho and Washington.

Called the Land of Winding Waters by the Indians, snow melt runoff keeps river waters high well into July. Living in the deep canyons and grassy plateaus, the Nez Perce built traps for salmon along the Way-lee-way, the Grande Ronde, hunted in the land of Big-Rocks-Lying-Scattered-Around, and dug plant roots at E-mi-ne-mah, Minam. A profusion of plants, aspen, maples, cottonwood, serviceberries, and grasses grow in the Wallowas. Rattlesnake Creek, Cougar Creek, Wildcat Creek, and at least three Bear Creeks reflect an abundance of wildlife in these mountains. Summer evenings are long. After the sun disappears behind the mountain peaks, the air turns chilly, and campfires are welcome.

Chief Joseph of the Nez Perce.

Early History of the Grande Ronde River

The Wallowa Mountains, home of Chief Joseph's band of Nez Perce Indians, cover 3,145 square miles and includes the Wallowa, Imnaha, and Grande Ronde river valleys of northeastern Oregon. In 1860 there were 60 men and 120 women and children under the leadership of Tu-eka-Kas, later known as Old Joseph. The Joseph Band of Nez Perce was the western branch of the powerful Nez Perce tribe occupying vast areas of Idaho and Montana to the east. (See Nez Perce chapter for culture and customs).

Prior to the 1860's, the Wallowa Mountains were still the homelands of these Indians with just a handful of whites making their way into this corner of the northwest. Fur traders found few beaver. Only Capt. Benjamin Bonneville, in the role of fur trader but rumored to be assessing Oregon territory for the U.S. government, crossed the Imnaha and Grande Ronde rivers in 1834. In his journal Bonneville reported travelling along a "considerable tributary of Snake River," the Grande Ronde, spending some amount of time among "these good people,

more and more impressed with the general excellence of their character."

The discovery of gold in neighboring areas and the establishment of missions at Lapwai, Idaho, and near Ft. Walla Walla, Washington, brought increasing numbers of white emigrants into these remote valleys. A sawmill company and buildings were built at La Grande in 1861, and the first white farmers in the valley settled nearby. Benjamin Brown, formerly a freighting agent between The Dalles, Oregon, and Walla Walla, Washington, decided to take up farming. He found "thousands of acres of just as desirable land (in the Grande Ronde valley) yet untaken" where he built the first private home. Other "esteemed pioneers" followed occupying lands "not held by any band of aborigines as a home . . . and visited annually during the summer months by hunting and fishing parties" (Hist. of Union Co., p.143).

Complaints were later made to the government that Indians visited the valley and "brought vast herds of ponies which sometimes trespassed on what the settlers conceived to be their rights" (Hist. of Union Co., p.143). Clearly

Benjamin Bonneville, adventurer and explorer.

these pioneers viewed the "vacant" land differently than Old Chief Joseph who had refused to sign any government treaties relinquishing his claim to the territory of his ancestors.

The only treaty signed in 1855 by the Joseph band and the U.S. government gave the Nez Perce the valleys of the Grande Ronde, Imnaha and Wallowa rivers. The Indians were satisfied with this treaty. However, continual civil pressure to remove the Indians to smaller reservations voided this treaty and created a new one in 1863. Old Joseph refused to sign. His son, Hin-mah-too-yah-lat-kekht, Young Joseph, returned from this treaty conference erecting boundary markers of three foot high piles of rock "to have all people understand how much land we owned . . . Inside is the home of my people. The white man may take the land outside. We will never give up these graves (of our fathers) to any man."

In 1871 Old Chief Joseph died and was buried with much ceremony at the confluence of the Wallowa and Lostine rivers. Some time later a permanent grave was dedicated at Wallowa Lake. His grave was vandalized by whites who removed the skull and put it on display in a dentist's office at Baker, Oregon. Young Joseph, born on the Grande Ronde at the mouth of Joseph Creek, took over leadership of his band. Over 6 feet tall, Joseph developed as a natural leader and orator.

White immigration and settlement on Indian lands continued to the point where it couldn't be ignored by Young Joseph. He insisted their land had never been signed away and eloquently and forcefully repeated this assertion to government representatives at Lapwai, Idaho, in 1873. These agents were creating yet another treaty and reservation ostensibly giving the Nez Perce the *upper* Wallowa Valley and Lake and giving white settlers the lower part of the valley. Unfortunately, bureaucratic ineptness further added to the confusion when the final treaty gave the Indians the *lower* portion of the Wallowas as well as the Grande Ronde. At that time, the lower portion was already settled by whites. Hostile confrontations arose. Again vascillating and responding to public pressure, President Grant rescinded this last treaty in 1875 throwing open all of the Wallowas to white settlement.

Indians held an angry council. Military troops were sent in under General Oliver Howard of Ft. Vancouver. The winter of 1876 passed without open violence, but in June the first skirmish took place. A. F. Findley and Wells McNall, who was notorious for disliking Indians, started out to look for some of Findley's horses missing near Minam. What happened exactly will never be known, but the white men came upon an Indian hunting camp. Ever tactful, they accused the Indians of stealing their horses. A scuffle ensued between McNall and one young Indian, Wilhautyah, a good friend of Chief Joseph. The Indian was killed and the two men fled back to the settlement for reinforcements. Settlers, fearful of the Nez Perce, felt Findley might have

General O. O. Howard who fought against the Nez Perce.

overreacted and were alarmed at Wilhautyah's death. Several days later Findlay discovered his missing horses near his ranch.

General Howard responded with customary military diplomacy in July by adopting a policy of "harsher measures" as suggested to him by local religious leaders and government agents. Joseph, on the other hand, took the position that a "quiet, well-disposed" man had been killed by whites who were "quarrelsome and aggressive," so that the valley in which he was buried was claimed as recompense for the life taken. All whites must be removed from the valley.

Howard promised McNall and Findley would be brought to trial, but when they were still at large that Fall, Joseph and sixty hostile Indians demanded these men be punished. Settlers again sent for military aid which arrived from Ft. Walla Walla under Lt. Forse. Forse requested the men turn themselves over to a court, which they did. No charge was filed against McNall who was "defending himself," and the Judge ruled Findlay's actions were justified.

A temporary peace descended on the region as everyone awaited five commissioners dispatched to settle the Nez Perce claims. The commission was composed of three easterners who knew nothing about the situation in Oregon as well as General Howard and Major Wood. At the outset the commission, which cared nothing for the niceties of Indian feelings toward their ancestral lands, was determined to settle the Indians on a reservation in Idaho. The decision, of course, was preordained, and probably the only ones surprised by the outcome were Chief Joseph and his people. They were given thirty days to leave their home lands.

Hurt, resentful and angry, they gathered their horses, cattle and belongings and departed across the Snake River at Dug Bar. Joseph was joined here in camp by Chief White Bird and his people who had just arrived from their homes along the Salmon River. They had also been displaced. The two were soon joined by Chief Looking Glass whose village had been burned. Angry men of these two bands hastened the Nez Perce War of 1877.

The overwhelming numbers and arms of the white soldiers forced the Indians into retreat. Over the Summer and into Fall there were five major encounters, and more than 1,300 miles were covered by Indian families on the move. Deciding to join Sitting Bull's non-treaty Sioux in Canada, the Nez Perce camped at Bear Paw, Montana. Forty miles south of the border they were attacked by troops under General Miles. Chief Looking Glass was killed, and Joseph surrendered with his band. Chief White Bird successfully led a party of approximately 200 Indians over the border where they received sanctuary from Sitting Bull. The Joseph band was never allowed to return to the Wallowas. Joseph lived for many years pleading for "only a small piece of land for my people in the Wallowa Valley." His request was never granted, and he died on September 21, 1905, and was buried at Nespelem, Washington.

Indian Life Along the River

Evidence of both prehistoric and historic man can be found in the Wallowas. Interestingly enough, more material from Indians has been recorded than from non-Indians. Campsites were nearly all located next to springs where tools, chipping stones, and arrowheads are scattered over the surface or found at shallow depths. Abandoned cabins, sawmills, and fences show where pioneers once lived. With the long use of the Wallowas by the Nez Perce, it

Chief Joseph's home at Nespelem.

is probable much remains to be uncovered.

Most of the Nez Perce villages were along rivers and small streams, not in the high mountains or on exposed plateaus. Along Wildcat Creek, before it enters the Grande Ronde, arrowheads, flakes, and basalt objects can be found. Cemeteries were placed near the villages along the riverbank but above flood level. Rounded river rocks were placed over the graves. Most graves have been completely disturbed by artifact hunters, so that it is virtually impossible to find one unexcavated. One grave near the mouth of the Grande Ronde had large piles of stone with pieces of cedar placed upright in them. Burial sites contained a variety of items including a metal given by Lewis and Clark, beads, knives, trade goods, guns, and even the skeleton of a horse buried over that of a man.

Lewis and Clark medallion

A number of rockshelters lie in the steep sides of Joseph Creek before it flows into the Grande Ronde. It is possible this creek was part of a route connecting different areas of food supplies. Basalt outcrops could have provided temporary shelter for the night. To support this idea, a very small amount of prehistoric material was contained in one of the shelters, and it is interesting to speculate on these early travelers.

The thick vegetation cover and exposure to weathering conditions make rock art difficult to locate. It is worthwhile mentioning the Wallula Monolith, or Umatilla Stone, the name given to a large rock covered with pictographs found to the northwest of the Grande Ronde.

In 1827 a railroad survey party under J. P. Newell was eating lunch while sitting atop a large rock. Glancing down they noticed carvings on the moss-covered surface. This six foot long, ten ton hexagonal rock, which came to be known as the Wallula Monolith, was transported to Portland by the Oregon-Washington Navigation Company, and in 1940 it was placed on a pedestal on the grounds at City Hall. Carvings on the two rock faces are numerous, and many theories have been put foreward to explain them. It has been suggested this was a slaughter stone for human sacrifices, but this seems unlikely.

On high rocky ledges it is possible to find rock cairns once erected by children of nine or ten years of age. These children were on a sacred vision quest for power. Pictographs can frequently be seen on the rocks as well. Other favorite rock art places are cliffs where paintings with yellow or red colors, highlighted by incising, have been drawn. Elements in the designs include mountain goats, deer, snakes and geometric symbols.

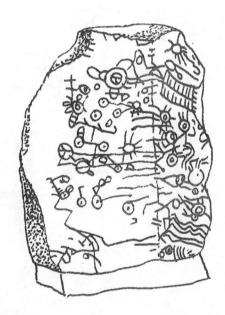

Wallula Monolith, an unusual stone covered with Indian art.

Geology of the Grande Ronde River

The Grande Ronde River run normally begins at Minam, Oregon, on the Wallowa River and ends at Hellers Bar on the Snake River less than one mile north of the confluence of the Grande Ronde and Snake. This is a run of 90 miles on the Grande Ronde itself. At first glance the Grande Ronde would seem a very pedestrian river with a limited stratigraphic sequence of rocks and real whitewater only in the last few miles of the river above the Snake. On the contrary, several aspects of Grande Ronde geology make it a fascinating canyon to raft.

As the river moves to the north and then northeast to meet the Snake on the Washington-Idaho border, its pathway is controlled predominantly by the geology. For most of the river's course, the channel dodges around hard spots like dikes or basalt outcrops, or it takes advantage of soft rocks such as the ground up rocks in faults. The names of several of the streams flowing into the Grande Ronde, as Big Canyon, Deep Canyon, and Elbow Canyon, provide clues to their fault origin.

In addition to the bedrock geology along the course generally run in rafts, one must not overlook the "Oregon Alps," the Wallowas, just upstream twenty miles from the put-in at Minam. This small range of mountains is glaciated and provides snow melt water to the Wallowa River and ultimately to the Grande Ronde.

The vast array of rock types in the Wallowas includes a core of 130 million year old granites located beneath sediments as old as 250 million years. These are among the oldest rocks in Oregon. Pebbles and cobbles of this diverse suite find their way into the Snake via the Grande Ronde. A quick inventory of any gravel bar on the Grande Ronde will provide representative rocks from the areas to the south which are drained by the river.

The stratigraphy of the watershed is relatively simple. Within the river's runnable section, the bed lies upon the basaltic lavas of the 15 million year old Columbia River group. These lavas are downwarped into a major southwest/northeast basin or trough. Lavas here are informally referred to as the "younger", upper flows (5 million year old Pliocene) on the mesa tops and the "older", lower flows (Miocene) down in the river valleys.

The two lava flow series are separated by a layer of sand and clay sediments 200 feet thick. These softer sedimentary rocks are not easily seen from the river because they are typically covered with thick vegetation and soils. Slumping and landsliding will occasionally open a raw exposure visible here and there along the river. Above some of the steep meander bends, bluffs high on the canyon walls show the sediments immediately below the "young" basalt flows that form the mesa tops. These sediments are part of a widespread series of bedded, stream-deposited volcanic ash and rock debris. The rare fossil mammals found in these sediments have been dated as 5 to 10 million years old (late Miocene to early Pliocene Age).

One of the foremost aspects of the Grande Ronde is the pervasive control by geologic features. Many of the major stream meanders are due to faults crossing the river. Here the course of the river briefly follows the fault trace to the northwest before coming back to the dominant northeast trend. The effect is a zigzag pattern of stream flow.

The structural "grain" or fabric of the Grande Ronde watershed is north by northwest. This fabric is expressed by both faults and dikes cut by the river. Most of the faults have shifted the local rocks only a few feet, but some show signs

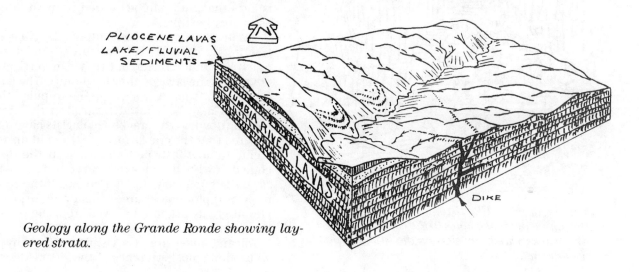

PLIOCENE LAVAS
LAKE/FLUVIAL SEDIMENTS
COLUMBIA RIVER LAVAS
DIKE

Geology along the Grande Ronde showing layered strata.

of having moved as much as a quarter of a mile. Within the faults, the broken rocks are separated by a clayey "gouge" or ground-up, smashed rock. This soft gouge is often washed out by streams leaving a well-marked trench like a moat.

Faults can often be spotted by the pronounced trenches cutting across solid rock. Usually the straight trend of the river itself is a good indicator that the canyon is following an old fault or group of smaller faults. If you can look up or downstream more than a mile, the stream is almost surely following an old fault.

The triple layer stratigraphy of the Grande Ronde, with soft sedimentary rocks sandwiched between two very hard basalt lavas, gives the canyons a typical mesa and butte topography. As the softer sediment erodes, it undermines the younger basalt above yielding flat-topped mesas. With continued erosion, these mesas will be chewed away to form smaller isolated buttes. One effect of the erosion of the soft sediments is to open up the canyon. Although there are some sheer walls in the Grande Ronde Valley, the usual topographic expression is of open canyons with mesas and buttes far away on the skyline. The lack of extensive sheer walls implies that the river has been freed to race past only a few obstructions or narrow spots. Because of this, very few real whitewater stretches develop.

Columnar basalt is an obvious feature in the lavas along most of the river. Over much of the route the columns are high above the river appearing as vertical, polygonal pillars in the flat horizontal lavas. At some of the rare rapids,

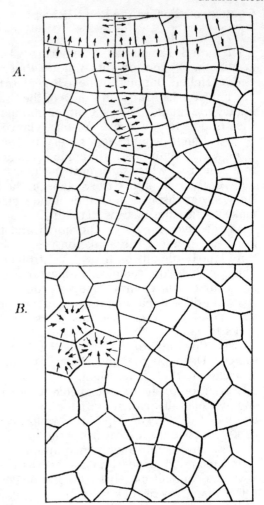

Basalt column (cooling cracks) viewed on end. Slow cooling (A) yields long curves and right angles. Rapid cooling (B) produces short, straight lines that meet at obtuse angles.

however, the columns are at water level, and it is here that they can be viewed from an interesting perspective.

Basalt columns are the result of the lava contracting as it cools. The fractures formed are much like mud cracks, but they run vertically down into the lava yielding a column as the lava is eroded from a wall. At the rapids, the columns may be viewed from atop the flow looking down at the cracks. Geologists have discovered that the cracks forming the column are a clue to how fast the lava cooled. If the flow cooled slowly, the cracks are curved and tend to form right angles with other cracks. If the cooling was rapid, cracks are straight and form angles at greater than 90 viewed from the top of the flow.

Volcanic dikes are also visible in the canyon walls along most stretches of the river. At sev-

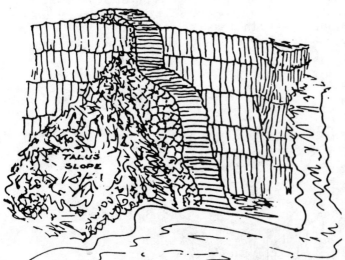

Three foot wide dike through columns of lava. Columns, at right angles to the cooling surface, are cracks which develop as the lava hardens and shrinks.

eral places these dikes appear in dense swarms occurring every few feet. Here broken zones in the basalts have been followed by rising lavas that fed the flows above. Most of the dikes are very thin and are measured in terms of feet. Dikes are often visible because they tend to stand out in erosional relief like walls above the more easily eroded surrounding rocks. The appearance of an erosionally resistant, hard dike traversing a landscape like a stone fence is the origin of the term "dike."

Some of the largest geologic features cut by the Grande Ronde are of unknown origin. Large scale, smooth lines trend east/west across the river canyon in the vicinity of Rondawa and Promise. These features are mapped only as "lineations" and are so large that they are only seen on ERTS satellite photographs. Ultimately the lineations may prove to be some previously unrecognized fault or even a buried river system under the lavas. In any case, the structures are so large as to have escaped notice by geologists mapping the canyon. It is indeed curious that structures could be so large as to be visible only to the cameras of an orbiting satellite. Clearly this is a case of not seeing the forest for the trees.

In defense of the field geologist on this point, it might be worthwhile to recount the history of the San Andreas fault in California. Although that great fault had been noticed and mapped by geologists at several separate points along its extent, it was not until after the disasterous San Francisco earthquake of 1906 that the full length of the fault from Mexico to beyond San Francisco was recognized.

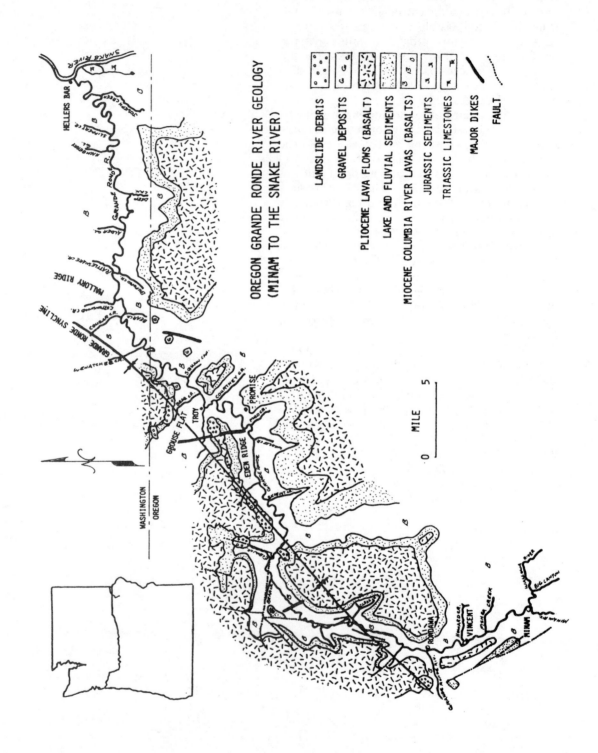

OREGON GRANDE RONDE RIVER GEOLOGY
(MINAM TO THE SNAKE RIVER)

LANDSLIDE DEBRIS

GRAVEL DEPOSITS

PLIOCENE LAVA FLOWS (BASALT)

LAKE AND FLUVIAL SEDIMENTS

MIOCENE COLUMBIA RIVER LAVAS (BASALTS)

JURASSIC SEDIMENTS

TRIASSIC LIMESTONES

MAJOR DIKES

FAULT

0   mile   1

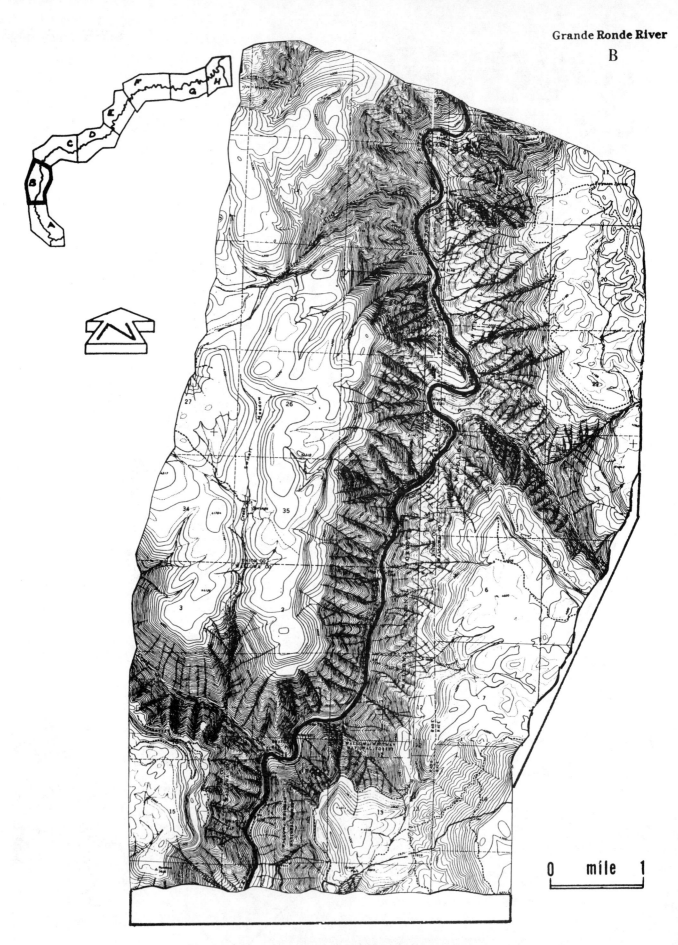

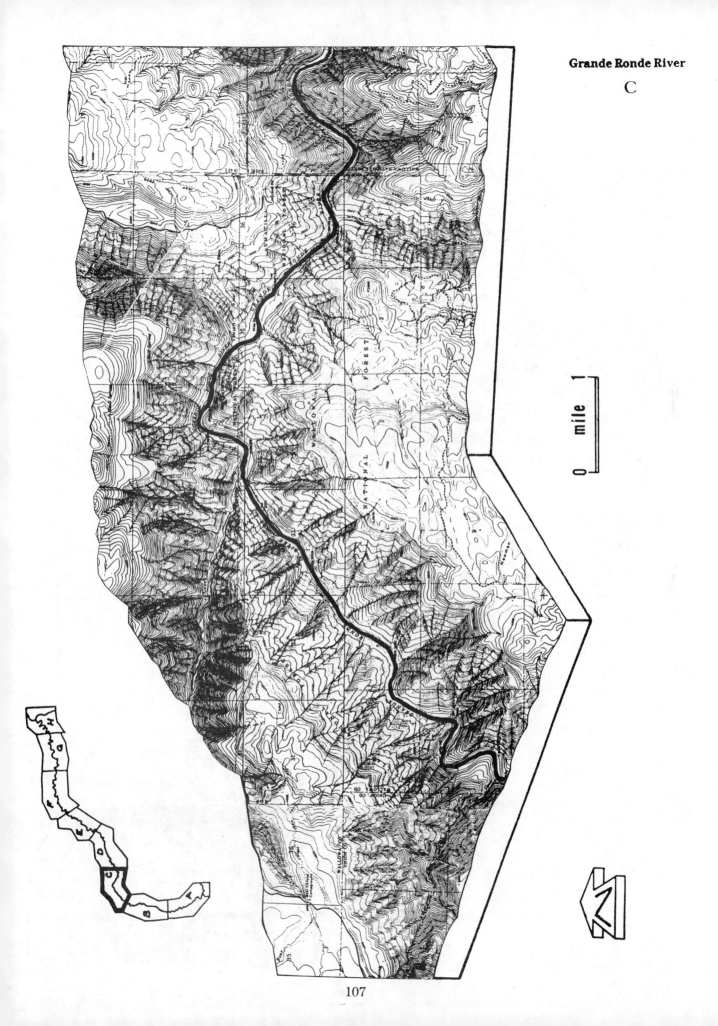

0   mile   1

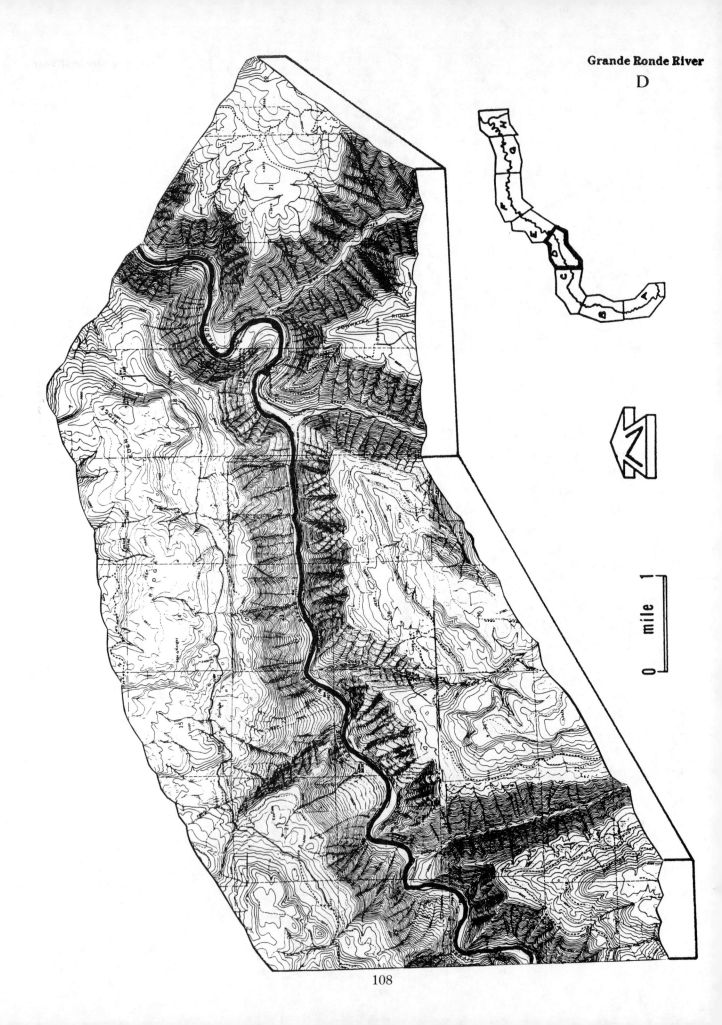

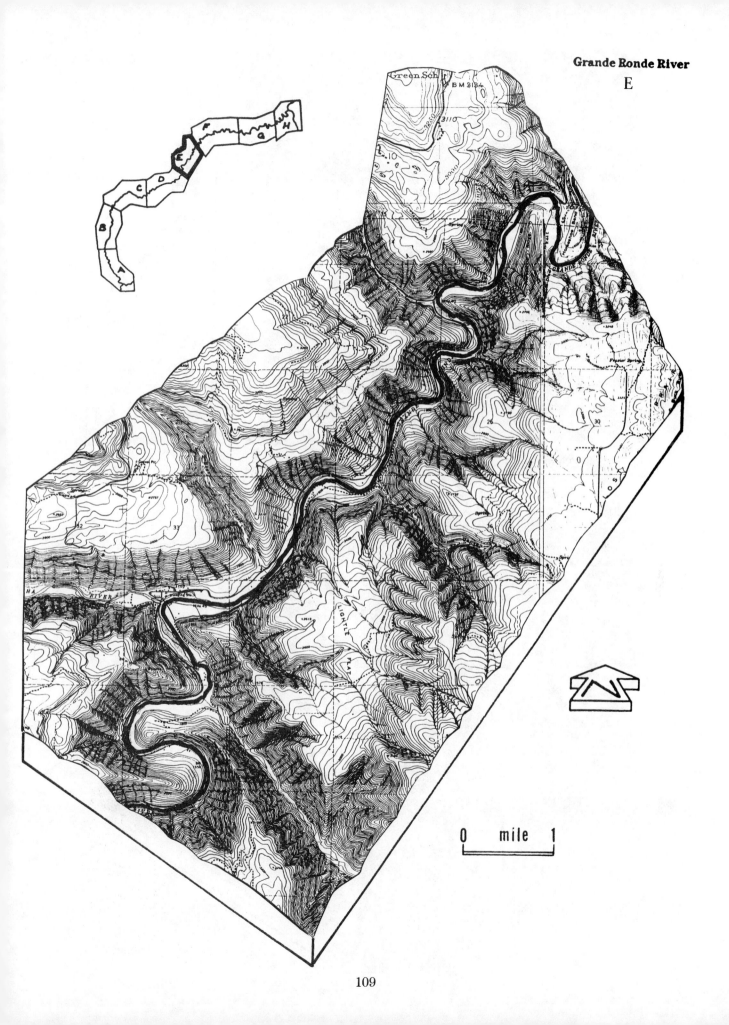

Grande Ronde River

E

0 mile 1

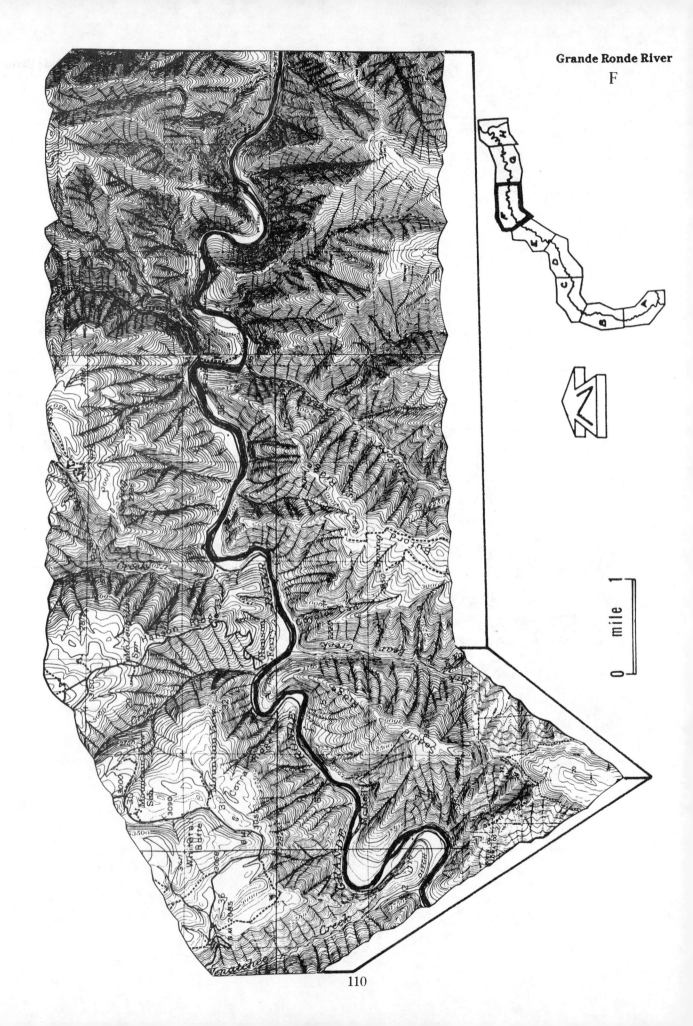

0 mile 1

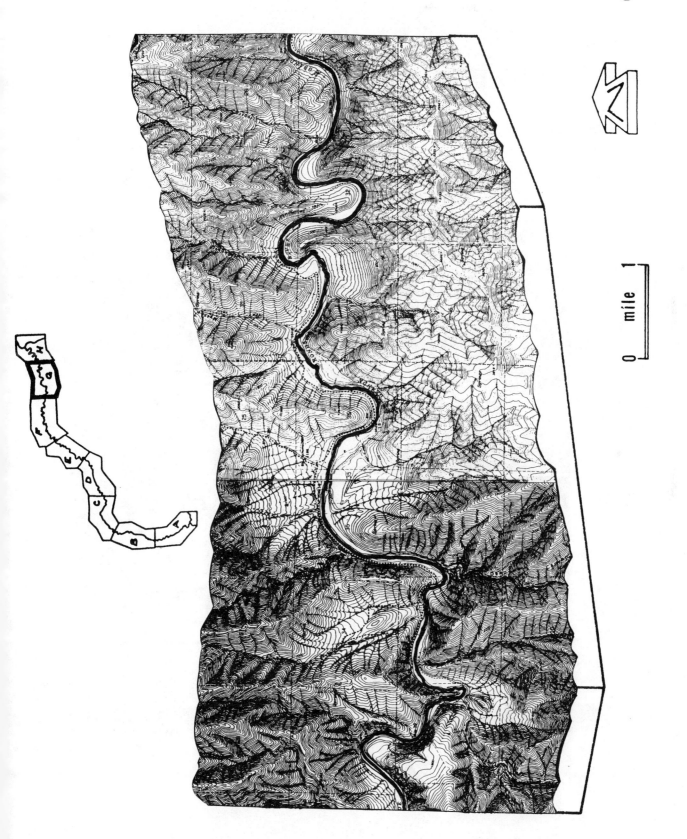

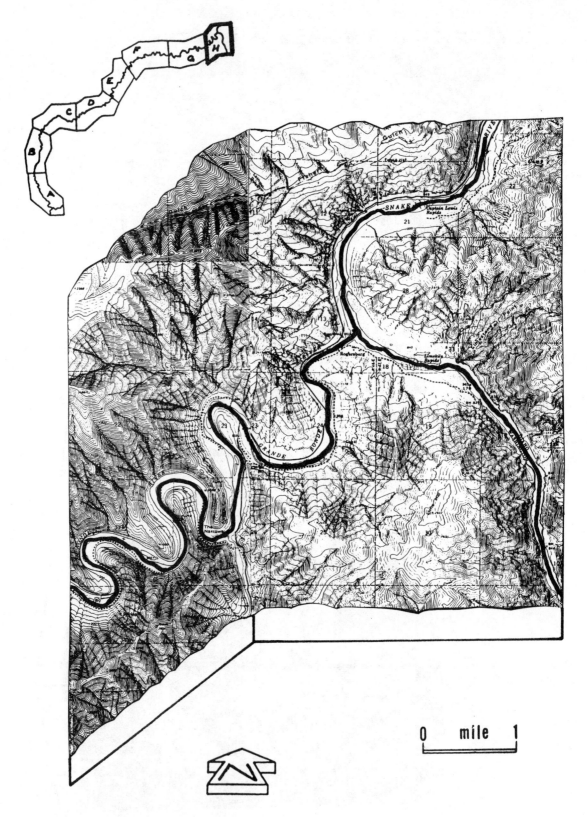

0 mile 1

John Day River

(Service Creek to Cottonwood Bridge)
Wheeler, Wasco, Sherman, Gilliam Counties, Oregon
Length of run: 114 miles
Number of days: 6-8

The John Day River, with its relatively calm waters, runs 280 miles from its headwaters in the Blue Mountains to where it joins the Columbia River. The river flows through seven counties in eastern Oregon and past an area particularly famous for its fossils and rock exposures. Today the river is even more forsaken than during the 1860's and 1870's when traffic along The Dalles-Canyon City Road here was heavy with gold miners, stagecoaches, and military troops.

Early History of the John Day River

In times past Tenino Indians lived along the south bank of the Columbia River as well as along the lower stretches of the John Day and Deschutes rivers. (For customs and history of the Tenino Indians see the Deschutes River chapter).

The John Day River was named after an early fur trapper who was on his way to The Dalles. Accompanied by a friend, both were captured by Indians near the river. John Day, a six foot tall Virginian, had started for Oregon with a hunting party when Day fell ill in eastern Idaho from exhaustion and lack of food. He and a companion, Ramsey Crooks, elected to separate from the others and continue on by themselves after Day recovered. It was their misfortune to meet with unfriendly Indians who stole everything they had including clothes before letting them go. As Day and his companion walked north toward Walla Walla, Washington, they were rescued by a trading party from Jacob Astor's fort on the coast. Day never recovered from his experience and died in less than a year.

As Day himself had discovered, eastern Oregon wasn't overly hospitable. Even in 1859, when Oregon was applying for statehood, a body of Senators felt only western Oregon should be included. It was obvious to the Senators no one would ever want to live east of the Cascades. Early explorers risked being attacked by Indians, poisoned by rattlesnakes, eaten by wolves or coyotes, frozen in wintertime or dehydrated during the summer. This is doubtless why the frontier didn't "push west" in Oregon, but instead lept over eastern Oregon into the

Peter Skene Ogden, who helped organize fur trappers in the Northwest.

lush, western Willamette Valley.

Fur trappers endured these conditions in their quest for new sources of animal skins along river valleys. Peter Skene Ogden, as an employee of the British based Hudson's Bay Company, led the 1825-1826 Snake River Expedition made up of French Canadians and probably several women and children. In search of beaver, they reached the John Day near Picture Gorge where they held a rendevous with other trappers. On the return trip along the now established route paralleling the John Day, Crooked, and Deschutes rivers, the party was reduced to eating their horses while almost starving and freezing to death.

Wagon trains with settlers anxious to reach their destination to the west, prudently followed the Oregon Trail across the Snake River to the Blue Mountains and Columbia River avoiding the eastern desert as much as possible. The few wagons which tried to take short cuts — guaranteed to save hundreds of miles — became lost in the desert, and history has recorded them as The Lost Wagon Train of 1845, The Clark Train Massacre, and The Lost Wagon Train of 1853.

The Blue Bucket Train in 1845 was led by Stephen Meek who had already done some trapping along the John Day. Leaving the Missouri River, he felt confident enough to guide his party of 300 wagons and nearly 1,000 men, women, and children through eastern Oregon. Leaving the Oregon Trail and following the Malheur River during the hot month of August, wagons were broken up on the rugged terrain. The only water to be found was alkaline which brought fever and death to the emigrants. Many perished in the heat. Threatened by several men who had lost family members, Meek and a companion, Reverend Elijah White, left the train and rode for help at The Dalles. The wagon party was ultimately rescued and travelled an additional week to reach the Columbia River and the Oregon Trail again.

Gold nuggets had been picked up by members of the Meek party during their wanderings in the desert, the spot marked with a wooden, blue bucket. The lost Blue Bucket Mine has yet to be found, but rumors of gold were enough to spur gold seekers from mines in California and Idaho. A large, well-equipped expedition left Portland in 1861, crossed the Cascades and followed the Crooked River to its headwaters in search of the misplaced blue bucket. Suffering the intense August heat and bad water, the party failed to find the treasure they were seeking.

Irregardless, prospectors came, set up tents, and built sturdy cabins to winter over. Eventually gold was found in 1861 up Canyon Creek, a tributary of the John Day — reportedly by a man wading into the river in his long johns — as well as along Griffin Creek just southwest of

Baker. The thousands of miners and prospectors who arrived at Griffin Creek held a formal meeting and called their settlement Auburn. Overnight Auburn grew to be the largest town in Oregon before disappearing again in 1903. Canyon City, two miles south of the John Day, had 10,000 persons including hundreds of Chinese who worked the mine tailings.

Miners and freight suppliers needed roads to their camps so the federal government, directed by Congress, sent several military parties through the eastern desert scouting out possible routes for roadways and railroads. The army hoped to accomplish what civilian wagon trains had failed to do, find a short cut to the Oregon Trail. With military thoroughness, this region was traversed from south to north, east to west, and back again through the years 1855 and into the 1860's. Finally five military wagon roads were laid out and generous grants of land were given to private contractors to build the routes.

Of these, The Dalles Military Road was constructed along old, existing trails from The Dalles, along the John Day River, through Burnt Ranch and Canyon City, and on to Boise, Idaho. The construction company was granted 556,627 acres, of which only 126,900 were actually used. An outraged public declared the company didn't make any real improvements, and eventually the outcry forced the government to withdraw the contract. However, in a court battle, company claims were upheld.

Travellers, miners, settlers, and Tenino Indians here suffered from the unwelcome atten-

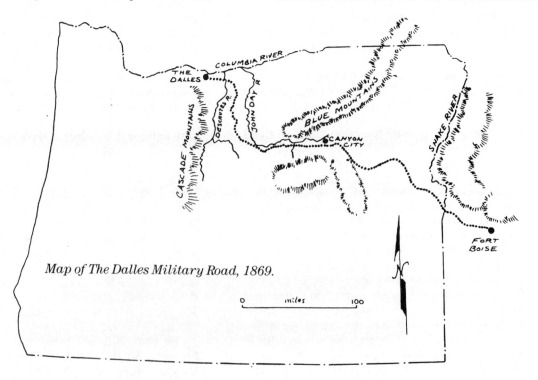

Map of The Dalles Military Road, 1869.

tions of the Northern Paiute who conducted raids to the north of their homelands. Led by the famous Chief Paulina, they captured livestock, burned buildings, and murdered whites. Settlers took refuge in mine tunnels at Canyon City or in hastily built fortifications. Prospectors who chose to remain near their claims were killed.

Some farmers were not scared off by the threat of Indians and remained on their farms. Burnt Ranch, south of Clarno, was the home of James Clark who ranched there with his brother-in-law. While cutting up drift wood along the river one sunny September morning, they were attacked by a band of armed Paiute Indians. Clark and Masterson fled, successfully escaping, only to find Clark's ranch house burned when they returned. Although another house was built on the spot, the name, Burnt Ranch, has been used since.

Camp Logan had been abandoned by the military some time back, and Camp Watson was to far away to provide much protection. The Dalles-Canyon City Road became too hazardous to use. Stagecoaches along the road were an easy target for the Northern Paiute, and Henry Wheeler lost over 80 horses and $20,000 worth of property to them.

The maraudings of Chief Paulina and his band came to an end in 1867. Stealing 25 head of livestock from Andrew Clarno's station, they were riding west toward the Deschutes River when spotted by a stage driver on The Dalles-Canyon City Road. He turned back to Maupin where a troop of men gathered and rode in pursuit. The Indians camped on Trout Creek where they were attacked, and the tall chief, distinctively dressed in an old, blue cavalry coat, was wounded in the thigh. The white men were reported to have shot Paulina three more times when he requested to be put to death. This removed any obstacles which stood in the way of white settlement, and towns could now develop without fear of Indian reprisals.

Clarno itself was never a large town. It was named for Andrew Clarno who settled there in 1866, operating one of the first post offices in the county. His son, Charles, ran a ferry close to where Clarno Bridge was built in 1897. Clarno, an enterprising person who admired steamboats, constructed a miniature model, the John Day Queen. The Queen, 40 feet long, 10 feet wide, and 10 feet high, navigated as a ferry and pleasure craft on a 10 mile stretch near Clarno. Passengers frequently chopped their own wood to fire the boiler.

One day Clarno and 20 men decided to pull the Queen downriver with ropes, when they were forced to let the ropes go at Clarno Rapids,

Pictographs at Butte Creek.

Pictographs at Muddy Creek .

a drop of 40 feet in one mile. The steamboat crashed on through the rapids to be destroyed against some river boulders.

One of the most unusual, British-sounding town names along the John Day is Twickenham. Established as a post office in 1896, the town was named by a local schoolteacher after Theophile Marzials' poem, Twickenham Ferry. The Twickenham referred to in the poem was a suburb of London. The early town included a store, ferry, hotel, blacksmith, as well as the post office.

Nearly ten years of white settlement passed before a brief, six month uprising, the Bannock War of 1878, was enacted in the southeastern desert. This was the last effort by northwest Indians to reestablish their old way of life. After a final defeat at the hands of the military, dispossessed Paiutes were forced back onto even smaller reservations in Oregon and Washington.

Indian Life Along the River

The sheltered valley of the John Day River was where Tenino located their houses during the wintertime for hunting and fishing. Over 150 sites on both banks have yielded artifacts.

Probably one of the most interesting archaeological spots in the John Day drainage is Butte Creek cave in Wheeler County. A skeleton buried in a grass-lined pit was uncovered by anthropologists from the University of Oregon. The skeleton had been buried with a cape made of

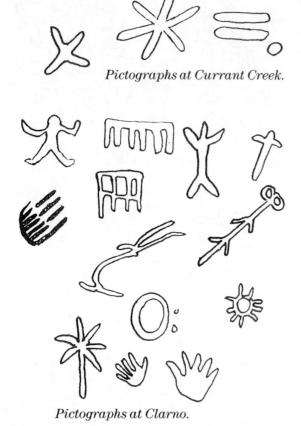

Pictographs at Currant Creek.

Pictographs at Clarno.

Pictographs at Cherry Creek.

include an elaborate series of geometric and realistic designs on the cliff above the river. The red pigments may have been applied at different times to the fine-grained absorbent rock. Although some colors are still clear, others are badly weathered. The designs include human figures, hands, floral designs, and many which can't be separated from the mass of superimposed lines.

South of Clarno near the northward bend of the river, cliffs east of Muddy Creek are covered with pictographs colored in dark red pigments. "Lizards" superimposed over grid patterns are in fairly good condition.

Red pictographs are painted on the cave walls in the same area at Currant Creek cave, and at nearby Cherry Creek red pigments display a variety of geomoetric designs on a cliff near the old road along the creek. Initials of early settlers have been also scratched into the rock. The Currant Creek Cave has a small opening which conceals a larger interior where ten small pictographs were painted.

Geology of the John Day River

The John Day River is run in two sections. The first is from Service Creek to Clarno in Wheeler and Wasco counties, and the second is from Clarno to Cottonwood Bridge in Gilliam and Sherman Counties. The stretch between Service Creek and Clarno Bridge easily makes up in scenery and excellent geology what it lacks in whitewater rapids.

Over this interval the river slices directly across a huge, elongate dome referred to as the Blue Mountain Anticline. This structure is a large up-fold that trends northeastward across eastern Oregon for almost 200 miles. As erosion proceeds to strip away the upper rock layers, the core of the dome has been increasingly exposed. Some of the core rocks are older than 250 million years. Traversing this huge dome on the river, we will move initially westward at Service Creek up the southeast side of the dome. Near Muddy Creek Ranch, the river trends northward to move off the dome to the northwest.

Working our way downriver we reach older rocks at the core of the dome. Cherry Creek marks a point near the center of the dome, and from there northwest the rocks will become progressively younger and plunge away to the west. The second section of the river from Clarno to Cottonwood is largely cut into lavas of the Columbia River group. Layers of these lavas once completely covered the dome but have been peeled off by erosion.

rabbit skin strips. A dog was placed alongside the body. Bone utensils, basketry, and horn spoons were recovered from different levels in the cave. An unusual, red pictograph with round eyes, two horns, and zigzag lines for the body was drawn at a Butte Creek location.

Just above Butte Creek, on Hoover Creek, a cremation pit was examined at the same time. Pipe fragments, copper and iron show the site was used fairly recently.

Pictographs just south of the Clarno Bridge

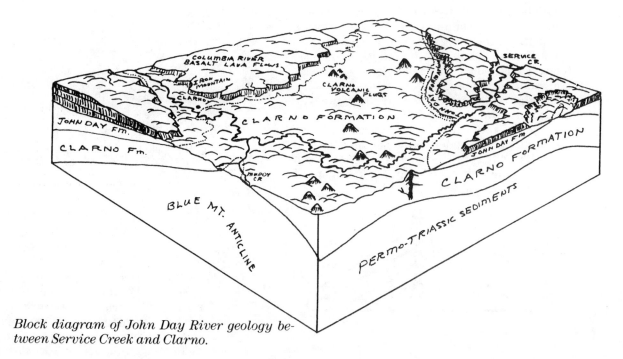

Block diagram of John Day River geology between Service Creek and Clarno.

Geology Along the Route

Beginning at Service Creek, the first half dozen river miles are in Columbia River lavas and parallel a trough-shaped basin (syncline) whose center is less than a mile to the southeast. About eight miles downriver, the first exposures of the 30 million year old John Day Formation (Oligocene/Miocene) are visible where they dip eastward under the Columbia River lavas. The John Day Formation is famous for its beautifully preserved fossil leaves and mammal bones.

This formation can be conveniently divided into three members on the basis of color. The lowermost, deep rust red member gives way upward to a pale, pea green (Turtle Cove) member which in turn is overlain by a cream/buff (Kimberly) member. Although the red interval, called the Big Basin Member, contains only a few fossils, the overlying, green Turtle Cove Member and the buff Kimberly Member are rich in fossil remains.

To find fossils within the Turtle Cove and Kimberly Members, the best strategy is to set aside about an hour or two, then comb fine gravels in the small streambeds which drain out of the rocks. Fossil teeth and bone show up as carmel to chocolate-brown colored fragments against the tan and green, lighter silts. Teeth are often best preserved as the phosphate composition of these fossils resists solution and weathering very well. Larger bones will fre-

quently show marks where rodent's gnawed the bone for its calcium after the mammal died but before it was buried.

Drifting downriver from Service Creek, the rafter is, in effect, moving backward in time or down section beginning with the youngest, buff John Day member and ending in the oldest, red member to the west. Within the green Turtle Cove Member, a volcanic, ash flow layer is especially prominent.

The John Day erodes into open valleys with rolling hills. Toward the middle of this formation a pronounced eastward plunging ridge, Kentucky Ridge, is visible on the skyline. This ridge is formed from the Picture Gorge Ignimbrite and is a classic cuesta ridge developed because of its resistance to erosion. An ignimbrite is a volcanic ash flow that was incandescent as it descended. The ash flow anneals itself as it settles to earth creating a glassy hard rock set with larger crystals.

The transition downward from the lowermost red, Big Basin Member of the John Day Formation to the underlying Clarno Formation is startling. The Clarno Formation is from 32 to 42 million years in age (Oligocene to Eocene) and is made up of volcanic ash, lavas and sediments. The lavas give the Clarno a highly colored appearance with reds to tans, buff, and deep green. Clarno volcanics reflect a particularly violent, eruptive cycle. Much of the ash was collected by streams, mixed with water, and recycled as mud. This mud often flowed like syrup when it became excessively water laden and fluid. Termed "lahars," the mudflows can be recognized by their larger pebbles set in a fine-grain mudstone matrix.

Spotted along the river throughout the Clarno interval are volcanic vent rocks. These vents are visible as hard erosional knobs above the more easily eroded Clarno sediments and rocks. The sites of these ancient volcanoes can be seen by the rafter as Craggy Peak, Sugarloaf Mountain and Amine Peak.

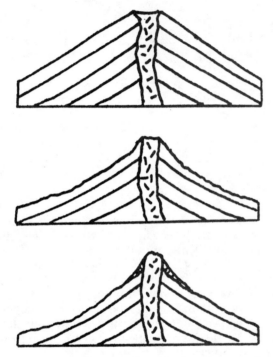

Development of an erosional volcanic vent plug in Clarno rocks.

About twenty miles into the Clarno interval, the rocks have been folded into a series of wrinkles, including domes (anticlines) and troughs (synclines). Near the crest of one of the domes and just south of the river along Muddy Creek are exposures of Permian/Triassic rocks in the Blue Mountain Anticline. These rocks, dated in excess of 200 million years, are much older than even those of the Clarno Formation. This older suite of rocks may even be foreign to North America having been rafted into this hemisphere from the western Pacific by the process of continental drift. At the center of this anticline lies the now famous Muddy Creek Ranch where the creek enters the John Day River.

As with the John Day Formation, sedimentary layers of the Clarno Formation yield fossils

Overhang, rockshelter cave development in soft ash or soils between lava flows.

although not in the abundance of the former. Clarno vertebrates include tropical forms such as crocodiles, rhinoceros, and tapirs. One particularly striking fossil-bearing section in the Clarno Formation is the so-called Nut Beds, where an unusual variety of beautifully preserved fossil seeds and nuts occur. In addition to these remarkable plant fossils, palm wood and fronds are further evidence that the Clarno climate was tropical.

Just at the raft take-out near the Clarno Bridge, the river again crosses the contact between the Clarno and John Day formations. Near the bridge, a volcanic mud flow can be seen as a rubbly broken-looking rock exposure.

The easily eroded nature of the John Day Formation is expressed well in the next ten river miles where much of it is mapped as an area of landslides. As the John Day decomposes and soaks up water, it will not support even slight slopes and begins to respond to gravity by sliding over itself. The landslide topography is recognized by its unusual hummocky pattern and poor drainage.

Beyond the landslide section, the river re-enters the Columbia River basalt lavas and remains in this rock type all the way to Cottonwood Bridge. This section of the river gives the rafter an excellent opportunity to see layered volcanic, lava and individual flows in detail.

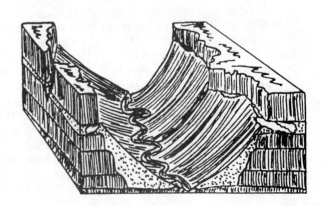

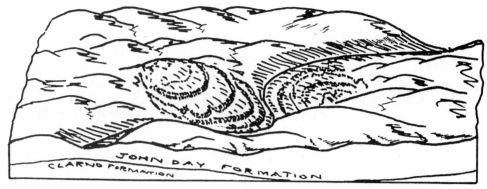

Landslide (debris flows) in the John Day Formation.

Some of the thicker flows of 10 to 20 feet can be followed for several miles along the river in the canyon walls. Gentle east/west anticlines (domes) and synclines (troughs) fold the Columbia River lavas in this interval into slightly tilted sections.

The great mass of Columbia River lavas erupted during the middle Miocene around 12 to 15 million years in the past. The tendency is to think of this lava as a single event, although it is actually a series of smaller flows. In the vicinity of the John Day River, the flows are called Yakima Basalt named for similar lavas near Yakima, Washington. Although some of these flows covered hundreds of square miles with fluid, hot lava, the flows were intermittent. Often sufficient time would lapse between successive flows so that soils could develop over the cooled surfaces before the next flow swept over the flat plains. These old soils between flows are visible today from the river as reddish clay layers at the base of a column series.

Looking downcanyon, the individual flows form a stair-step effect on the slopes. The softer, old soils as well as volcanic ash layers tend to weather out into rockshelters and shallow caves along the canyon walls. Humans as well as animals have lived in many of these overhangs.

Multiple lava flows in the John Day Canyon between Butte Creek and Cottonwood Bridge.

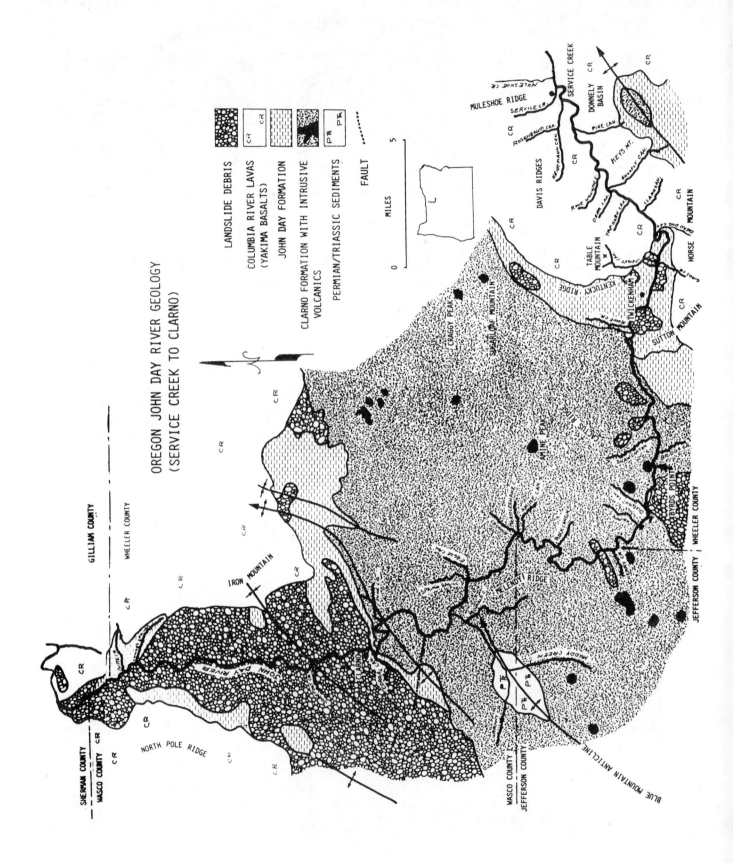

OREGON JOHN DAY RIVER GEOLOGY
(SERVICE CREEK TO CLARNO)

LANDSLIDE DEBRIS

COLUMBIA RIVER LAVAS
(YAKIMA BASALTS)

JOHN DAY FORMATION

CLARNO FORMATION WITH INTRUSIVE
VOLCANICS

PERMIAN/TRIASSIC SEDIMENTS

FAULT

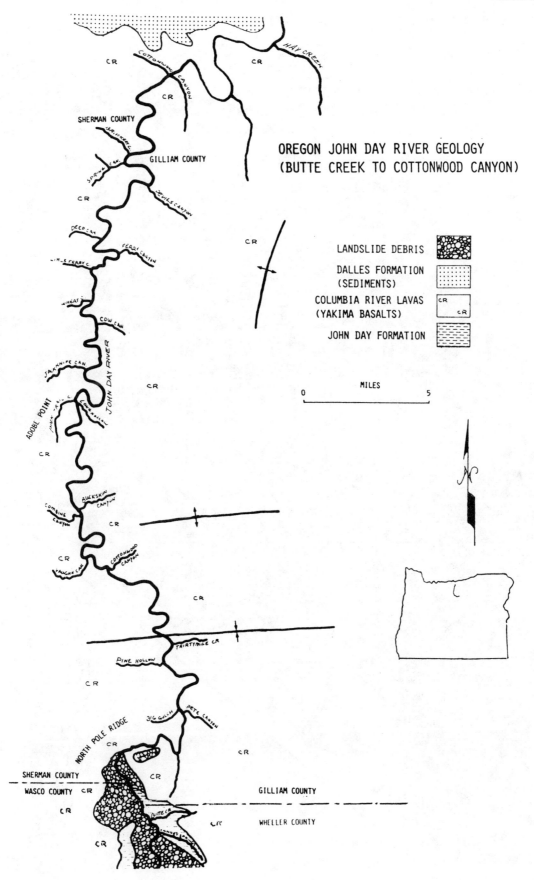

OREGON JOHN DAY RIVER GEOLOGY
(BUTTE CREEK TO COTTONWOOD CANYON)

LANDSLIDE DEBRIS

DALLES FORMATION
(SEDIMENTS)

COLUMBIA RIVER LAVAS
(YAKIMA BASALTS)

JOHN DAY FORMATION

MILES

0   5

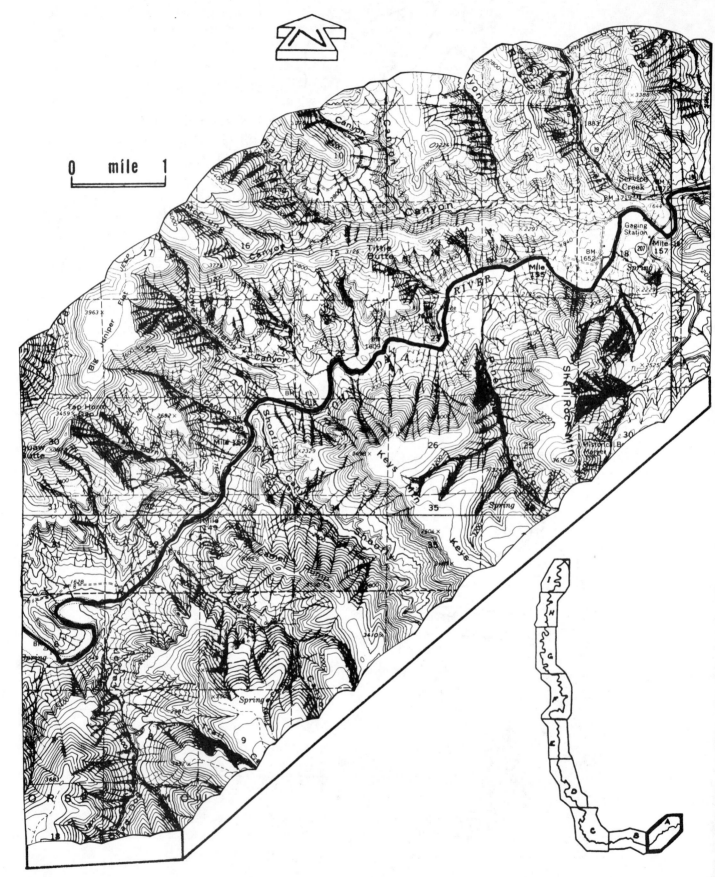

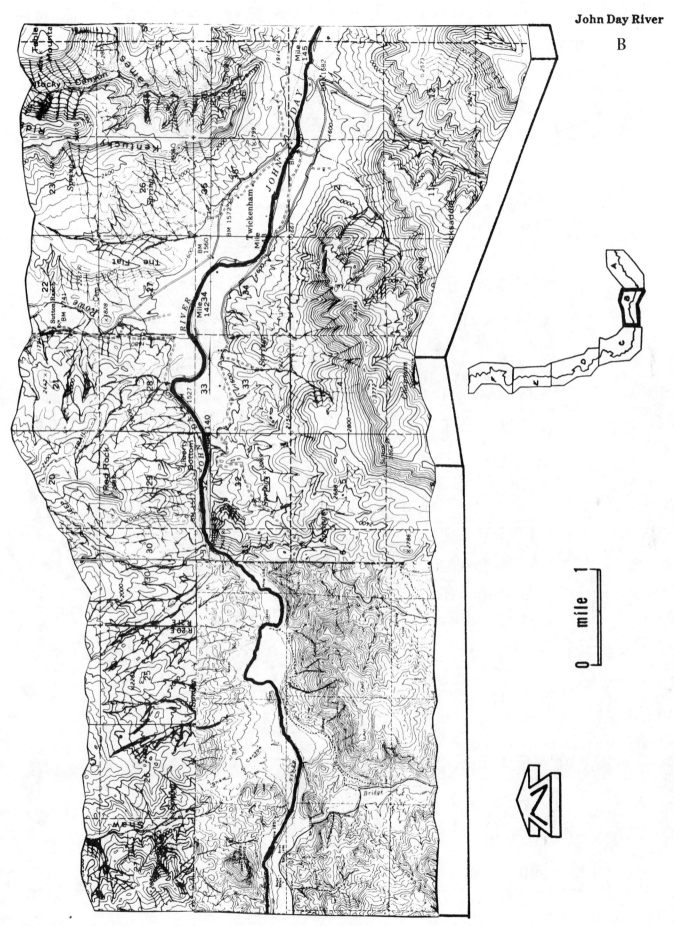

0   mile   1

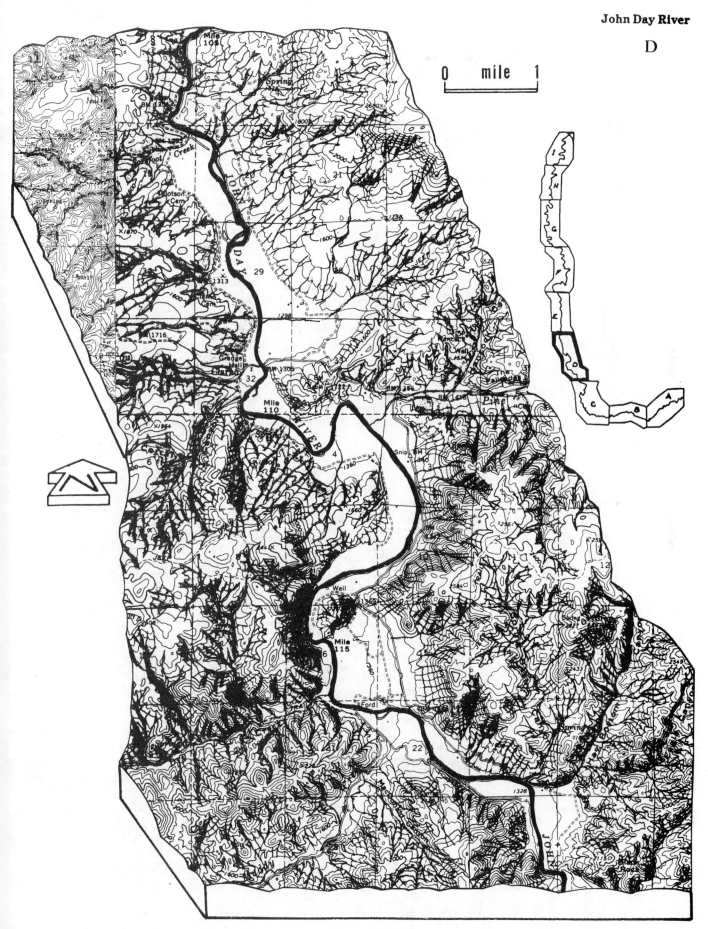

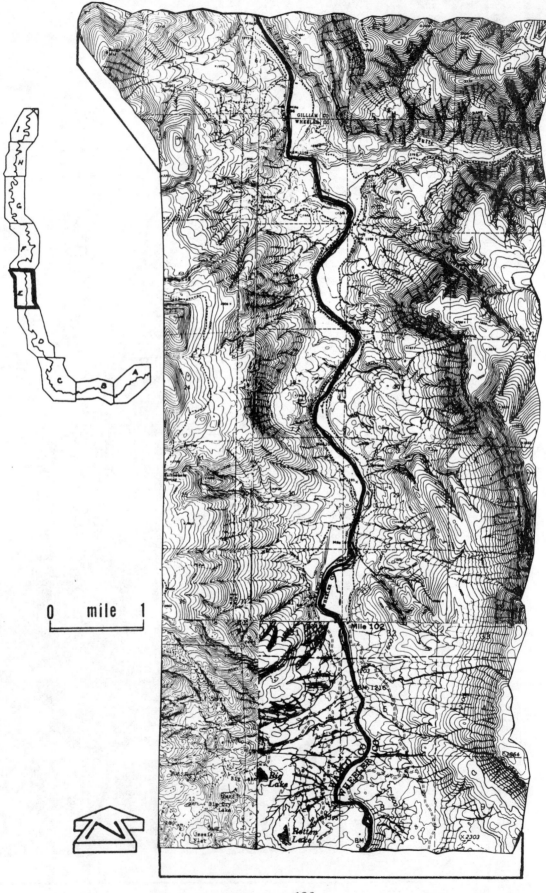

0 mile 1

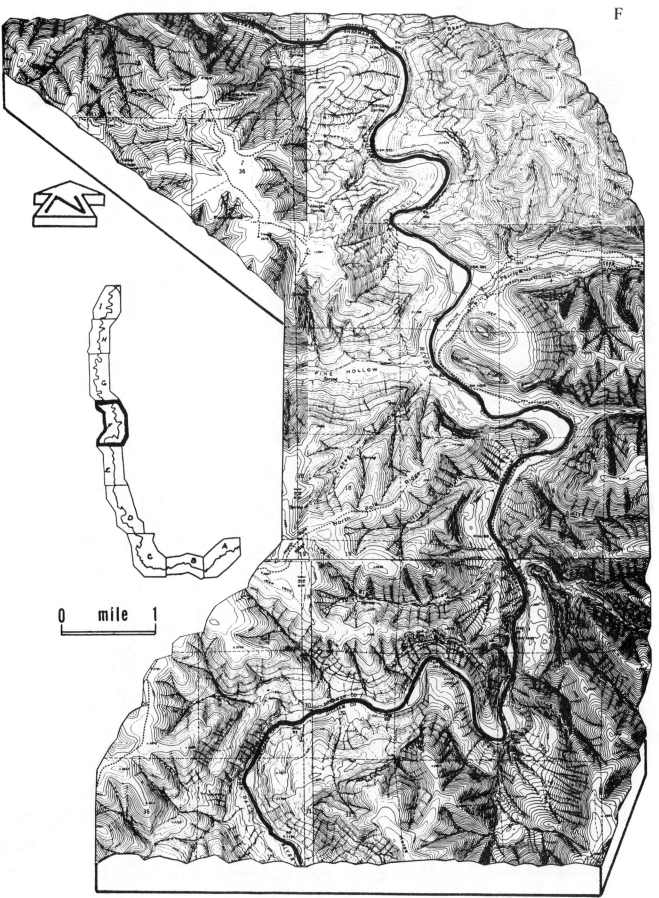

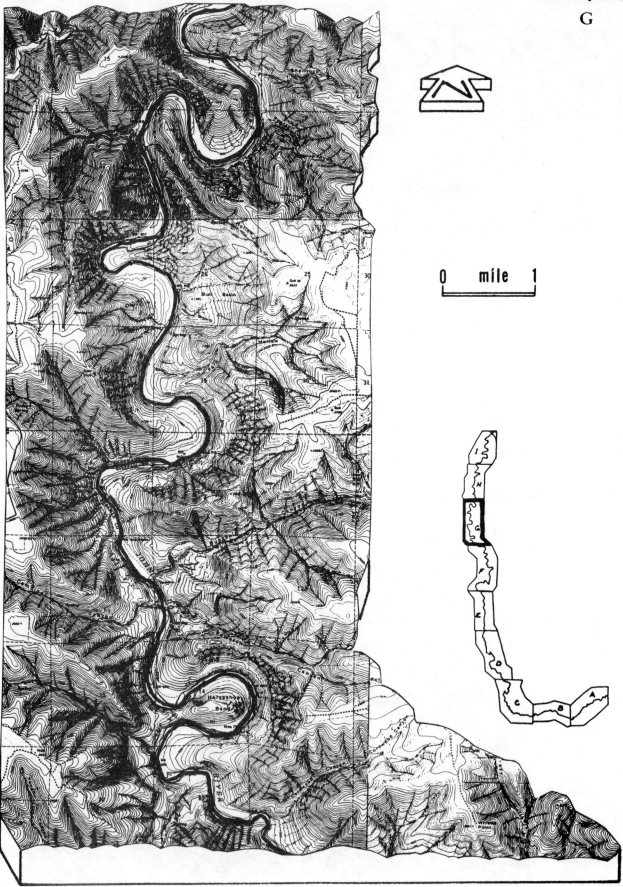

0   mile   1

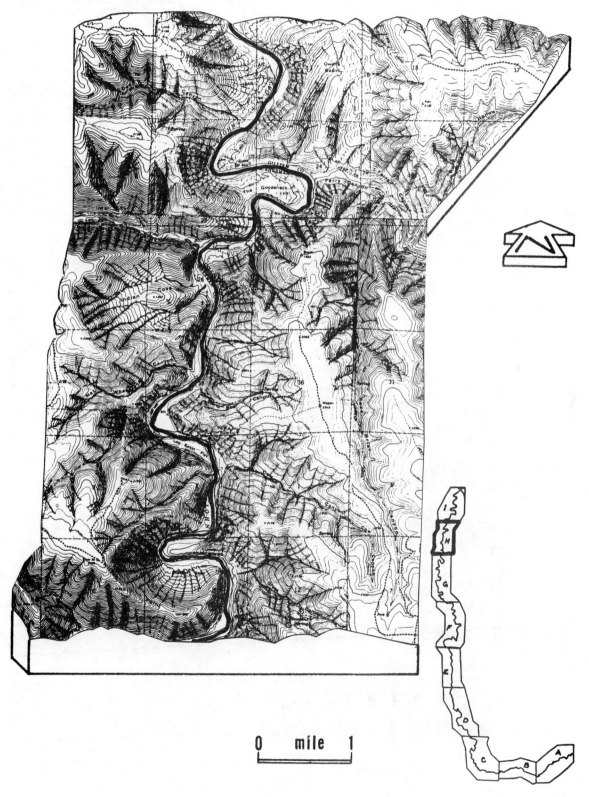

0 mile 1

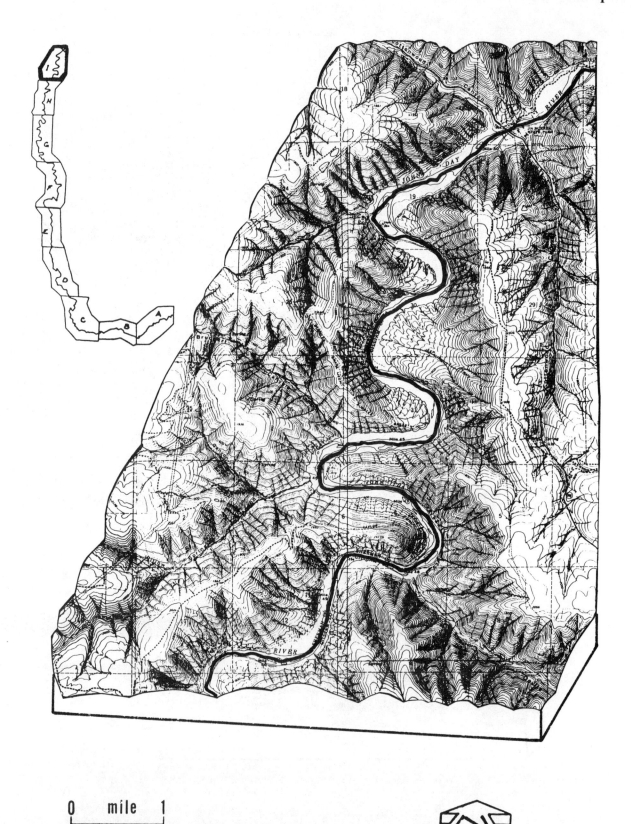

0 mile 1

Klamath River

(Keno, Oregon to Copco Lake, California)
Klamath County, Oregon to Siskiyou County,
California
Length of run: 35 miles
Number of days: 1-2

It is curious that this runnable section of the
Klamath river is suspended between two lakes,
Klamath Lake and Copco Lake as well as crossing through major physiographic regions from
the Great Basin into the Cascade Mountains.
Because of these physical features, sections of
the river are completely different from each
other. During a history going back 7,000 years,
Indians who lived here developed a life style
not found elsewhere.

This is one of the rivers in the West facing a
crisis at present because of plans to construct a
dam near Salt Cave. This beautiful river offers a
good deal more than just a potential for hydroelectric power. It is significant that this area already has Oregon's only active source of geo-
thermal power.

Early History of the Klamath River

In prehistoric times Klamath Indians occupied an area around upper Klamath Lake and
Klamath Marsh, Oregon, south along the Klamath River to where it crosses the Oregon-California border and west into the area of present
day Copco Lake, California. At this point, the
river was on the fringe of four tribes, the Takelma, the Klamaths, the Modocs, and the
Shasta. These Indians cooperatively used the
area for hunting, fishing and gathering. Of the
four groups, the Klamaths and Modocs had a
similar culture and language. The name "Modoc" is derived from Klamath Indian words
"moa" meaning south and "takhi" meaning
native or inhabitant.

The name Klamath was first used in 1826 by
trapper Peter Ogden of Hudson's Bay Company
after the Columbia River Indians told him about
the "Clammitte" to the south. The Klamaths
called themselves *maklaks*, meaning "men".

Population estimates are difficult as most
groups were greatly reduced by epidemics before any counts were made. Estimates are
based on size of villages, number of houses, and
number of individuals thought to occupy a
household. The Modoc-Klamaths were considered to number 1,200 to 2,000 persons. Of those,

800 to 1,400 lived around the lakes and marshes
in southern Oregon where they were called
"lake people" by their neighbors.

The Klamath tribe had a unique life style because they relied heavily on large shallow lakes
and marshes for food. As a result, the tools and
customs they developed were specifically
adapted to this environment.

Fish runs in March signaled the move from
winter to summer villages. Some villages were
named ceremoniously, as Sucker Fishery Village, where up to 50 tons of fish could be caught
in dipnets or lines set out during several weeks
of fishing. Torches and canoes were used for
night fishing. A distinctively triangular-shaped
fishnet weight was thought to have been attached to a net dragged after the canoe.

As Summer progressed, camas, arrowroot,
and tule were harvested. Tule seeds were eaten
and the stems were woven into baskets, mats

Distribution of Klamath and Modoc Indians in Oregon and California.

for houses, shoes, and other items. Wokas, a yellow-flowered, water lily covered 10,000 acres and was the most important food staple in the Klamath diet. The seed pods were harvested in August, baked, and pounded to separate the seeds from the pod. The seeds were then ground with a mano. The horned mano was a uniquely Klamath development and found only in the marsh area where wokas grew. One person in a canoe might gather 4 to 7 bushels of pods in a day. Berries, seeds, and nuts were harvested in Fall.

Edward Cope, a geologist travelling in the Klamath area in 1880, reported his Klamath guide Chaloquin, carried a bag of wokas on their trip. Chaloquin would mix the wokas with water and wait for it to swell before eating. Cope reported it made an "agreeable mush tasting like farina and coffee". Cope himself ate heated canned tomatoes mixed with crumbled biscuits.

Deer and antelope drives took place by herding the animals into the lake to be speared by women in canoes. Marsh birds and their eggs were also eaten. Birds were snared at night when they flew toward a fire built in the front of a canoe.

Fishing through the ice during December fish runs as well as winter hunting added to their store of food. During particularly long winters, patches of bark were stripped from pine trees, and the tender inside was eaten along with a stringy black lichen. The Klamaths rarely lacked for food living as they did in an area of vast and varied resources.

Places along lakes and rivers where food

plants and animals were plentiful were revisited each year. Permanent villages were build nearby for convenience. Larger conical winter houses, covered with tule mats and earth, had an opening in the top for entry. These houses might be set up in a continuous line — as reported along the Williamson River — for 5 to 6 miles. More than one family would occupy a winter house. Summer houses were above ground and were smaller brush hut structures.

A house and contents were burned after the occupant died. The body, along with some possessions, was placed atop a traditional cremation pile, one of which was said to be 12 feet high and 30 feet in diameter. Horses were frequently burned at the same time.

Most Klamath tribes had three leaders, one for religion, warfare, and domestic problems. Duties didn't overlap, and each leader advanced himself politically by wealth, oratory, or experience. Initially these positions weren't hereditary and depended on popularity and good will in the tribe. However, this may have changed after the arrival of Americans with a more noticeable accumulation of wealth among the leaders.

Manos for grinding seeds and acorns.

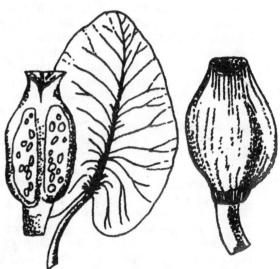

Wokas seed pod and leaf.

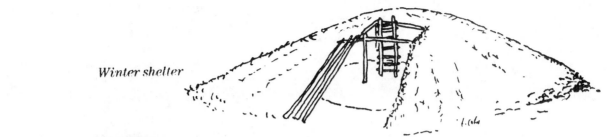

Winter shelter

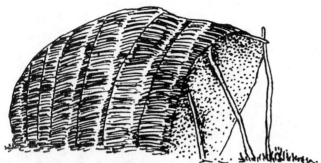

Summer mat house

Klamath and Modoc were aggressive and maintained actively hostile relationships with many Indian neighbors. The Molallas to the north were an exception. Raids on groups some distance away were conducted to obtain slaves and goods to trade at The Dalles in northern Oregon.

"When the Snakes made war on us that made us keen to fight other Indians and we made war without provocation . . . Those wars lasted a great many years. We found we could make money by war, for we sold the provisions and property captured for horses and other things we needed. It was like soldiers nowdays who fight for money. We made war because we made money by it and we rather got to like it anyhow" (Minor, Beckham, Toepel, p.117).

The acquisition of the horse enabled the Klamaths to range further afield in their raids, bringing on a period of prosperity which lasted until white settlement. After Euro-American contact, guns and horses were the main items obtained through trade.

Fur traders explored the area in 1825 only to find hostile Indians and not many fur bearing animals. According to trapper, Peter Ogden, the "scantiness of the local beaver made their territory of little interest . . . and they (Klamaths) came to enjoy a reputation as fierce warriors." Ogden and his men, who were in a starving state on this journey, obtained roots, fish and plentiful dogs to eat from the Indians in the vicinity of what Ogden came to name Dog Lake, present day Klamath Lake. Dog flesh served as emergency fare for early explorers, and Ogden recorded "We traded 10 dogs this day . . . at the following rate — 4 rings for one, the same number of buttons or thimbles — and for a scraper too."

Romanticized explorer, John Fremont, trekked back and forth through the area several times in the middle 1840's. On something of a scientific expedition, his cartographer produced maps naming geographical features. He was accompanied by scientists who collected plants, recorded geology and kept lengthy journals. Several of Fremont's men were ambushed and killed by the Klamaths in 1845.

Settlers, in their ever persistant search for an easy or "better" route to the West Coast, were the next to penetrate Klamath/Modoc lands. In 1846, Levi Scott, along with Jesse and Lindsay Applegate, started south through the Rogue and Umpqua Valleys of Oregon, turned east to-

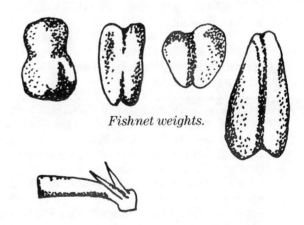

Fishnet weights.

Fish hook from the tail bone of a sucker.

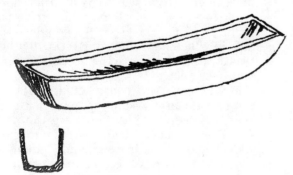

Canoe for deep water, about 25 feet in length.

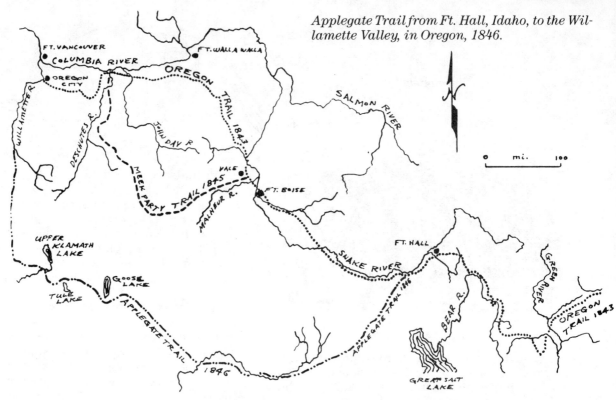

Applegate Trail from Ft. Hall, Idaho, to the Willamette Valley, in Oregon, 1846.

ward Klamath Lake, through the Klamath River basin to Tule Lake, ending their trip in Ft. Hall, Idaho. This was the new southern route, or Applegate Trail, as this alternate to the Oregon Trail came to be called. Although never as popular as the Oregon Trail, this route was used heavily in 1849 by gold miners.

Additional emigrant traffic brought clashes between white settlers and the Klamath-Modocs. Each side blamed the other for starting the trouble. Mr. T. Carter relates that hunting and shooting Indians for amusement was a favorite pastime irregardless of whether they were friendly or not. Emigrants, about to hang an Indian along the trail, were chagrined to find it was a woman.

Bloody Point, where the Applegate Trail crossed near Tule Lake was the most popular spot for Indians to seek revenge upon passing wagon trains. It was reported that 150 persons had been killed and 300 wounded by Indians near here. In 1852 packers who had their horses stolen followed the river to near Keno, Oregon, where they found and killed four Indians. Later, in retaliation an entire wagon train was destroyed at Bloody Point.

This action precipitated what came to be known as the Ben Wright affair. A group of volunteers from Yreka, California, seeking vengeance, were led by the infamous Indian hunter, Benjamin Wright. Wright provides an example

of what was no doubt a sadistic and unbalanced personality whose actions were permitted by the society in which he lived. He was killed at a later date by the Rogue River Indians.

Early one morning Wright and his men deliberately lured the Modocs with promises of food and peaceful council into camp before killing and mutilating 40 out of the 47. California newspaper accounts of the incident weren't more humane in reflecting the American attitude that Indian killing was just retribution for whites killed.

Increased hostilities led to the abandonment of the southern route. A fort was needed to protect the trail, but proponents couldn't agree on its location. Once the Klamath region was selected as the most practical spot, one group favored the present day Ashland area, the other Jacksonville to the west. The final decision was made by Col. Charles Drew who selected a site in favor of the Jacksonville proponents, the side to which he belonged. Eventually the fort was located 35 miles to the north of the road it was to protect.

Ft. Klamath was constructed in 1853 and, with customary wisdom, manned with volunteers aggressively hostile toward the natives. Col. Drew, a man with an obsession to exterminate Indians, was in command. Indians complained "Col. Drew has taken possession without their having any assurance that they would ever get anything for their country, and that it would shortly be overrun by settlers."

Such proved to be the case. Settlements increased, and in 1864 the Klamaths had signed a treaty giving most of their land to the U.S. government creating the Klamath Reservation. The reservation confined them to the Sprague and Williamson river basins and lands adjacent to Agency Lake.

Cattlemen with herds numbering 2,000 spread around Keno. Keno, originally the site of a ferry operated by Robert Whittle, went through a series of name changes. First called Whittles Ferry, it was renamed Plevna (a name in the 1878 Russo-Turkish war). This was changed to Keno after a Capt. D. J. Ferree's dog. The bird dog was named after a popular card game.

In 1866 the first cabins were built near present day Klamath Falls. Formerly called Linkville, George Nurse, a merchant from Ft. Klamath, opened his one room store there with his hotel across the street. The store and hotel did a large amount of business during the Modoc War of 1872-1873.

That war was the only major conflict between Indians and whites in this part of the country, and the year long conflict took place to the southeast of the Klamath River in the lava beds of Siskiyou County, California. Modoc Chief, Captain Jack, after several brief unhappy tries at life on the Klamath Reservation, left taking with him a small band of 150 men, women, and children. In 1872 the Army sent out troops to forceably return this band to the reservation.

Captain Jack, Modoc leader during the war of 1872-1873.

The Indians weren't in the mood to be persuaded, and instead Captain Jack took up a position in the natural fortress created by the lava fields in northeastern California. The ensuing war lasted until Captain Jack and the remainder of his band left the stronghold, demoralized by continual mortar shelling, lack of food and water, and internal dissention.

After a few skirmishes, Jack along with three of his men was captured and hanged. Their graves at Ft. Klamath were visited by geologist E.D. Cope a short time later. All had been rifled at the time of his visit, and the bodies removed. Cope afterwards obtained the skeleton of one of the Indians.

Education of the Klamaths was about to begin. White settlers and soldiers anticipated a civilizing influence on the "barbaric . . . wild and high spirited savages" (Bryant, p.77). Buildings, sawmills and schools were built. In subsequent years lumber and natural resources proved to be substantial making the Klamaths a wealthy tribe. White settlers and their livestock encroached onto the reservation, and lumber was poached. The reservation was abolished in 1954 by the U.S. government thereby opening this land to white take-over.

Indian Life Along the River

Villages, quarry sites, rock shelters, rock cairns, and hunting blinds along the river testify to the presence of Indians dwelling from Keno south in ancient times.

Only one house pit was discovered at Keno, but on nearby Bogus Creek, three human burials, two horse burials, a dog, a wagon, as well as beads and trade goods, were excavated. Obsid-

Bridge at Keno, Oregon (Klamath County Historical Society).

ian and stone artifacts were among the items found.

Just south of Keno on the river, during excavations for Big Bend Dam, two rockshelters and one open site provided an abundance of artifacts — pottery, points, and faunal remains — used by hunters and gatherers around the years 1,000 to 1,800.

Eleven sites near Salt Cave on both sides of the Klamath River where it crosses the Oregon-California border indicate occupation as far back as 7,000 years ago. Rockshelters, villages, a shell midden and open sites were excavated. Bone tools, scrapers and points, along with many other artifacts, show two or three cultural groups used the cave on a year round basis. The dwellers were adapted to a diversified hunting, fishing and gathering life. Some artifacts were obtained in trade from California and from the Oregon coast a distance away. A wooden shovel, partly buried near a salt spring at one of the caves, indicates this area was used fairly recently as a source of salt.

Mortars, basalt tools and projectile points have been found south of Copco Lake on Indian Creek and on a river terrace downstream from Beswick.

At Iron Gate Dam, a village with three house pits used as late as the year 1,500 was excavated. Perhaps 30 people lived in this village.

It is worthwhile mentioning a petroglyph site in the Klamath area, although it is north of Keno about 60 miles. Klamath/Modoc petroglyphs are unusual in style and coloring. Near the remnants of a cremation pit and old village site, red and copper, blue-green petroglyphs have been drawn to represent lizards, circles, bird tracks, human figures, and a unique "shield" design. The yellow center and red design rarely occur in Oregon petroglyphs.

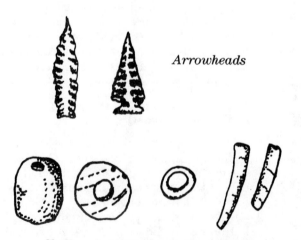

Arrowheads

Shell and steatite beads from Salt Cave.

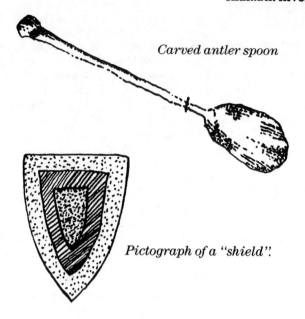

Carved antler spoon

Pictograph of a "shield".

Geology of the Klamath River

The Klamath River crosses through both Oregon and California as well as traversing two major physiographic provinces — the Great Basin and Western Cascades. The geology of each of these provinces exerts a clear and distinctive control over the river's pathway.

Upper Klamath Lake, north of Keno, Oregon, once extended over a much wider area than its present limit. The presence of ice age, aquatic clams and snails has been used to reconstruct shorelines far beyond the reaches of even the marshes today. During the ice ages (Pleistocene), in excess of 10,000 years ago, all of the lakes in this vicinity of the Great Basin were considerably larger. Today swamps and marshes are all that remain of many of these lakes.

Just north of Keno, on the Klamath River, the stream sluggishly meanders through a marsh more than 10 miles in length. The levels of the lakes and marshes have been dropping steadily for the past hundreds of years as the climate has progressively become warmer and drier. At Keno the bedrock is volcanic lava flows, ash, dust, and cinders, much of which is less than 10,000 years old. Geologically referred to as the Klamath basin, this area is the northwest corner of the Great Basin.

Drainage in the Great Basin is dominantly internal, but the Klamath drains west to the Pacific. Fossil salmon preserved in old lake sediments suggest that this drainage had an outlet to the ocean well over a million years in the past. This area of the Klamath basin is cut by hundreds of small and large faults. Most of the

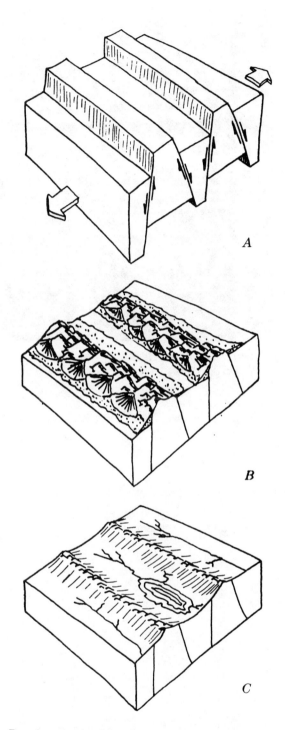

A

B

C

Development of basin and range topography from large-scale block faulting.

Geology Along the Route

The course of the river for the first ten miles west of the put-in at Keno meanders back and forth across pronounced fault blocks. Typically the river will flow westward then change course sharply to the north or south as it intersects the softer rocks of a fault. Even a cursory look at the topographic map of this area reveals a distinctive northwest/southeast "grain" to the physiography.

Only a mile or so out of Keno, sediments are evident as deposits exposed in the canyon walls. These sediments are locally referred to as the "Yonna Formation", and they are volcanic material of sand and silt which has been redeposited by water. The perfect stratification visible as even, banded layers here attests to a lacustrine or lake deposit origin for most of these rocks.

Mammal fossils including peccary or pig and fish collected from these sediments have been dated at just over 5 million years old. This period of time, the Pliocene epoch, is predominantly one of erosion in much of Oregon, and this area is one of the very few in the state where fossils of this age are to be found.

As one rafts downriver along this stretch, the faults are very evident as straight valleys with few meanders. From the river you can see as much as a mile or two up into these remarkably straight fault controlled canyons. After only about 5 miles of exposed Yonna Formation lake sediments, the unit disappears from the canyon walls, and the river moves back into younger volcanic rocks again.

Six to seven miles north of the Oregon/California line, the river exposes an older series of volcanic rocks. These are a thick sequence of volcanics ranging in age from 10 million (Miocene) down to 40 million (Eocene). They are part of the Western or older Cascades which occur in a broad north-south strip lying just west of the volcanic peaks of the high Cascades. The river remains in this western Cascades province within what is called the Rhododendron Formation all the way to Copco Lake in California.

Rocks in this interval are hard and resist erosion. Consequently the canyon walls close in to form a deep, narrow channel. On the skyline near the border, mesas capped by lavas overlying the older flows dot the landscape to the southeast. Within this stretch from the border to Copco Lake the older volcanics tilt gently toward the east. Traversing downriver toward the west, we are in effect going down section or into older rocks.

The Klamath River along this stretch has been

faults are of the vertical type with an upthrown and down-thrown side. On the down-thrown side of the fault, a shallow basin with a lake or playa is often developed. Round Lake and Aspen Lake are good examples of such "sag" basins.

referred to as an antecedent stream. This means its course was established prior to the uplift of the underlying rocks. Stream erosion was able to keep pace with the uplift process. Similarly, the Rogue and Umpqua rivers to the west are antecedent streams.

A series of thermal springs are located in a meadow bordering the Klamath River near where Shovel Creek enters. Called Beswick Hot Springs, the five springs are both clear-water and hot-mud baths where the temperature reaches 152 degrees fahrenheit.

The take-out at Copco Lake is a fascinating area geologically. Historically many dams for large reservoirs were the sites of original natural dams and small lakes. Copco Lake is one such example, but here the natural lake far exceeded the present manmade reservoir. Shorelines and beaches of the old lake are still visible high up on the canyon walls. This lake developed and drained within the past few thousands of years, but instead of concrete, nature used basalt (lava) from nearby volcanic cones.

This natural engineering process began with three volcanic cinder cones 200 to 300 feet high. These are visible today just below the present dam powerhouse in a north-south line less than a mile apart. Lava flows from the cones poured into the river valley for a distance of two miles downstream. Later flows piled up in the vicinity of the cones, and the combined flows built a dam up to 120 feet high in the gorge. The huge dam was later eroded, breached and drained by the river.

Beautifully preserved volcanic cones dot the landscape in the Klamath River Valley below the state line. Dozens of small flows as well as eroded volcanic necks and related features such as dikes impart a rugged topography. Just to the south toward Mt. Shasta, less than 5 miles away, even larger volcanic structures are scattered over the landscape. `

The climate here has prevented the development of thick mature soils common to wetter areas to the west in the coast ranges and Klamaths. One consequence of this is good exposures of rocks of the particular lava flows and ash falls from cinder cones. It is usually possible to trace individual lava flows right back to their source at a fissure or volcanic cone by carefully walking out the exposures.

A series of hydroelectric dams near the border interrupt the normal stream flow and cut off salmon spawning on the upper Klamath. Copco Dam No. 1 was completed by 1922, the name Copco formed from the words California-Oregon Power Company.

Waterwheel at Beswick used for irrigation (Siskiyou County Historical Society).

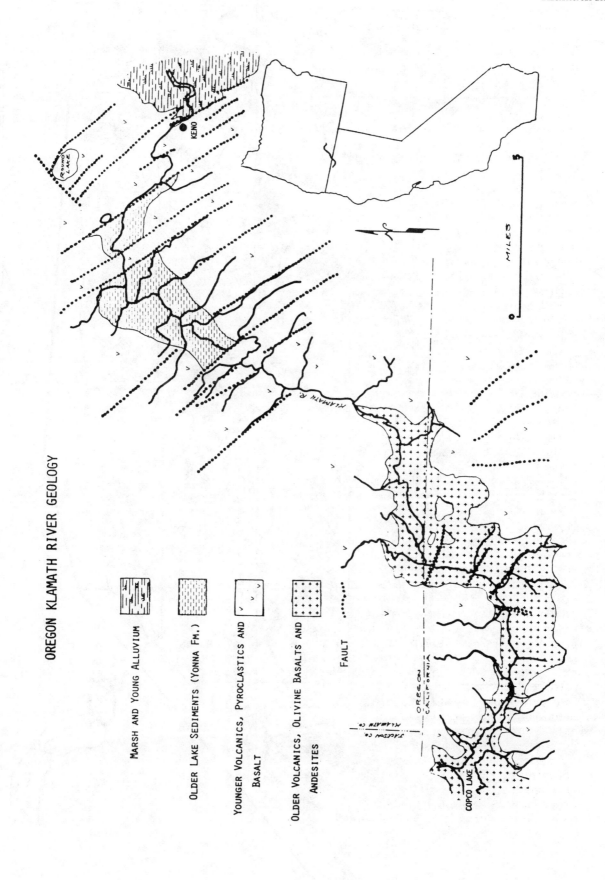

OREGON KLAMATH RIVER GEOLOGY

MARSH AND YOUNG ALLUVIUM

OLDER LAKE SEDIMENTS (YONNA FM.)

YOUNGER VOLCANICS, PYROCLASTICS AND BASALT

OLDER VOLCANICS, OLIVINE BASALTS AND ANDESITES

FAULT

KENO

ROUND LAKE

KLAMATH R.

MILES

OREGON CALIFORNIA

JACKSON CO. KLAMATH CO.

COPCO LAKE

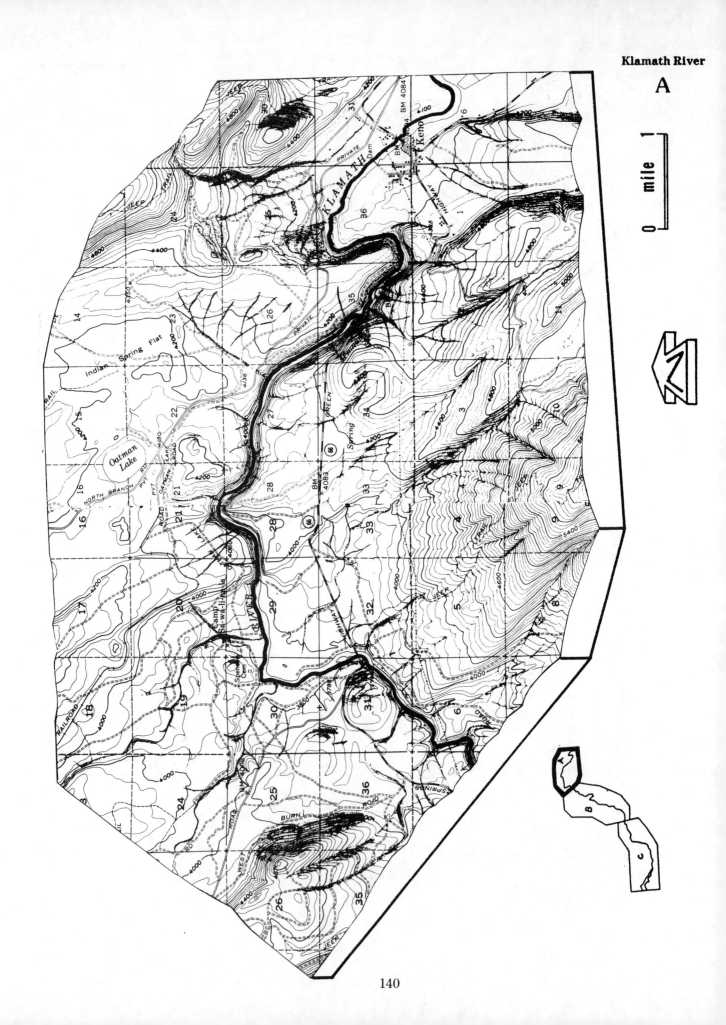

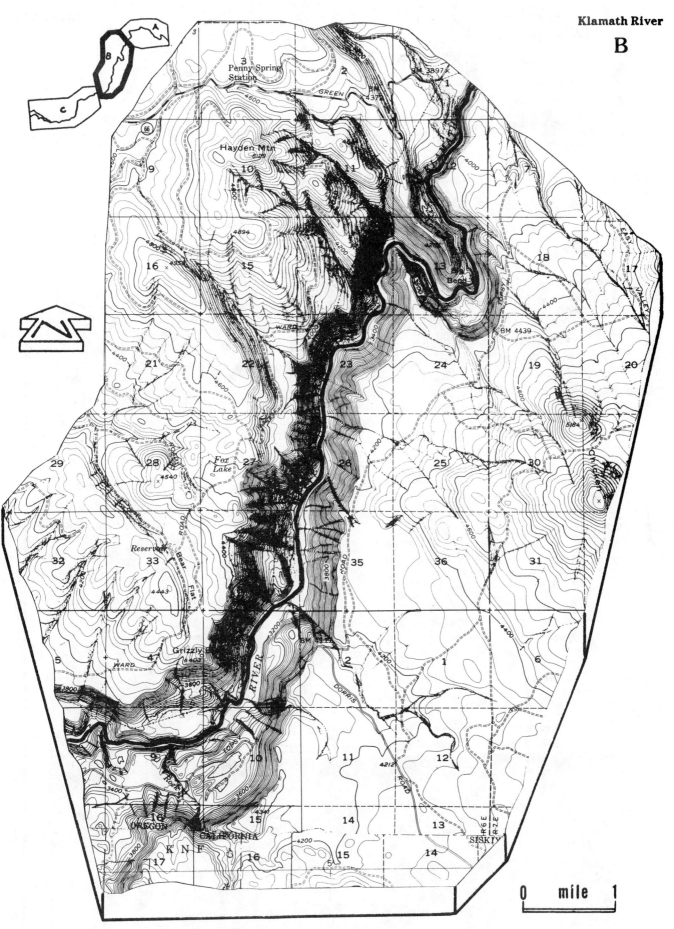

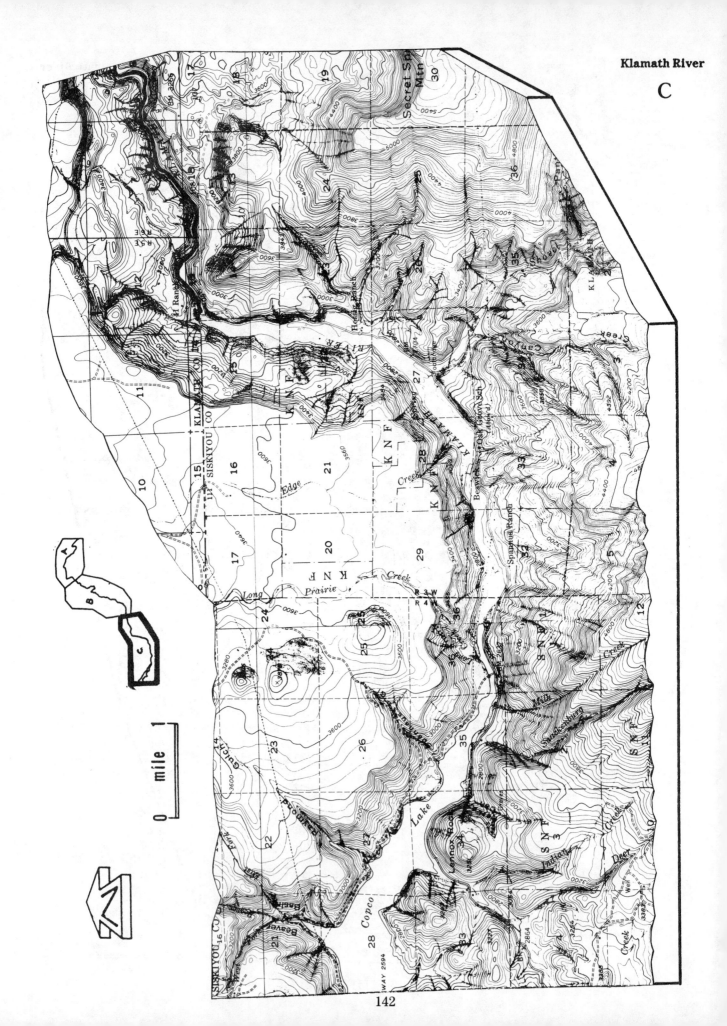

Owyhee River

(Rome to Leslie Gulch, Lake Owyhee)
Malheur County, Oregon
Length of run: 63 miles
Number of days: 3-5

Flowing through what is still one of the most inaccessible areas of the West, the lengthy Owyhee River covers a distance of 400 miles through steep, narrow canyons before merging with the Snake River to form the border of Oregon and Idaho. A long geologic history of large inland lakes and episodes of volcanic eruptions ended with the formation of the Owyhee Canyon more than 10,000 years ago.

Since that time, little has changed along the Owyhee. Throughout prehistoric times Indians lived on the plateau, and it was here the last Indian battle in the Northwest was fought. Rafters and hikers can see eagles nesting on lava ledges hundreds of feet above the valley floor. Wild horses, donkeys, and deer compete for grasslands. Birds and an unusually high number of rattlesnakes may prove to be the only companions seen by travellers in this region.

Early History of the Owyhee River

Northern Paiute belonged to a larger group of Indians living in the desert regions of Oregon, Idaho, northwestern Nevada, and a small area of northeastern California. The Owyhee River basin formed only a small part of their territory.

Early travellers through the Owyhee Plateau called the Northern Paiute by a variety of

Chief Buffalo Horn of the Bannocks.

names. Some names as "Snakes" and "Bannocks" referred to other Indians as well as to the Paiutes, whereas, "Paviotso" and "Digger Indians" referred only to the Northern Paiute themselves. "Paiute" means water (pa) and direction (ute).

The particularly harsh climate where they lived kept their numbers to a minimum, and only one person is estimated for every 10 to 20 miles. Life expectancy was low, and babies or old people would be abandoned when there wasn't enough food or they could no longer travel. The Paiute were loosely organized into bands of perhaps 40 to 50 persons, although families could move from band to band.

Paiutes didn't feel land was owned by any one group, but each band kept to its own region. Bands were, in turn, named after a distinctive food or landmark. Twenty-one bands lived in the region with such names as the Root Eaters, the Seed Eaters, or the Elk Eaters. The tribe occupying the Owyhee and parts of the Snake River basin was the Tagu or Tuber Eaters.

Food was so scarce that families spent most of their time travelling from place to place, searching for what the desert offered. One disaster, as lack of rainfall, might bring starvation. Plants were vital to Paiute survival. After using up winter supplies, Paiute sought fresh green thistle or squaw cabbage around streams and lakes in the Spring. As plants ripened and produced seeds throughout the Summer, Paiute families might travel up to 40 miles to gather

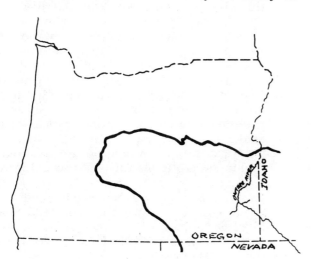

Distribution of Northern Paiute in Oregon, Idaho and Nevada.

143

seeds at a particularly productive location. Grass seeds — fescue, wheatgrass, Indian rice grass, bluegrass, cattail, rushes, and sunflowers — were collected, winnowed, and ground into flour. By adding water a mush was produced. Seeds were also stored in pits, baskets, or caves near winter villages. Later in the Summer, Paiute travelled again to moist areas where berries were picked and bulbs of camas, lily, arrowroot, and wild onion were dug. With the onset of winter, Paiute bands moved to villages near their stored food supplies where they stayed during the cold months.

Mat-covered house

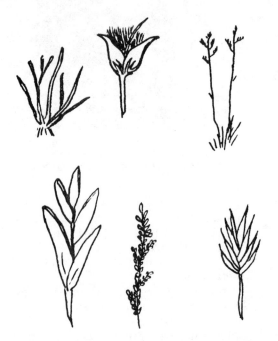

Grasses gathered and winnowed by the Paiute.

Winnowing basket

Game animals and fish were caught, but large scale, communal hunts could only take place on the average of every 11 years because of the scarcity of large desert animals. During Fall, animal hunts and round-ups would be organized to catch rabbit and antelope. Dogs herded the antelope into corrals. Spring pole traps, nets and clubs were used for rabbits, birds, rodents and other smaller game. Caterpillars, bee eggs, ants, and insects were eaten when in season.

Different parts of plants were useful as well. Serviceberry and juniper were made into bows. Baskets were woven tightly enough to carry water since the Paiute didn't make pottery. Bags, blankets, ropes, and sandals were woven from sagebrush bark, and plants were mixed for both poisons and medicines.

Before the introduction of horses, the Northern Paiute had to carry all their own possessions. Many implements were made and abandoned on the spot or stored in the winter camp to be reused later. Horses straying into Paiute territory were killed and eaten as a practical solution to life in the desert where the vegetation couldn't support a population of grazing animals. However, by the middle 1800's the Paiute were using the horse for riding, trading, and carrying goods. This greatly expanded the amount of territory they could cover and made life much easier.

In 1819, Donald McKenzie of the North West Fur Company bestowed the name "Owyhee" on the river. In his search for more profitable regions of fur bearing animals, McKenzie sent three Hawaiian Islanders down the river to explore. They never returned, and the river was named after them. It was first called the Sandwich Island River, but the name was later changed to Owyhee after their homeland.

As an agent of the Canadian fur company, McKenzie was following the company policy of discouraging serious American competitors by exterminating all fur bearing animals in the Snake River watershed. The Canadians felt American trappers would be so frustrated at finding no profitable furs, they would retreat once again east of the Rockies. To carry out this policy, Director John McLoughlin and Peter Skene Ogden provided the 1824-1825 Snake River Expedition with "2 gentlemen, 2 interpreters, 71 men and lads, 80 guns, 364 beaver traps, 372 horses."

Americans weren't completely dissuaded from exploring the Owyhee Plateau, but once the area was depleted of fur bearing animals, trappers had no further interest in remaining. After their departure the Owyhee wasn't disturbed again for 30 years. When gold was discovered on Jordan Creek in 1863, 2,500 miners overflowed from the Boise, Idaho, gold area into the Owyhee basin. Instantly towns arose. By 1865 the Poorman Mine was producing $4,000 to $5,000 dollars a ton for gold. Five hundred feet down in the Poorman, a beautiful 300 pound mass of ruby silver was discovered. The block of reddish, argentite crystals was displayed at the 1866 Paris exhibition and won a gold metal.

Traffic for gold brought on the development of wagon roads. These so-called roads usually followed Indian trails and on occasion were only slashes on trees or broken sagebrush. In 1864 two military parties scouting routes for roads left the first wagon wheel tracks in many parts of this desert region. These two parties descended into the southern Owyhee canyon. Lt. George B. Curry described the water carved rocks as . . . "an archway curiously shaped,

yonder is a tunnel running the face of a sandstone ledge hundreds of feet from the bottom."

The first passable east-west road from Idaho, known as the Oregon Central Military Road, traversed the Owhyee near Rome — so named because its white, chalk-like cliffs reminded travellers of pillars in ancient Rome. Near the Owyhee crossing, this road was developed by Sam Skinner who arrived to look for gold only to become disillusioned with the hard work and small returns. He opened his road in 1866 and charged $3 for a wagon and team, .25 cents for a horse and rider, and $1 for a buggy. Miners had hoped these improved roads would reduce food prices, but such was not the case. Flour reached a high of $32.20 for one hundred pounds, eggs were $3 per dozen, picks were $8 each, and boots cost $12 a pair. A stage ride from the Owyhee area to Chico, California, was $50.

Traffic was so heavy along the roadway Skinner and his partners, Mike Jordan and Peter Donnelly, had to inspect the route continually for damage to the bed and grading. The partners were constantly on the lookout for hostile Indians who were determined to keep the whites out. During one of these inspection tours, Jor-

Poorman mine in the Owyhee territory, 1865 (Idaho Historical Society).

dan and his brother were killed, scalped and so mutilated, one of his would-be-rescuers, David Shirk, said "they were so cut to pieces that we had to gather them up in a blanket to bury them."

Once aroused, Paiute proved tenacious and fierce in defending themselves and their territory. Beginning in 1857 they raided the Tenino Indians living to the north on the Warm Springs Reservation. About 250 Northern Paiutes ran off 150 livestock and killed 13 women and children. A later raid only involved the theft of horses. Miners and travellers were harassed continually by the Paiute. Elusive Indians easily avoided volunteer military expeditions sent out to protect road travel in 1864-1865. Francis F. Victor wrote of these campaigns "Few skirmishes here and not a dozen Indians killed from April to August."

The Oregon Volunteers had been formed from miners and settlers who found the regular income better than could be earned elsewhere. As soldiers their performance was less than spectacular. Their attitude was to "clean out the murdering savages". They even managed to become irate over the murder of 50 Chinese on their way to the mines, in spite of the fact Chinese themselves were regarded as heathen in other circumstances. The hapless Chinese had been armed with wooden rifles to scare away the Indians. The Owyhee *Avalanche* reported the "bodies were mutilated in the most shocking manner." The newspaper went on to dismiss

Camp Lyon, 1866, headquarters of the Owyhee District.

Easterners who expressed concern for the welfare of the Indians as "skimmed milk philanthropists."

Regular military troops arrived in the Owyhee area after 1865. Initial encounters were won by the Paiute who had perfected a guerrilla warfare. From the sidelines anxiously watching miners didn't know whether to laugh or take up arms themselves again after hearing stories of the army attacking a group of strangely shaped rocks in the night. In 1866 disgusted miners organized for a second time to go "Indian hunting" as they defined their purpose. Military and volunteers unsuccessfully chased the Paiute around the desert, while the Indians continued to raid settlers. However, on one occasion when the volunteers were almost eliminated in an ambush, they prudently retired to mining saloons and let the Army subdue the Indians.

By 1868 the increased establishment of military camps and forts manned by soldiers compelled the Northern Paiute to surrender. Paiute bands were placed on the Malheur Reservation, created in 1871 by President U.S. Grant. Not happy as reservation farmers, an alien way of life to them, the Paiute left when a clerical error in 1878 opened the Camas Prairie Reservation in Idaho — written by the clerk as Kansas Prairie — to whites. This mistake precipitated the Bannock War.

The Paiute were led by Chief Buffalo Horn and Chief Egan who were protesting conditions on the Malheur Reservtion. General George Crook who had been on a tour of inspection just prior to the outbreak of hostilities reported "with the Bannocks and Shoshones our Indian policy has resolved itself into a question of warpath or starvation . . ." Egan, a Cayuse, whose parents were killed in a massacre, had been raised by the Paiute.

During the short war from April to September, settlers panicked and fled to Ft. Harney or Pendleton abandoning their houses and possessions, many of which were destroyed. "Our route was marked on every hand by evidences

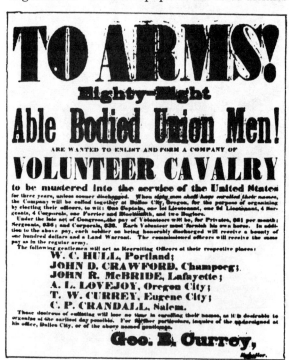

Recruiting poster for the Oregon Volunteers.

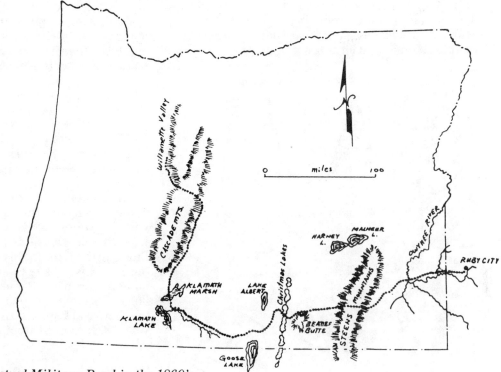

Oregon Central Military Road in the 1860's.

of destruction and carnage, sometimes we found ranch buildings razed to the ground by fire, again, simply ransacked, furniture destroyed, clothes, carpets, etc., stolen, windows broken and goods scattered around. Occasionally we found a white man murdered by the savages,'' was the comment of an early pioneer.

Buffalo Horn was shot in June by a volunteer on his way toward the Owyhee from Idaho. Egan crossed the river near Rome with a substantial number of Army and volunteers in pursuit. Moving north he captured horses along the route. Near La Grande, Oregon, he was tricked into a meeting and killed by Umatilla Indians who cut off his head and displayed it to the Army as proof of his death. During this last Indian uprising in the Northwest, perhaps 33 whites were killed along with 150 Indians.

Mining activity on the Owyhee Plateau all but ceased by the late 1800's, and once flourishing towns were abandoned. In their place, cattle and sheep spread over summer grasslands. Five hundred and ten Paiute Indians were moved to the Yakima Reservation in Washington. Some scattered to the Warm Springs Reservation or took up residence near small towns.

Indian Life Along the River

Evidence left by Indians who lived in the Owyhee desert is scarce. Petroglyphs have been drawn along the tortuous Owyhee canyon near Hole-in-the-Ground. Designs here include circles, human figures, bird tracks, ladders, and rain symbols. To the south along Jordan Creek, several sites display a series of petroglyphs in the canyon walls and on boulders near springs. The drawings on boulders have been exposed to weathering processes, and the lines and rake pattern are almost illegible. At the second locality, designs include zigzag lines and stars.

Rock art at Hole-in-the-Ground.

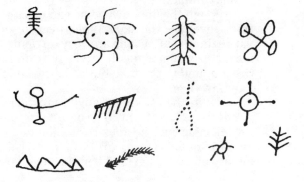

The Dirty Shame Rockshelter, about 30 miles south of Rome, is located on Antelope Creek, a tributary to the Owyhee. This shelter was occupied for a long period of time, from 9500 to 400 years before the present. Investigations have turned up arrowheads, leaf-shaped knives, milling stones, and large spear points.

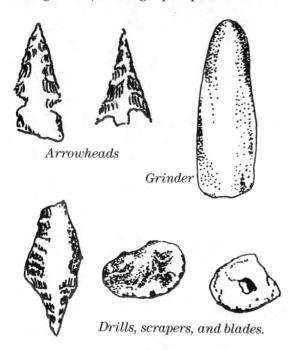

Arrowheads

Grinder

Drills, scrapers, and blades.

Geology of the Owyhee River

If the scenic Owyhee has any drawbacks, it is that the river is normally runnable with rafts for only a short interval in the Spring when there is sufficient water. This six week window of runnability is often a period of rain and even light snow. Some years, on the other hand, are a pleasant surprise. As late as the middle of June, 1984, the river was run with no difficulty, and it appeared that it would be runnable for sometime after that. Snowmelt on the Owyhee can cause the river level to rise and fall rapidly changing the complexion of many of the larger rapids completely.

Although the upper Owyhee can be run with rafts, several of the major rapids over this interval beginning in Nevada are very difficult. This narrative summarizes the geology from Rome, Oregon, and to the take-out at Leolio Gulch on Lake Owyhee.

Rocks in the Owyhee Canyon are relatively young in geologic terms. The oldest are those of the Miocene Sucker Creek Formation deposited about 16 million years ago. The general geologic history of the region has been the development of shallow lake basins and flood plains which are in turn filled with volcanic debris ranging from ash to lava. Volcanic eruptions were of variable composition and occurred intermittently. Fossil animals preserved here include mammals as well as fish. Ancient floodplain and lake deposits in between the lavas contain plant fossils such as wood, leaves, and pollen. The rocks of the canyon have been subject to only slight disturbances and are predominantly flat lying. In some areas the rocks have yielded to regional earth pressures by faulting. The breaks or faults locally have broken the landscape up into a series of blocks which give southeast Oregon its characteristic rugged topography.

The river between Rome and the Owyhee Reservoir cuts neatly through the strata beginning with the youngest, the Rome Beds, and ending with the oldest, the Sucker Creek Formation. Running the river in this interval is in effect going progressively back in time from rocks 10 million years old at Rome to strata dating 16 million years of age at the reservoir. This amounts to a backward step in time of roughly 100,000 years with each river mile.

Geology Along the Route

Beginning at Rome, Oregon, the buff to tan Rome Beds are visible on the skyline in every direction. The evenly layered (stratified) nature of these sediments is the hallmark of an ancient lake of considerable size. Between the mud/clay layers one sees coarse sands and gravels of beaches and bars of the old lake shore. The soft Rome sediments erode easily and impart to the river its typical mud brown color.

The fossils preserved in this old lake include a diverse fauna of mammals including peccaries, rodents (rats, mice, beaver), rabbits, antelope, otters, rhinoceros, horse, camels, bears, moles, and others. In addition, the bones and scales of many species of fish are preserved in the laminae of the lake muds like leaves in a book. All of these fossils have been dated as late Miocene in age or about 10 million years old. The abundance of huge lake fish and the many aquatic types of rodents and otters confirms the presence and size of the prehistoric lake.

Toward the northern extent of these lake sediments, five miles north of Rome, the river cuts directly across the axis of a gentle basin or "syncline". Here the Rome Beds have been slightly folded to form a shallow north/south trough.

Less than six miles downriver from the put-in, the river valley changes abruptly from the characteristic wide open valley with rolling hills to a narrow canyon with steep walls. The canyon walls here mark an exposure of a volcanic rock called rhyolite. Although the rhyolite weathers to a black or grey color, fresh exposures are an attractive pink. When they are hot, rhyolite lavas are characteristically stiff, viscous low temperature flows. The dough-like consistency, evident as swirled flow banding like a marble cake, can be easily seen in the sheer canyon walls. The rhyolite is called the Jump Creek Formation. That rock is well-exposed around Owyhee Dam where it forms prominent ridges.

From here the river moves into a series of very narrow canyons cut into lavas of the Deer Butte Formation. The Deer Butte consists of alternating thin lake and stream sediments and thicker lava flows. Near Jordan Valley large collections of fossil rodents have been extracted from the Deer Butte sediments. These rodents as well as carnivores and hoofed mammals have been dated at 15 million years. For the next 6 to 10 miles through the basalts of the Deer Butte Formation, the river also traverses a series of fractures or faults in the rock. These faults must have been accompanied by catastrophic earthquakes.

The river canyon opens up gradually as more and more soft sediment is exposed along the banks. Lambert Rocks is the name originally given to lava flows in the river valley. Over the years, however, the name has been applied to a spectacular monolith series in the badlands topography developed on the west side of the river. Interestingly enough, these badlands, or chalk basin, are the eroding sediments of another large shallow, ancient lake.

The most striking aspect of this badlands is a series of lava flows that poured out at intervals over the lake muds. These lavas are dramatic black horizontal layers in sharp contrast to the buff to tan lake sediments. In addition to the dark black parallel bands of the lava flows, the volcanic rocks also impart other colors to the landscape. As the lavas rapidly flowed over the moist lake flats, clay in the muds were baked to a natural red brick by the intense heat. This brick layer is still visible below each of the lava flows, and it is particularly resistant to erosion. Each of the picturesque columnar rock formations in the badlands is capped or armored by by a layer of the brick. This hard brick retards erosion and is responsible for the characteristic columnar topography.

In the dry washes emerging from Chalk Basin, quantities of opal are found. The opal is white with a glassy fracture and pearly luster. The silica (quartz) in the opal is derived from dissolved ash of the volcanics.

Another fascinating geologic feature visible in this stretch of the river is intercanyon lava flows. About 10,000 years ago the Owyhee Canyon as we now know it was already well developed. Then a series of very hot, runny lava flows from volcanoes up on the plains poured out over the landscape and into the river canyon. Although these flows did not entirely fill the canyon, they did disrupt the flow of water for a time. Faced with a solid rock (lava) plug blocking its path, the river began the slow process of cutting around the blockage. The easiest new route was invariably alongside the old canyon wall. These "intercanyon flows" are obvious today because in these intervals the valley walls don't match. On one canyon wall, lake or

Owyhee Canyon between Rome and Chalk Basin cutting through successive flood basalts of the Deer Butte Formation.

Lambert Rocks, badlands of the Owyhee.

stream sediments are exposed in gentle slopes, and on the opposite wall blocky lavas form sheer cliffs.

Volcanic breccias are common in this stretch. The breccia develops as volcanic episodes turn explosive with the sudden release of water and gas. These breccias have a shattered appearance where small blocks of lava stud a yellow volcanic ash. Faults are also common here. Whistling Bird Rapids has developed where a large block of rock and debris slid down a steep fault plane on the east side of the river meander. The slide scattered debris over the riverbed, and much of the large block is still visible on the east side of the river channel. Here it forms an extreme rafting hazard where it lies up against an old fault.

At this point the river takes an abrupt eastward turn into Green Dragon Canyon. The incredibly sheer walls make this canyon one of the most picturesque sections of any western river. The valley itself is a straight three mile stretch directly cut into the rocks of an ancient volcanic vent. These rocks are predominantly banded rhyolite, and the varicolored pinks and greys characteristic of this rock are spectacular. Close inspection of the wall rock in the steep canyons reveals the swirled and banded "marble-cake" effect of the rhyolite.

Equally profound is the east/west fault the river is following. This break or fault can be traced for several miles beyond the canyon itself. The steep walls and straight river course are another expression of the hardness and ero-sional resistance of the rhyolites. Just as the river turns north again upon leaving Green Dragon Canyon, the rafter encounters Montgomery rapids. A hard meander series first left then right make this a difficult rapid for even experienced rafters.

North of Montgomery, the river exposes older lake and fluvial (river) sediments many of which are coarse sands and gravels. The multiple colors here, including tan to buff, greens and reds, are imparted by small amounts of oxidized iron in the sediments.

For six to ten miles, prior to entering the waters of Lake Owyhee, the rock exposures in the valley are of the Sucker Creek Formation. This brightly colored rock unit is an ancient stream and lake environment which also experienced a succession of volcanic ash flows. The Sucker Creek contains scattered mammalian fossil remains of early Miocene age (16 million years old), but it is most famous for its locally rich deposits of fossil leaves. Plant fossils have been used to estimate the local rainfall back in the Miocene here at over 43 inches per year. The regional altitude, latitude, minimum temperature, and effect of local topography have been calculated from the time when the fossil plants lived there.

At the take-out at Leslie Gulch on Owyhee Reservoir, the road out winds back up to the plain through the Gulch. The canyon out is a beautiful cross-section through the rugged and colorful erosional landscape typical of the Sucker Creek Formation.

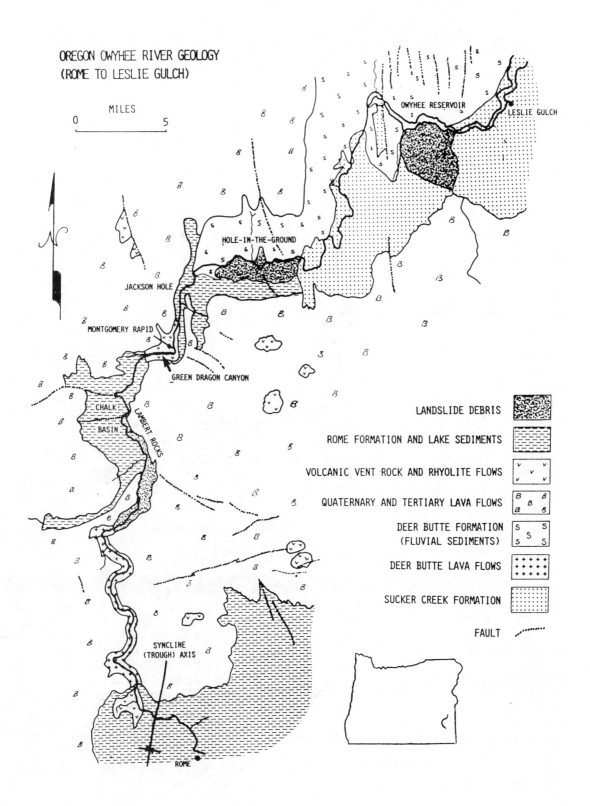

OREGON OWYHEE RIVER GEOLOGY
(ROME TO LESLIE GULCH)

MILES

0   5

OWYHEE RESERVOIR

LESLIE GULCH

HOLE-IN-THE-GROUND

JACKSON HOLE

MONTGOMERY RAPID

GREEN DRAGON CANYON

CHALK
BASIN

LANBERT ROCKS

SYNCLINE
(TROUGH) AXIS

ROME

LANDSLIDE DEBRIS

ROME FORMATION AND LAKE SEDIMENTS

VOLCANIC VENT ROCK AND RHYOLITE FLOWS

QUATERNARY AND TERTIARY LAVA FLOWS

DEER BUTTE FORMATION
(FLUVIAL SEDIMENTS)

DEER BUTTE LAVA FLOWS

SUCKER CREEK FORMATION

FAULT

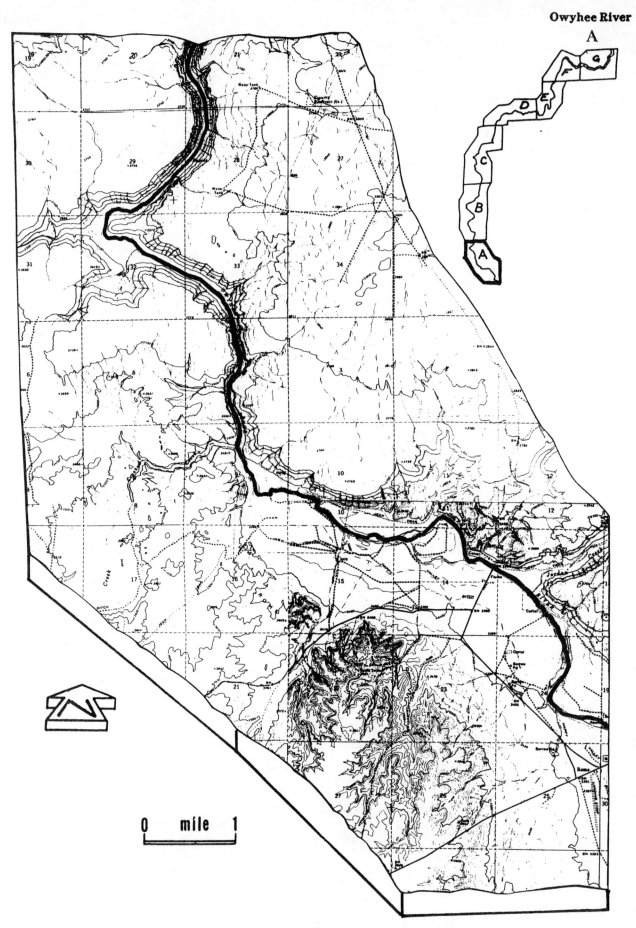

0   mile   1

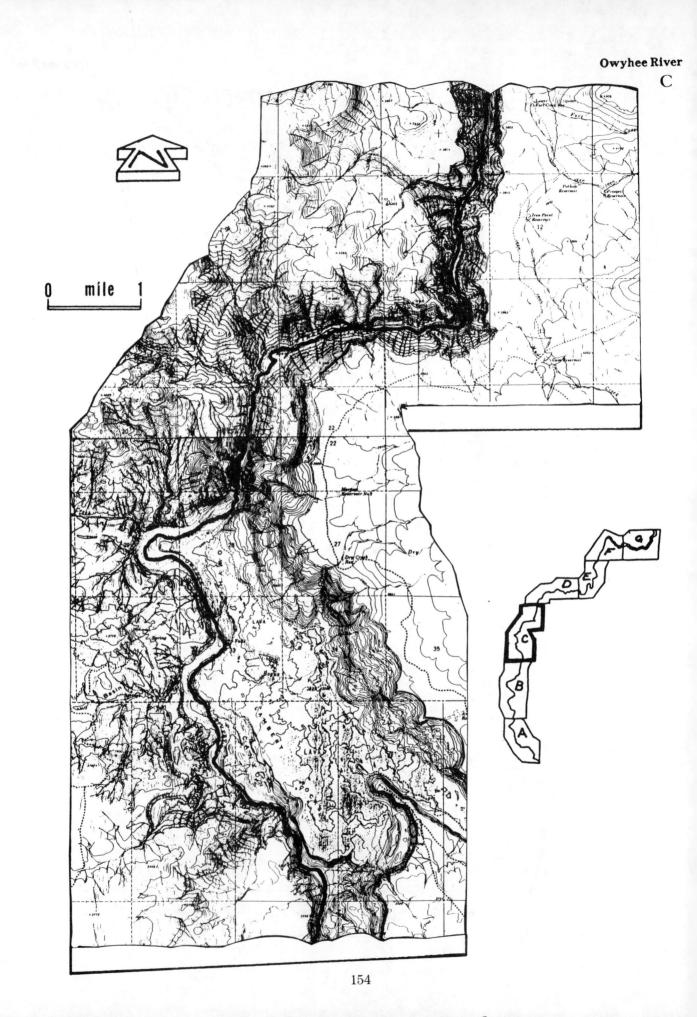

0 mile 1

Owyhee River

D

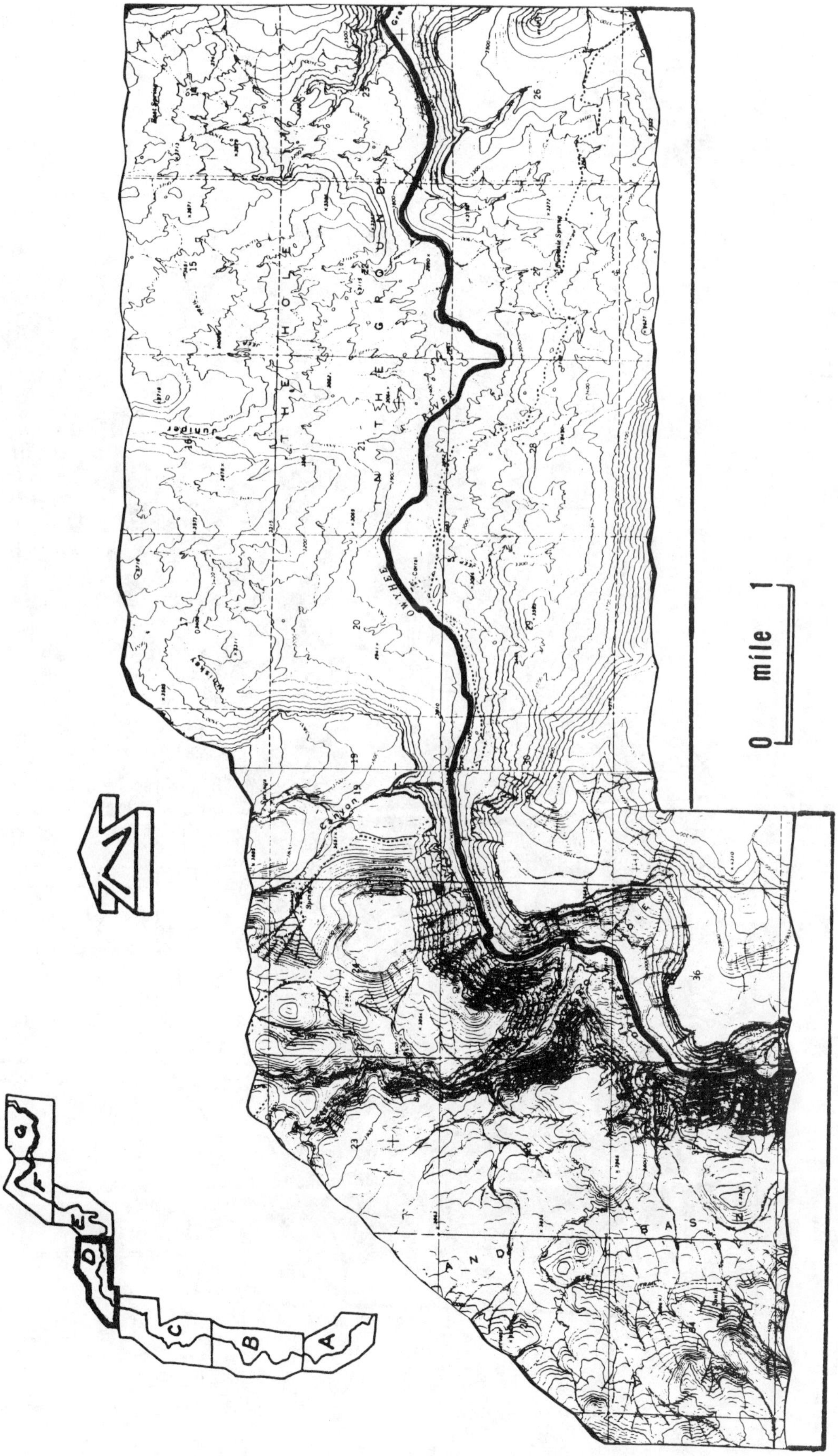

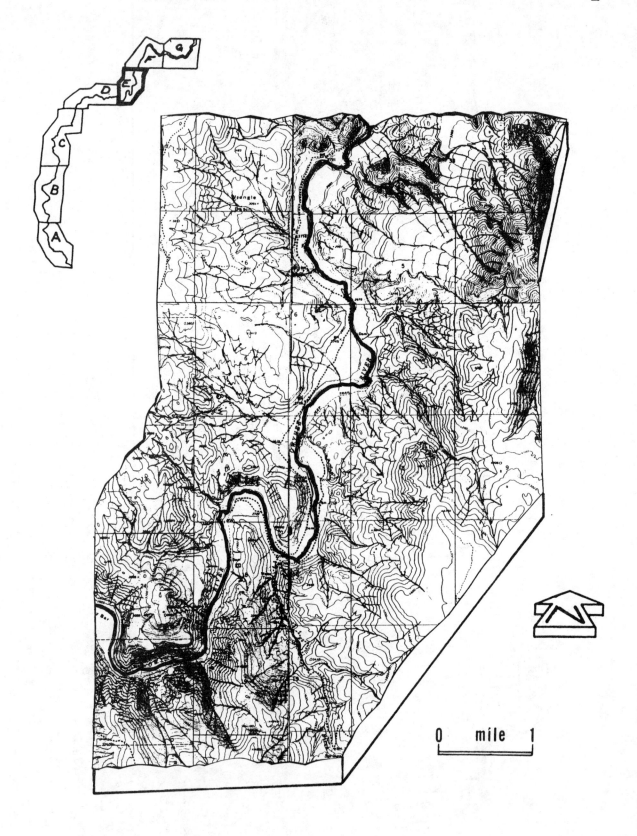

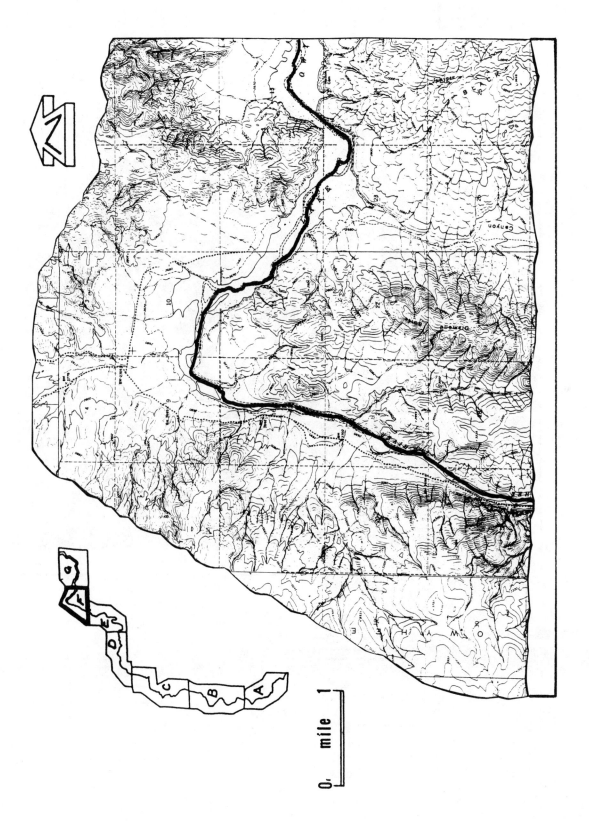

0, mile 1

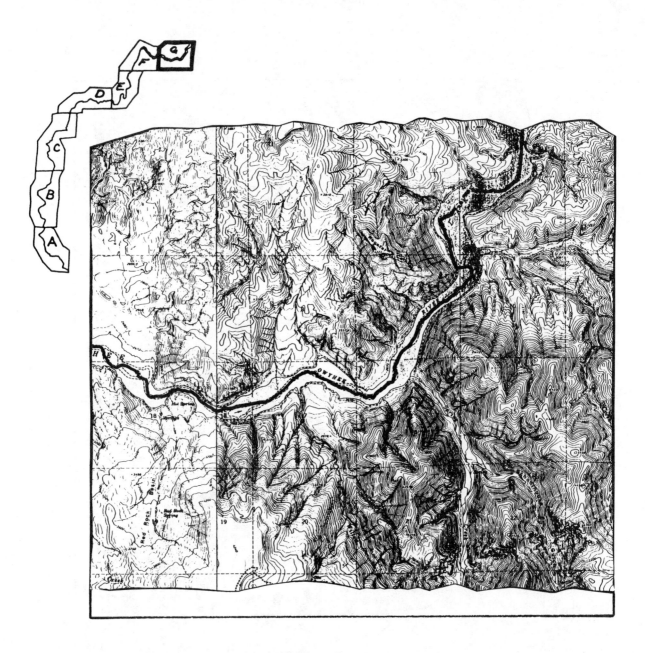

0 mile 1

Rogue River

(Grave Creek to Foster Bar)
Josephine, Curry Counties, Oregon
Length of run: 38 miles
Number of days: 2-4

The wild Rogue River, in its 215 mile run from Crater Lake in southern Oregon to the Pacific Ocean, is joined by waters of the Applegate and Illinois rivers. Ten million years ago when the Klamath Mountains were being uplifted, waters of the Rogue resisted being forced from their canyon and maintained their course across the mountains to reach the ocean. The unique physical geography of the Rogue formed the background for an unusually violent period of human history where Indians took refuge in the deep, narrow canyons to escape pursuit by soldiers and settlers.

Eighty-four miles along the Rogue River Valley have been protected for the future by the National Wild and Scenic Rivers Act of 1968.

"Old Doctor", 1874.

Early History of the Rogue River

The Rogue River changed names many times in its history. Trappers called it Tutuni, with various spellings, meaning rogue or rascal Indians. For the same reason, French traders called it La Riviere aux Coquins, meaning river of the rogues, and Henry Eld of the Wilkes Exploring Expedition in 1841 talked about the Rascally River in his journal. The term "Rogue", similar to the French word for "red", was erroneously thought to describe the color of the water when heavy rains washed minerals downstream. However, an early settler took issue with this, pointing out the water of the Rogue is one of the clearest in the state. The river was also known for a short time as McLeod's River, after an early explorer, and by the territorial legislature as "Gold River".

In prehistoric times, tribes speaking two different languages occupied the Rogue River Valley and surrounding areas. The Tutuni lived along the coast and at the mouth of the river extending a way inland and south toward California. In 1854 the Tutuni population was estimated at 1,311 persons, with 383 reported in 1910. The Takelma tribe lived in the middle Rogue River Valley and up the Illinois River. Takelma is an Indian word meaning "those who dwell along the river." In 1780 the entire population was estimated at 500, with only one person remaining in 1910.

The Tutuni were able to set up numerous large permanent villages because of a plentiful year round supply of food. They relied most heavily on the large salmon besides shellfish, seaweed, and occasionally whale meat. Salmon were split, and dried before storage. Hunting of deer, elk and smaller game provided additional meat. Berries, seeds, and root plants were available most of the year in the mild coastal climate.

The Takelma, on the other hand, were seminomadic moving when the food of one area had become exhausted. They relied more on plant food such as camas root or dried berries combined with meat from deer or other animals. Camas was found in the moist meadows in the Spring with celery and wild onions gathered during the Summer. Camas, with its blue, lily-like flower, grew in wet areas, and women spent many hours digging the bulbs. Bulbs were covered with leaves and baked for two to three days until they were ready to be eaten or stored. Berries and acorns were collected in the Fall, and a variety of tobacco was cultivated.

Deer and elk were hunted by men accompanied by half wild dogs. A hunting party set out in the early morning to surround and trap the game in a canyon. As the deer were driven toward the mouth of the canyon they would become entangled in nets strung across the opening. Deer-head disguises were worn on occasion. During times of hardship, even crows and ant eggs were eaten.

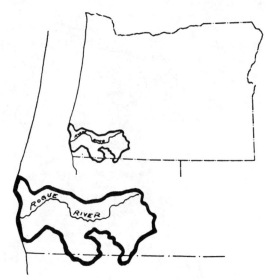

Distribution of Indians along the Rogue River.

Houses, made of bark slabs were partially excavated four to five feet in the ground and covered with a thatch roof. Dirt was piled up along the sides, and an open fire was built for warmth in the center. The result was a very smokey interior, and many women had sore eyes from working inside constantly. Canoes were only built along the coast but were used inland where the river wasn't too rocky or swift. Takelma upriver used log rafts instead of canoes.

Wealthy men of the tribe were distinctively clothed in buckskins decorated with redheaded woodpecker scalps and dentalia shells from the coast. Necklaces, nose and ear ornaments and wrist bands were favored by the wealthy who topped off their wardrobe with a basket hat. Fur lined clothing was worn by both men and women in wintertime.

The Takelma were more aggressive and warlike than the Tutuni. Takelma raided their neighbors to the west for food, slaves, or other necessities. Slaves were, in turn, traded to the Klamath Indians to the east.

Spanish and English explorers sailing up the West Coast were the first to discover the mouth of the Rogue River, but trappers of Hudson's Bay Company in 1826 were the earliest overland party to reach the river at Gold Beach. This group was led by Alexamder R. McLeod who felt the Indians were annoying. "McLeod . . . had (no) quarrel with the Indians, except at Rogue's River, which owes its name to the conduct of the Natives who were very impudent and troublesome, and went so far as to take the people's kettles from the fire and help themselves to the contents" (Peterson and Powers, p.10).

In July, 1828, Jedediah Smith, having trapped in California, was on his way north with furs and gold. He camped with his 18 men and 300 horses near the confluence of the Smith and Umpqua rivers. The trappers had been following the coast streams and trails on their way to Ft. Vancouver. The camp was attacked reportedly as a result of the trappers trying to carry off some Indian women from a nearby village. Smith and three of his men were the only survivors to eventually reach Ft. Vancouver.

Throughout 1830 and early 1840 the Rogue River Valley was untroubled. The aggressive reputation of the Indians as well as a report by McLeod that "the country was destitute of beaver" discouraged interest from outsiders. Travellers through the valley on their way north or south were not anxious to linger among these unfriendly Indians.

Two events in the early 1850's combined to bring many outsiders into southern Oregon. One was the Land Act passed by Congress in September, 1850, promising 320 acres of land in Oregon to settlers over 18 years of age. Enthusiastic whites were given 2,500,000 acres of Indian land. The other event was the discovery of gold on Josephine Creek along the Illinois river in 1850 and near Jacksonville in 1851.

A rush of miners to the the Rogue Valley made it one of the most heavily populated areas in Oregon for a short time. In August over one thousand miners and settlers had arrived only to depart by 1854 leaving behind a mere 52 persons. By 1853 most of the river and creek beds of southwest Oregon were being prospected for gold. Gold, discovered in sands at the mouth of the Rogue, hastened the development of per-

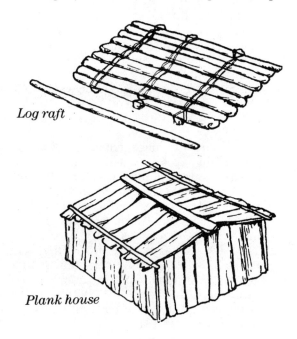

Log raft

Plank house

manent towns. Gold Beach was the largest along the coast.

Miners were responsible for naming Grave Creek. In the Summer of 1848 gold seekers found the desecrated grave of Martha Leland Crowley who had been buried along the creek. They reinterred the body and named the creek in her honor.

Miners cut trees, killed animals, and took over the river valleys. Displaced natives resorted to harassing wagon trains and disrupting mining operations. Whites, on the other hand, complained that Indians had stolen their cattle. No doubt this was true. Starving Indians raided food supplies in isolated cabins killing settlers who happened to be in the way.

In September, 1855, a wagon train was attacked in the Siskiyou Mountains, and two men were killed. Angry whites gathered, and a volunteer army was formed from among the white population. "Extermination" was the motto on a banner carried by volunteers arriving for army duty in Jacksonville. Local agents felt "self preservation here dictates these savages being killed off as soon as possible".

Vowing to destroy all Indians, the supporters rode out at night and attacked a village on Butte Creek, near Table Rock in southern Oregon. In the darkness Indian men, women, and children were killed indiscriminately. The darkness was used as justification for the killings. "None were killed after it became light enough to distinguish between the sexes," was the self-righteous statement made by one of the vounteers.

A series of similar attacks by the volunteers brought about the Rogue Indian Wars of 1855-1856. The wars lasted eight months with a series of battles beginning with the Battle of Hungry Hill near Grave Creek on October, 1855, and ending in May, 1856, with a battle south of Marial at Big Bend. Most of the war was fought in the remote, winding canyons of the Rogue.

Toward the coast, the Tutuni became increasingly alarmed with the aggressive attitude of

Benjamin Wright, Indian agent in Oregon and California.

white settlers and with tales they heard of Indian massacres inland. They didn't expect much help from local Indian agent, Benjamin Wright. Wright, a reputed drunkard responsible for causing the Modoc War, molested Indian women and mutilated Indians he had murdered, collecting fingers and noses of his victims. Although warned of a Tutuni uprising, Wright laughed at the idea. He was described as a "brave and successful fighter, and skilled . . . in a practical knowledge of their Satanic treachery, yet in this instance he would not be warned" (Dodge, p.72).

In February the Tutuni attacked, burning every house along the Rogue and up and down the coast. Agent Wright and a friend, trusting "the words of the Indians, fell as babes, and became the first victims of the outbreak. Their bodies were flung into the river and drifted away to the great cemetery of the sea" (Dodge, p.73).

Port Orford, 1859, showing the harbor, Battle Rock, and the citizen's fort on the hillside (Coos-Curry Pioneer Museum, North Bend, Oregon).

Settlers took what refuge they could find in the partially finished Miners Fort at the mouth of the Rogue. They remained inside for 30 days until an Army contingent marched up from California to rescue them. Twenty-three white settlers were killed during the raid. With the arrival of troups, the Indians fled upriver where they joined the Takelma.

On March 18th soldiers under Capt. C. C. Augur burned a large Indian village on the Rogue near Agness while the Takelma fired at them from a nearby hillside. In all, five villages were burned and their sleeping inhabitants massacred during the war. Near Solitude Bar, soldiers were ambushed and driven back by Indians rolling rocks down on them. During the battle at Grave Creek, white soldiers and volunteers were surrounded by Indians for hours of fighting until the soldiers retreated loosing ll men with 27 wounded.

The final major conflict, the Battle of the Meadows, occurred in late May, 1856, in the open meadow of the Rogue at Big Bend. The Army, under Col. Buchanan, accompanied by volunteers, marched upriver on both banks successfully trapping the Indians. Despondent, Chiefs George and Limpy agreed to surrender although Chief John vowed to fight on.

While small groups of Indians came down from the canyons to where the Army was camped, confusion arose as to whether the Indians were friendly or aggressive. Prematurely, zealous volunteers began to shoot at Indians all along the river. Hearing the shots, the remaining Indians under Chief John attacked the soldiers who fled to a nearby brushy ridge. Trapped in a crossfire, at least ten soldiers were killed and 19 wounded. After 12 hours of fighting, the battle ended with reinforcements arriving to help in defeating the Indians. Chiefs George and Limpy surrendered, but Chief John and his remaining 34 warriors didn't come in until June 25.

By Summer 1,400 Indians had been transported by ship, at a cost of $10 per person, or had walked to the Grande Ronde and Siletz Reservations in Oregon. They were heavily guarded along the way by the Army. One, Lt. Ord, noted in his diary "It almost makes me shed tears to listen to them wailing as they tot-

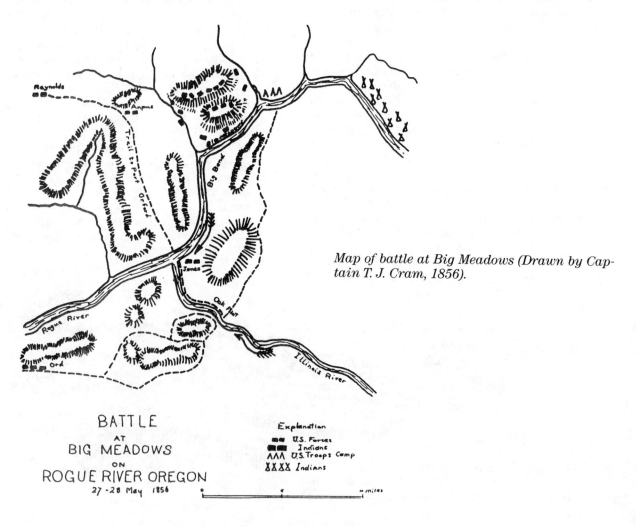

Map of battle at Big Meadows (Drawn by Captain T. J. Cram, 1856).

ter along . . . one old woman bringing up the rear, her nakedness barely covered with a few tatters — and barely able to walk . . . had been a long time getting here . . . ''

Indians hiding in the hills and canyons were tracked down by bounty hunters and killed by agents with government contracts to make the murder official. By 1857 the Rogue River Indians no longer inhabited their homelands in the river valleys.

Even after removing the Indian threat, the ruggedness of the Rogue Valley prevented rapid settlement. Trails into the upper Rogue were the only early means of communication, and mail was carried by mule or horse. The only boat transportation was an occasional ferry across the river. Letters might be carried in a pack for months before being delivered to the receiving party at a time convenient to the trapper.

A mail route up the Rogue was proposed in 1883 by an early settler, E. H. Price, who had a cabin at Illahe. The government in Washington, D.C. was unenthusiastic about backing the project financially when told it would serve only ll families. Price persevered and a temporary weekly round trip route from Gold Beach to Big Bend was set up. The mail boat was operated by pushing with a pole or pulling a tow rope. Price's 15 year old son carried the first mail on June 14, 1895, and throughout the summer, allowing one week for the trip. During the winter, with the increased river flow, the mail had to be carried on a trail overland. When the route was officially approved, Price received $600 a year, and his cabin became the first post office on the Rogue.

Harry Price's cabin and post office at Illahe.

Shells from the Gold Hill burial.

Obsidian blades from the Gold Hill burial.

Indian Life Along the River

The locations of many villages along the upper Rogue have been lost in the past. Of the seventeen previously known Takelma villages, most are at the confluence of rivers or streams. Kashtata was near the Rogue and Galice Creek, Hudedut at the forks of the Rogue and Applegate River, and Salwahka near the mouth of the Illinois. Most of the 30 or more Tutuni villages were along the coast although Mikonotunne, for example, was located along the north side of the Rogue, 14 miles inland.

Upriver from Grave Creek, a burial site with close to 22 skeletons was unearthed near Gold Hill. All the skeletons were lying on their left side facing west with their knees pulled up against their chest. Obsidian blades, serpentine pipes, and shells were also uncovered in several of the graves. Flint chips, arrowheads and some worked tools were found scattered throughout the area. An eight year old child's grave had a large collection of shells.

Geology of the Rogue River

Of all the Pacific Northwest rivers, the Rogue displays some of the most exotic rocks and minerals in its valleys. The Klamath Mountain province is, itself, one of the most structurally complex in the Oregon. This complexity is due to the Klamath's marginal position on the continent and the fascinating role that continental drift played in the evolution of the area. This history, including episodes of faulting, folding, and intrusion by volcanic rocks, has left unmistakable evidence here in the form of valuable ore minerals. The exploration and settlement of the Rogue Valley owes much to the discovery and exploitation of these ore minerals, most notably gold.

Against the background of shifting earth plates, the Klamaths and indeed the valley of the Rogue show many signs of this history. Our record goes back to 150 million years ago to what is called the Jurassic Period. Rocks of this age were first recognized and named for the Jura Mountains in Europe. During that period rocks were being deposited along an ancestral North American coast line.

Simultaneously the earth's crustal plate comprising the Pacific floor was moving toward the ancestral mainland of North American where they collided. As the Pacific plate began to slide beneath the mainland, volcanoes generated by this collision process developed in an offshore chain. With the continued collision, pieces of the sea floor were scraped off and pushed up against the mainland rather than carried under it. Today in the Klamaths we see an ancient sea floor at altitudes of 2,000 to 3,000 feet above sea level that once rested under three miles of water.

The Oregon Klamaths were originally situated 50 miles southeast of their present position. About 100 million years ago they were broken loose from the mainland along a fault

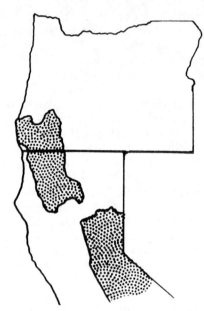

Fifty mile separation and possible displacement of Klamaths from the Sierras.

which slowly shifted and moved them toward the northwest and into their present position. Finally, 50 million years ago the area underwent uplift and distortion before being eroded to give it the present rugged topography. In addition to mineralization in the Klamaths, the crustal distortion and shortening suffered here in the later Jurassic show up today as large blocks thrust under each other in an eastward direction.

The Rogue River Valley represents an intermediate stage of erosion. In the erosional cycle, when the flat uplands of a terrain are completely dissected, the landscape has passed from youth to maturity. While this is the case in the valley of the Rogue, many youthful features appear which include steep, V-shaped valleys, waterfalls, and rapids. Recognizing that the Rogue is between youth and maturity, it is easy

Subduction zone with accretionary wedge.

PACIFIC PLATE

MAINLAND

Oncoming plate breaks up into a series of overlapping, smaller plates.

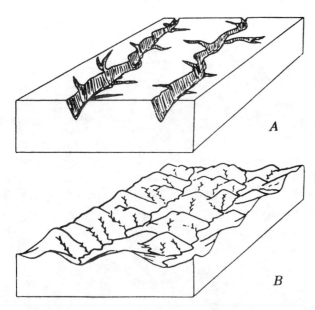

Evolution of a landscape from youth (A) to maturity (B).

to see signs of rapid erosion even in the harder rocks. ·

The earliest geologic reports of this area described the Klamaths as an eroded plane tilting from 4,000 feet inland down to 2,000 feet near the coast. The area has been eroded in two phases. An initial phase saw it dissected nearly to sea level only to be followed by an uplift. The later uplift has proceeded faster than the erosion and has produced the rugged steep canyons characteristic of the area today.

It is worth noting that of the west slope Oregon rivers, only the Rogue and Umpqua were able to resist this uplift and maintain their westward course. As a consequence of this, the Rogue Valley forms a natural cross-section of the regional geology. Other rivers, during this interval, were forced to change course to adapt to the topography.

Landslides and rockslides in the Rogue River Valley are common. Some of these are because of old gravel bars in terraces above the stream that are weakened by the river and eventually fall. Others are due to the collapse of weak serpentine masses which are part of the volcanic/metamorphic rocks. The steep valley walls, of course, contribute heavily to these mass movements of rock material.

On occasion, the slides may dam the river or temporarily muddy its course on a grand scale. Whiskey Creek, only a mile and a half downstream from Rainie Falls, experienced a slide in the late 1800's that dammed the river bank 15 miles upstream to Hellsgate. The steep canyon walls of the Rogue River as well as its narrow channel also contribute to periodic flooding. In recent times, the catastrophic floods of 1964 and 1974 sent incredible volumes of material down to scour the canyon thoroughly.

During and after their formation on the sea floor, rocks of the Klamaths were further deformed by the metamorphic process of heat and pressure. Of the rocks in this stretch of the river, only the Rogue and Galice Formations show potential for ore minerals. Many of the minerals are in volcanic rocks within these formations.

The process by which this enrichment with ore minerals took place has been poorly understood until recently. The discovery of volcanic vents on the sea floor in the equatorial Pacific has shed much light on this process. Hot ocean water circulating through the volcanic ocean-crust carries with it in solution a vast array of dissolved minerals. As the superheated water emerges along the mid-ocean ridges, many of these minerals are precipitated into porous volcanic rocks there. Primary economic minerals produced by this process include gold and silver, as well as several lesser minerals such as platinum.

Evidence of mining operations is visible on just about any stretch of the river either as rusting equipment or old tailings and dump piles. Large mining operations on the Rogue included the Almeda Mine with over 1,400 feet of subterranean workings and the Old Channel placer mine just south of Galice which yielded over one million dollars in gold. This placer mine was one of the largest hydraulic operations in the

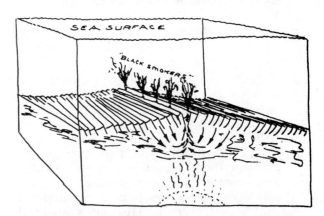

Mineralization along volcanic vents ("black smokers") in the deep sea. Heat from the magmatic source below sets up a convection current drawing seawater down through broken ocean crust to be collected and forced up through the mid-ocean rift area. Here the superheated, mineral-laden seawater deposits sulfides along the rift as the water cools.

Greenback Mine, No. 115 in 1905 (Ramp, 1979).

United States from 1860 to 1940. Tyee Bar, along the river west of Galice, was also claimed to be the source of one million dollars in gold dust removed by Chinese workers.

Chinese workers came to Oregon with the first gold discoveries and at one time may have numbered as high as 3,000. Chinese were industrious workers willing to work claims long abandoned by others. Resentment resulted when the Chinese became rich from their prospecting efforts. To put a stop to this, the Oregon Constitution was revised in 1857 to ban Chinese from owning real estate or mining claims. As if this were not enough, a poll tax was levied against them five years later.

Doubtless one of the most spectacular aspects of early history was the transportation of heavy equipment for mining and processing the ores.

Some of these came overland, but a few of the larger pieces came downriver on huge rafts. The heaviest equipment included stamp mills for processing the ore. A stamp mill is a cast iron, ore crushing device consisting of multiple trip hammers to pulverize ore for easy extraction of the gold. Some of the larger stamp mills weighed several thousand pounds.

Mining activity along Oregon rivers has been closely tied to the price of gold and the state of the U.S. economy. Increasing prices of gold and the poor economic climate of the past five years have both triggered renewed activity in gold mining. Unfortunately for the would-be gold seekers, the accelerated costs of mining itself have directed that these operations remain small.

Geology Along the Route

Perhaps the most convenient method of introducing layered rocks in a given area is to record them in the observed sequence of appearance proceeding down river. Quite often this strategy presents a sequence from bottom to top or oldest to youngest. Rocks being eroded in a mountainous area will usually have the oldest exposed at the core of the mountains and progressively younger rocks toward the mountain periphery. The complexity of the Klamaths dictates that this sequence will not be in order by age.

In the vicinity of Galice at the put-in, the Galice Formation is exposed in the river bed and banks. The measured thickness of the Galice is in excess of 15,000 feet, but we only see a short sequence of it along the river route. Rocks of the Galice include mudstones and some black shales. More rarely the Galice displays thin sandy beds, but the dominant rocks are fine-grained. In addition to the mudstones, the Galice bears a thick sequence of volcanic rocks some of which are relatively coarse-grained. Locally the Galice mudstones and shales have been altered by heat and pressure (metamorphism) to form hard platy slates as well as another platy rock called phyllite. It is not difficult to imagine the Galice mudstones as a deep ocean floor setting.

As the river winds northward out of Galice we very quickly pass from the Galice Formation into the Rogue Formation. We have moved downriver from younger rocks (Galice) into older ones of the Rogue Formation which originated from lava flows and ash. These rocks represent an ocean bottom environment in and around the chain of volcanic islands that developed 150 million years ago (Jurassic period). Much of the Rogue volcanics have been considerably altered by metamorphic heat and pressure to produce "metavolcanics." Like the Galice, the thickness of the Rogue Formation is measured in miles. Rainey Falls displays a nice representation of a variety of these metamorphic Rogue Formation rocks.

Rainey Falls itself is a good example of the hardness of these rocks. In and around the falls a characteristically banded metamorphic rock called gneiss is abundant. Black crystals of the mineral amphibole are common in the gneiss as well as in veins of a pale, pea green mineral called epidote.

Downriver from the Rogue Formation, the Dothan Formation dominates the next 20 miles along the river. This formation is up to 18,000 feet thick and is composed of sandstone with some siltstones and mudstones. In several places along the river, Dothan rocks have been distorted into tightly bunched folds attesting to high pressures and temperatures that accompanied the formation of the Klamath Mountains.

The change from Dothan back into Rogue Formation rocks occurs just below China Bar. The path of Mule Creek Canyon follows a fault for over a mile with Dothan Formation in the east wall and Rogue Formation on the west. The straight pathway and sheer walls of Mule Creek Canyon are an indication that the river is following an ancient fault. Here the Rogue turns up on its edge making rafting conditions hazardous with high turbulence and whirlpools as at Coffee Pot.

A very short interval of the 80 million year old Riddle Formation is geographically sandwiched between the Rogue Formation and the Flournoy Formation. The Riddle Formation is composed of coarse conglomerates and sandstones which must have been deposited in shallow water.

At Blossom Bar huge boulders of the Riddle Formation litter the riverbed making boat passage hazardous going downstream and impossible upstream. These boulders were eroded out of the bedrock from an earlier phase when the stream channel was much higher than at present. Gold was mined in the late 1800's at Blossom Bar as well as 4 miles downriver at Solitude Bar where old mining equipment still lies rusting on the banks and up tributaries along the river.

The remainder of the river bed downstream from Clay Hill Creek to the take-out at Foster Bar lies in the Flournoy Formation. Rocks of this marine (oceanic) formation are evenly bedded layers of sands and silstones with local thin conglomerates. Conglomerates are seen as rounded gravel pebbles which have been cemented together. At 50 million years of age, it is the youngest geologic formation along the river. On the skyline only two miles north of the river, Panther Ridge stands out as a pronounced resistant rock wall of even younger sandstones.

Rogue River gold company, Grave Creek, 1936 (Ramp, 1979).

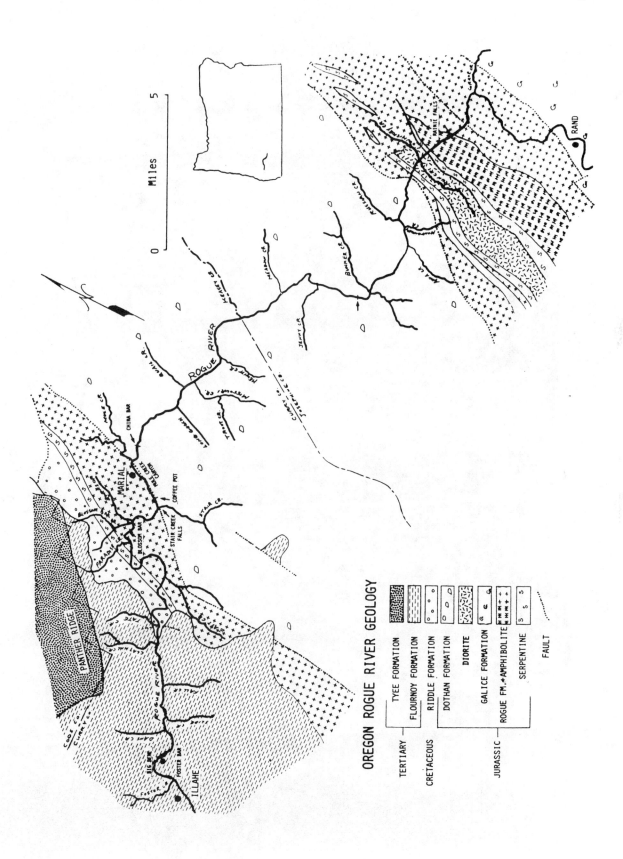

OREGON ROGUE RIVER GEOLOGY

TERTIARY — TYEE FORMATION
          FLOURNOY FORMATION

CRETACEOUS — RIDDLE FORMATION
             DOTHAN FORMATION

             DIORITE

             GALICE FORMATION

JURASSIC — ROGUE FM. + AMPHIBOLITE

           SERPENTINE

           FAULT

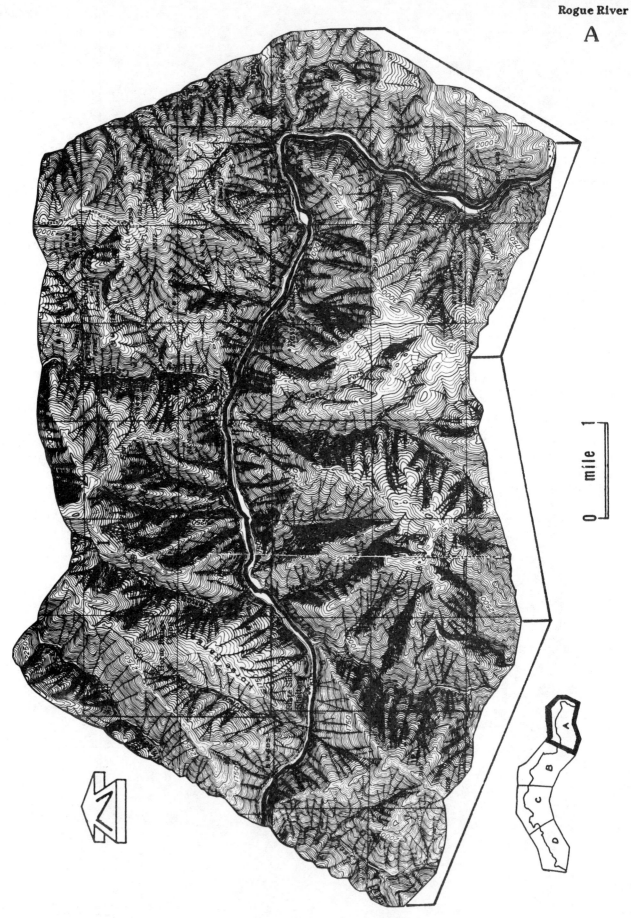

mile 1

0

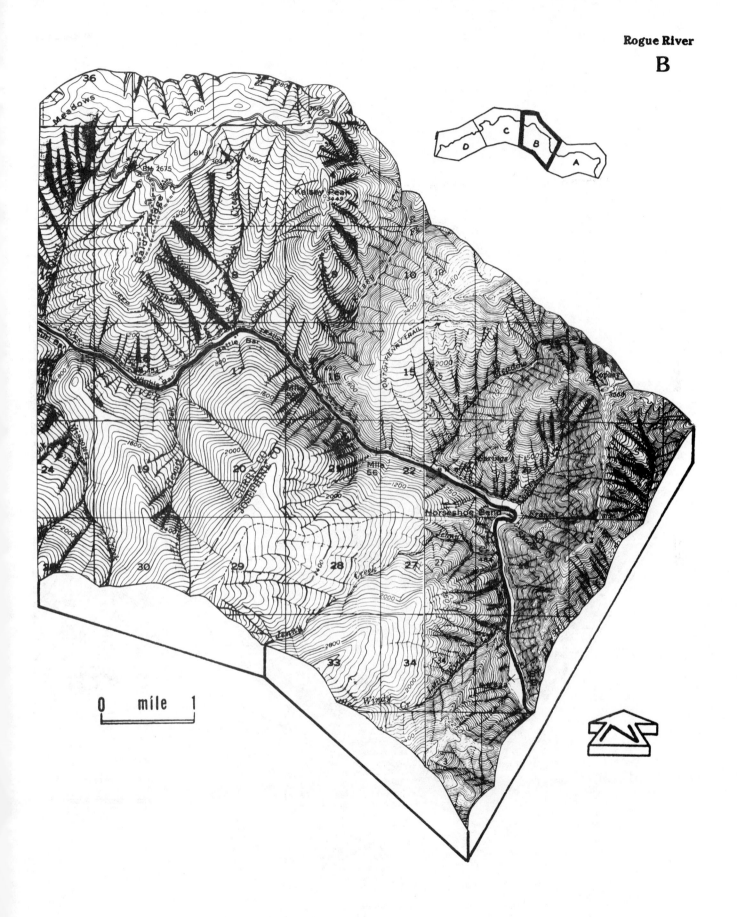

0   mile   1

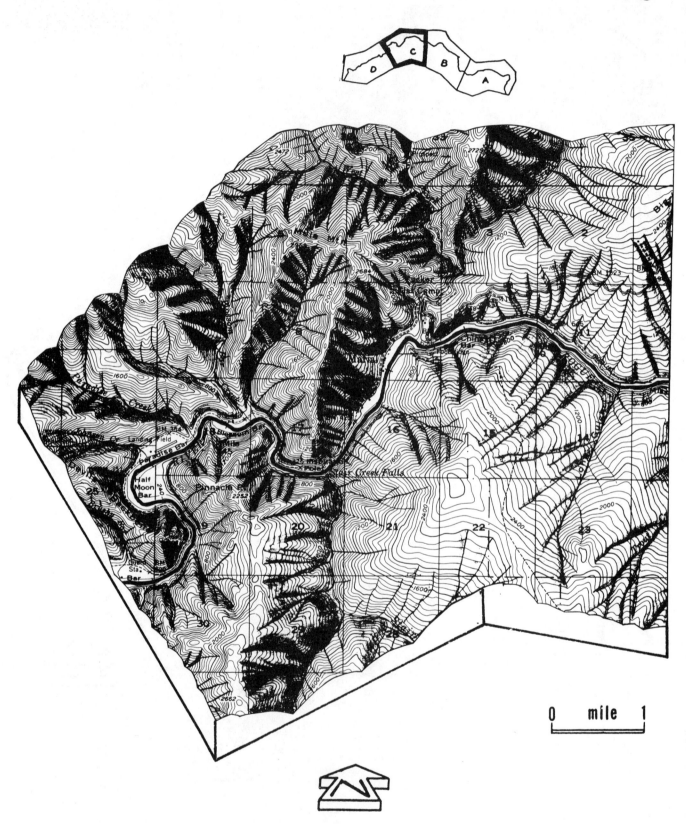

0 mile 1

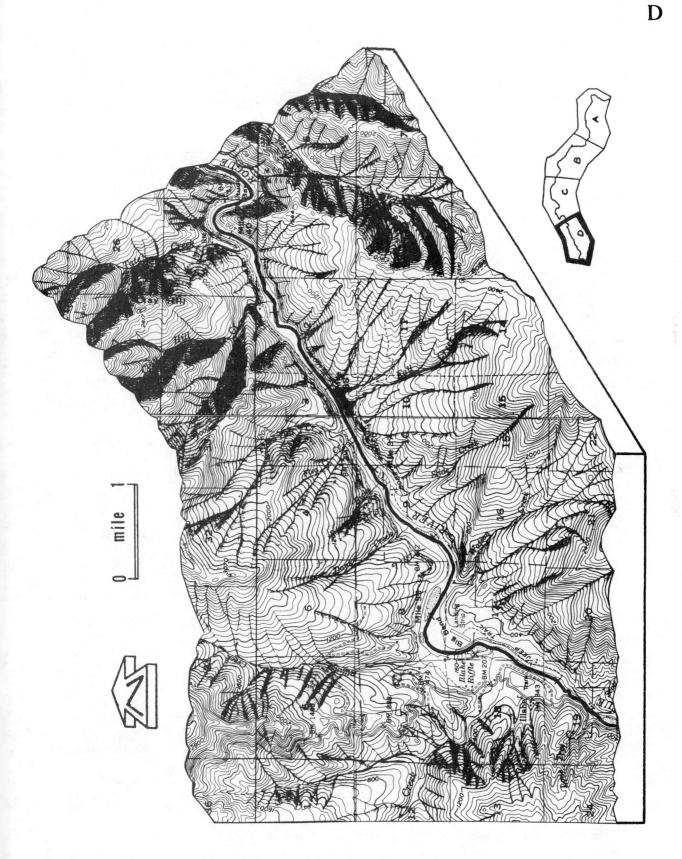

Santiam River

South Santiam River
(Foster to Jefferson)
Linn County, Oregon
Length or run: 38 miles
Number of days: 1-2

North Santiam River
(Packsaddle to Stayton)
Linn to Marion County, Oregon
Length of run: 25 miles
Number of days: 1-2

The Santiam is a good river to wet your feet and test the water. Calm waters are interspersed with mild rapids, and usually problems only arise from rocks in the riverbed scraping along the bottom of the raft in shallow stretches. Drifting is pleasant and takes the rafter past scenes of early Oregon history where lumber and grist mills were placed along the riverbanks. The community of Foster grew up around P. J. Foster's grist mill, John Shaw's lumber mill started at Mill City, and even Waterloo was a mill town. Here, after extensive court litigation involving heirs of a grist mill owner, a local wit suggested the name Waterloo after the case was settled.

Even though it's close to a large metropolitan center, immense Douglas Fir, ferns, moss, and the damp smell of decaying leaves along the Santiam take the visitor back into the distant past and into what can only be a pleasant experience.

Early History of the Santiam River

The Upper Santiam band of Indians who lived along the Santiam River long ago were only a small part of a larger group which included everyone who spoke the Molalla language. The vast Molalla territory covered the western Cascades from Mt. Hood in the north to the Rogue River in the south. The Upper Santiam band of the Molallas was restricted to the central Cascade region. Little is known of them as a separate tribe, and most information is derived from our knowledge about the Molallas. References to the "Santiam Moolalla" are few. They were mentioned with other Oregon Indians in an 1851 treaty to create reservations, however, they weren't named in a later treaty in 1855.

The meaning of the word "Santiam" is unknown. In 1906 there were 23 of the Santiam

Joseph Meek, mountain man and explorer in the Willamette Valley.

Molallas living on the Grande Ronde Reservation, and only nine remained by 1910.

There has been much speculation on the history of the Molallas themselves and how long they occupied the Oregon Cascade region. They were thought to have lived in the area east of the Cascades around the Deschutes River at one time. Perhaps they were pushed out by a more aggressive group, but for at least the past century and up to the time of the first explorers they lived in the Cascades.

While the Molallas may have numbered only 500 at one time, they ranged over a vast territory to hunt. "All they did was hunt," related an early pioneer. Deer, elk, and other game were plentiful higher in the mountains, and during Summer and Fall most of a family's time was spent there hunting. Salmon, trout and suckers abounded in mountain lakes and streams. Much of the food, including fish and berries, was smoke dried and stored. Families banded together for gathering food, but the Molallas weren't organized into strong tribal groups. Leadership was loosely given to the person responsible for getting the job at hand done.

Temporary summer houses in the cooler mountains were made from available brush, whereas permanent winter houses were built on the lower slopes of the Cascades and were partly underground.

175

Distribution of Molalla and Santiam Indians in Oregon.

On the whole the Molallas were friendly with their neighbors often engaging in trade with them. They traded with the Klamath Indians to the south for beads, goods, and slaves although the fate and number of these slaves is not known.

A scattering of facts about their life style has come down to us. Both Molalla men and women decorated their bodies with tattoos, and noses were pierced for decorations. The heads of female babies were flattened at birth. Cremation was practiced, but nothing else was known of their religion.

The great demand for furs on the American and European markets during the 1800's brought fur trappers into the Northwest rumored to contain vast resources of beaver. Jacob Astor's fur post at the mouth of the Columbia River was the first permanent northwestern settlement, and trappers from his Pacific Fur Company entered the Willamette Valley in 1811. One trapper, Gabriel Franchere, wrote of the valley that "our guide informed us that up this river about a day's journey there was a large waterfall and beyond it the country abounded in beaver, otter, deer, and other wild animals."

The lush animal wealth of the Willamette Valley enabled trappers to surpass themselves in fulfilling their quotas. During a good year a fur trapper might bring back as many as three or four hundred pelts. The American, Jedediah Smith, in 1825, brought in an impressive 668 skins.

Frequently Indians themselves were willing participants of the fur trade, procuring skins and bargaining for higher prices in trade goods. Alexander McLeod of Hudson's Bay Fur Company found the Oregon region to his liking. He reports trading cheap goods in return whatever the Indians brought into camp.

"Traders introduced a wealth of trade goods: metal, tools, kettles, cotton . . . and liquor. This trade had major impact on their technology, ability to acquire wealth items, and their relationships with each other. Some groups were pitted against others in the scramble to become middlemen in the trade. The mariners also introduced virulent new diseases . . . " (Beckham, Minor, Toepel, p.189).

While trade items changed the Indian way of life, diseases killed off the Indians themselves. Smallpox, measles, and cholera, for example, were brought in with the first non-Indians. Indians had no resistance to them and would invariably die from a disease which might only cause a white settler to be ill for a short time. Many Indians would be killed in a wildfire epidemic never having seen a white person. The Indian method for curing smallpox and measles by sitting in a partially underground sweathouse and then jumping into cold river water often brought a quick death.

Mat-covered long house.

House made from bark slabs.

Fort Astoria before it was owned by the British.

A smallpox epidemic swept Oregon territory from 1781 to 1782, and a second epidemic of unknown cause appeared in 1829 and sporatically through the next ten years. Hundreds of Indians from the Columbia River south to California were killed. Symptoms of the epidemic of 1829-1840 were fever, chills, and pain. Speculation was that the disease may have been malaria or scarlet fever. Whatever the disese, thousands died.

"A gentleman told me that only four years ago, as he wandered near what had formerly been a thickly populated village, he counted no less than sixteen dead men and women, lying unburied and festering in the sun in front of their habitations. Within the houses all were sick; not one escaped the contagion; upwards of a hundred individuals, men, women, and children, were writhing in agony on the floors of the houses . . . some were in the dying struggle . . . shrieked and howled in the last sharp agony." (Beckham, p.109) Diseases in these years had reduced the Indian population by as much as 70 percent.

Never very numerous to begin with, Indians had died leaving vast unpopulated tracts of land. Settlers, farmers, and roadways filled in the region, and the government spent $20,000 on oyster and champagne dinners persuading the Indians to part with the remainder of their lands. Treaties and reservations were made and voided with equal rapidity while the area owned by the Indians shrank with each new negotiation. After signing reservation treaties in 1855, the Molalla groups, perhaps numbering 50 persons, moved to the Grande Ronde Reservation on the Yamhill River. As late as 1927 descendants of the Molallas were still found among the Klamath Indians to the south.

A beaver drawn by a French traveler in 1703.

Indian Life Along the River

Relatively little material evidence of prehistoric Molallas has been found in the Cascades. There has only been limited construction activity to attract archaeologists, and the heavy forest and soil cover discourage discovery.

Of the three Molalla settlements recorded, two could possibly be of the Santiam group. One was at the headwaters of the Santiam River, and a second location was only given as the western Cascade slope.

Cascadia Cave, near Cascadia in Linn County, was a seasonal camp along the South Santiam River. The cave was a main stopping place on an east-west Indian route as far back as 7,000 years ago. This is one of the earliest occupation sites in western Oregon. The cave contained a particularly large number of hunting implements, domestic tools, bones of deer, rabbit, and other animals. Larger mammal bones were found in deeper levels. Lack of fish bones, even though the river was nearby, suggests the Willamette Falls to the north formed a barrier to most migrating fish.

The cave walls were decorated with a series of petroglyphs incised into the cave wall to a depth of 1/2 inch. Most designs are geometric lines. The most humanistic elements are what might be called bear tracks or human footprints.

Near Lebanon, on a bank of the Santiam River, a mammoth tooth, section of a tusk, and a small chisel of bone were found together. Early speculation was hopeful that an early man tool had been found in association with a Pleistocene mammoth. However, close examination could not confirm this, and to date no manmade artifacts have been found in Oregon associated with these early mammals.

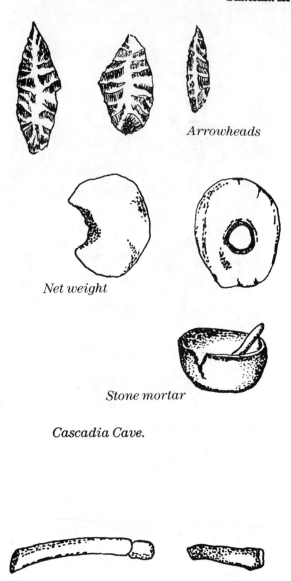

Arrowheads

Net weight

Stone mortar

Cascadia Cave.

Worked stone and bone tools from along the Santiam.

Rock art from Cascadia Cave.

Geology of the Santiam River

The Santiam drainage is normally run in two intervals along the north and south branches most of which lie within Linn County. The river cuts through thick sequences of volcanic rocks commonly known as the "older" or western Cascades. Lava flows here are several miles to the west of the better known Cascade peaks forming the chain of volcanos known as the "younger" Cascades. Rock formations of the western Cascades are made up of successive lava flows and ash deposits thousands of feet thick.

A rough, mature erosional landscape was well-developed on the older surface prior to the out pouring of the Columbia River and Sardine Formation lavas. This uneven landscape resulted in the development of intercanyon flows. Molten lava flowed into and very often filled river valleys after racing downstream for miles. The wide extent of some of these 20 to 40 million year old (Tertiary) lava flows suggests a very hot, runny fluid.

Because of extensive rainfall, soils of the western Cascades are well-developed and thick. Often the best or only exposures of rocks to be seen are either in the river canyons or in fresh highway roadcuts. Sweet Home, Leba-

non, and Lyons are situated along an old ocean shoreline that can be traced from Eugene in Lane County all the way past Silverton to Mollala in Clackamas County and beyond. Buried beaches and offshore sand bars of this old seaway hold some promise for oil and gas production.

Geology Along the Route of the South Santiam

The South Santiam is run between the community of Foster with the take-out at either Lebanon or Jefferson well into the Willamette Valley. The upper reaches of the South Santiam watershed primarily drain the Little Butte Formation. This poorly understood geologic unit is a complex rock mass of marine and non-marine sediments and lava flows. These lavas are among the oldest in the Cascades and consequently appear much the worse for wear. A very common rock variety among the lavas here is andesite. Andesite usually appears as a gray to light maroon color with imbedded, elongate white and gray crystals.

The Little Butte is predominantly flat-lying, but a subtle arch (anticline) in the rocks appears between Sodaville and Lebanon. North of the

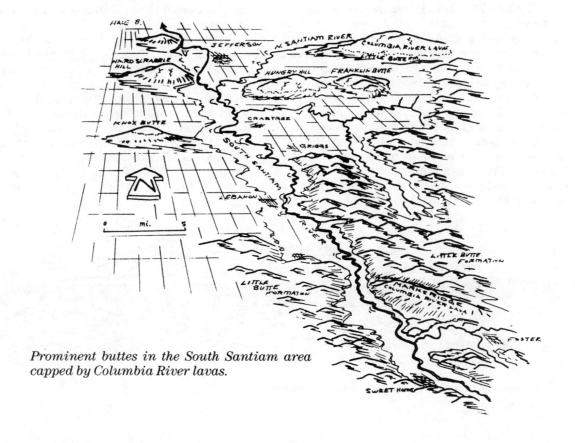

Prominent buttes in the South Santiam area capped by Columbia River lavas.

river the arch dives under the regional geology to reappear in the wall rocks of the North Santiam. Along the latter drainage, the arch cuts the river at right angles between Lyons and Fox Valley. The arch is one of a series of gentle north-south folds which lightly wrinkle the rocks of the western Cascades.

Stream valleys in the upper reaches of the South Santiam are sharp V-shaped chasms typical of a youthful stage of erosion. Steep valleys in the Foster/Sweet Home area are paved on the bottom by extensive gravels of "Quaternary" Age. Much of this river sediment is less than a million years old.

Out in the valley beyond Lebanon, there is very little bedrock exposed, and most of the riverbank is in gravels until it reaches Jefferson. In the Sweet Home vicinity, a prominant mesa, Marks Ridge, forms a northwest/southeast hill just north of the river valley. This mesa is capped by a thick layer of the resistant, 15 million year old (Miocene) Columbia River lavas. This is a small remnant of what was once a huge, flat sheet of basalt lava covering thousands of square miles in the Cascades. The flat top of the mesa is the cooled surface of the lava.

A series of winding meanders just above Sweet Home is a signal that the stream valley is in a mature stage of erosion. Passing through Sweet Home, the river cuts through an older elevated, fluvial stream terrace of gravel, sand, and clay before it breaks out into the Willamette Valley at Lebanon.

At Lebanon the gradient of the river flattens considerably. Here complex meanders, oxbow lakes, and flat gravel/sand deposits reflect a late maturity to old age stage of erosion. At the confluence of the North and South Santiam rivers, a prominent range of hills stretches out to the east. Hungry Hill, Franklin Butte, and Rogers Mountain have a foundation of Little Butte Formation rocks overlain by a hard caprock of Columbia River lava. Immediately west and south of Jefferson, the hills of Knox Butte, Hardscrabble Hill, and Hale Butte are also ero-

sional remnants of a much larger flow sheet of these lavas. Many of these local buttes of basalt have been quarried for construction rock.

Shallow petroleum wells have been drilled near Albany and Jefferson. These are probes for gas reserves in the 45 million year old (Eocene age) Spencer Formation lying only 2 to 3 thousand feet below the surface. Gas "shows" were discovered, but no commercial quantities were located.

Little Butte rocks exposed near Scio and Franklin Butte include ancient stream and lake deposits about 30 million years old (Oligocene age). These gravels and claystones contain rich accumulations of petrified wood. Tree species identified from this wood were from a mild, humid tropical climate along the coastal plain of that period.

Geology Along the Route of the North Santiam

The North Santiam is run between Packsaddle Park just east of Gates down to Stayton. Although this section is shorter than the run on the South Santiam, the geology is more variable. In the vicinity of Niagara, volcanic rocks exposed in the streambed are largely andesites of the Sardine Formation. Lavas of this formation cover vast areas in the western Cascades just west of the Cascade peaks. Steep canyons, rapid water and few meanders suggest a youthful stage.

Shortly into the run, the river breaks out of the steep canyons and into a flat valley with a floodplain over a mile wide. Six miles into the valley, Mill City straddles the river. On the north facing hills adjacent to Mill City, the Little Butte Formation is visible below the Sardine Formation which forms the skyline. About 1,000 feet above the river, there is an obvious break in the topographic slope between the steeper hills of the Sardine lavas capping the peaks and the more rolling topography of the older Little Butte below.

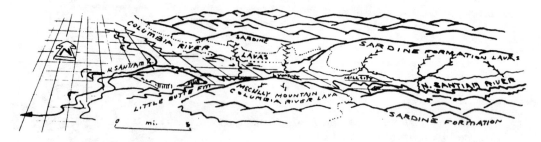

North Santiam River showing layered volcanic sequence.

For a distance of four to five miles beyond Mill City, the Sardine Formation lavas form the mountain crests on the skyline. Below that point at Lyons the Columbia River lavas are visible for the first time as the crest of McCully Mountain to the south.

The Little Butte Formation near Lyons is famous for beautifully preserved leaf fossils of the Lyons Flora. This tropical Oligocene flora reflects the mild temperatures of the coastal plain here 30 million years ago. At the take-out near Stayton, the river has altered its style into a late mature stage with a number of meanders, oxbow lakes, and flat open floodplains.

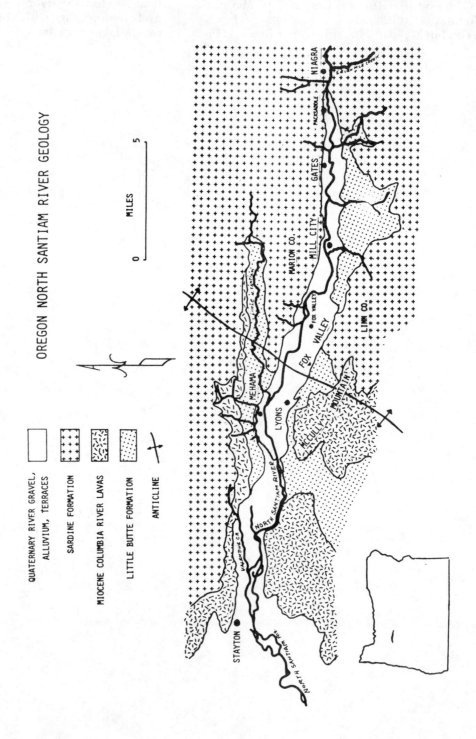

OREGON NORTH SANTIAM RIVER GEOLOGY

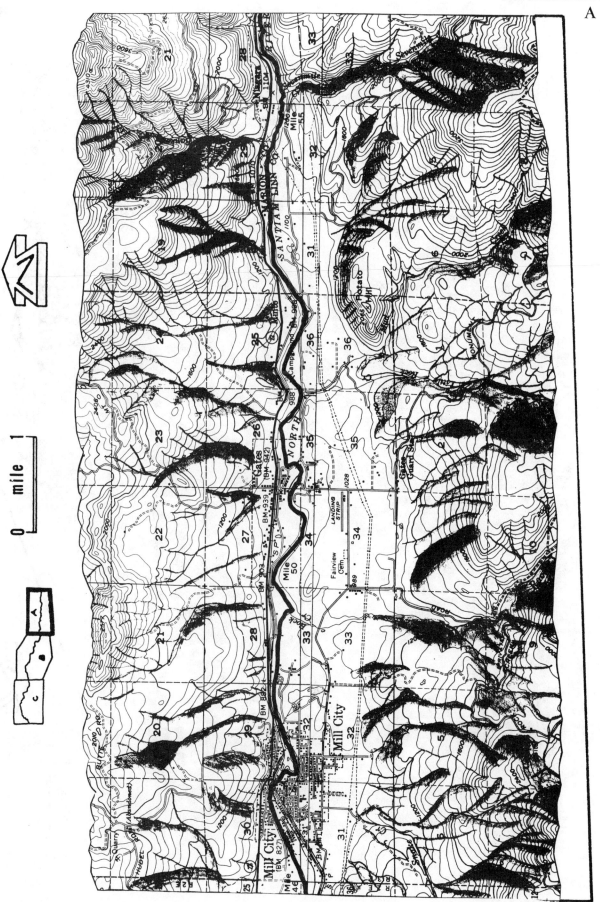

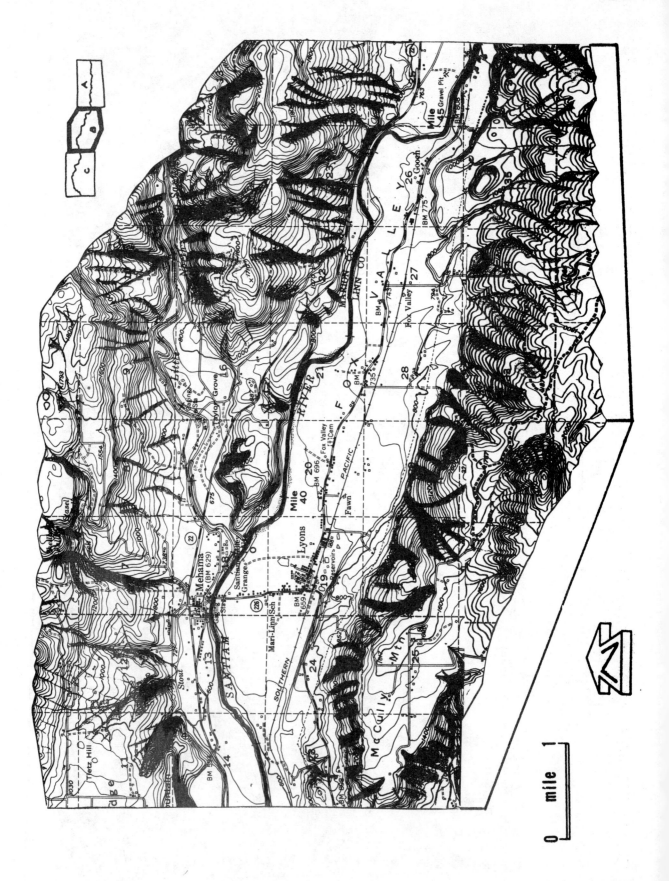

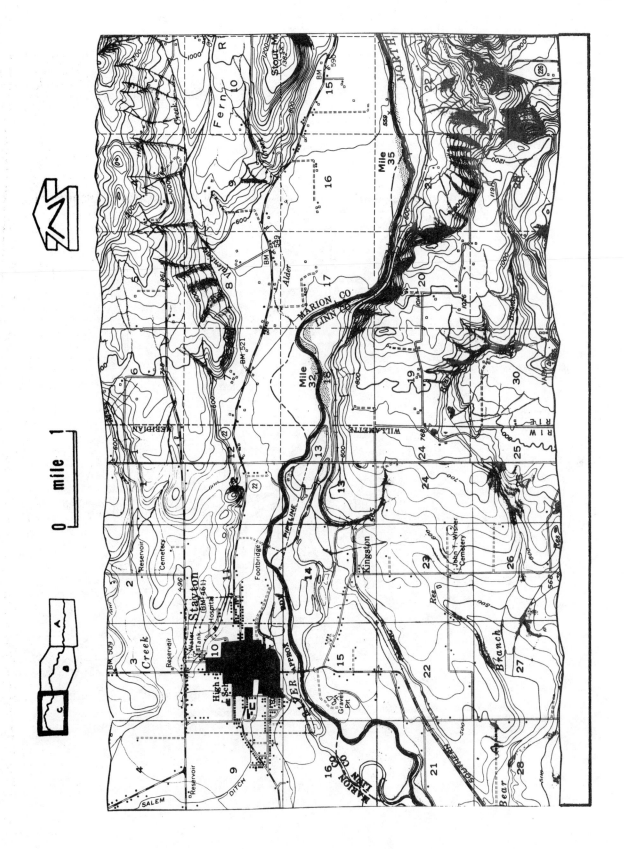

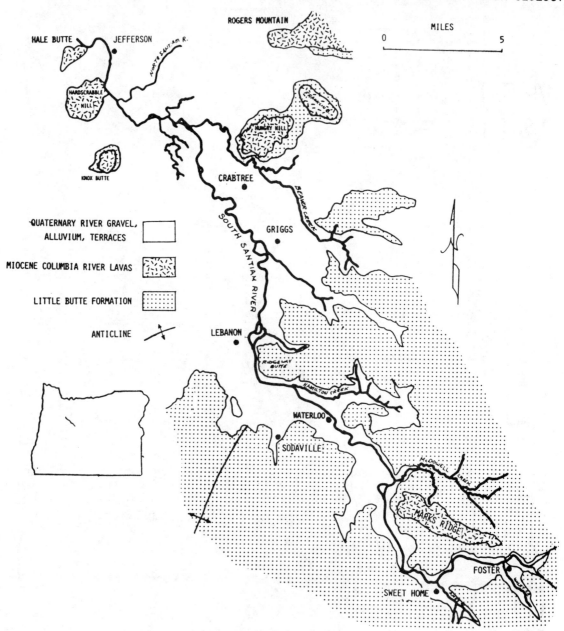

OREGON SOUTH SANTIAM RIVER GEOLOGY

MILES

0   5

ROGERS MOUNTAIN

HALE BUTTE   JEFFERSON

NORTH SANTIAM R.

HARDSCRABBLE HILL

KNOX BUTTE

HUNGRY HILL

CRABTREE

BEAVER CREEK

QUATERNARY RIVER GRAVEL,
ALLUVIUM, TERRACES

GRIGGS

MIOCENE COLUMBIA RIVER LAVAS

SOUTH SANTIAM RIVER

LITTLE BUTTE FORMATION

ANTICLINE

LEBANON

RIDGEWAY BUTTE

HAMILTON CREEK

WATERLOO

SODAVILLE

McDOWELL CREEK

MARKS RIDGE

FOSTER

SWEET HOME

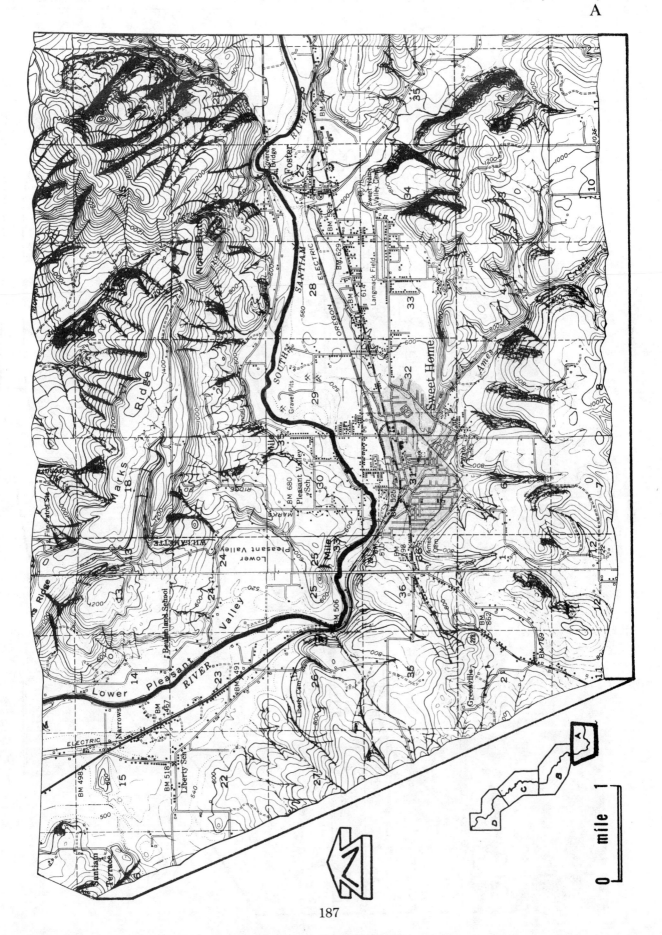

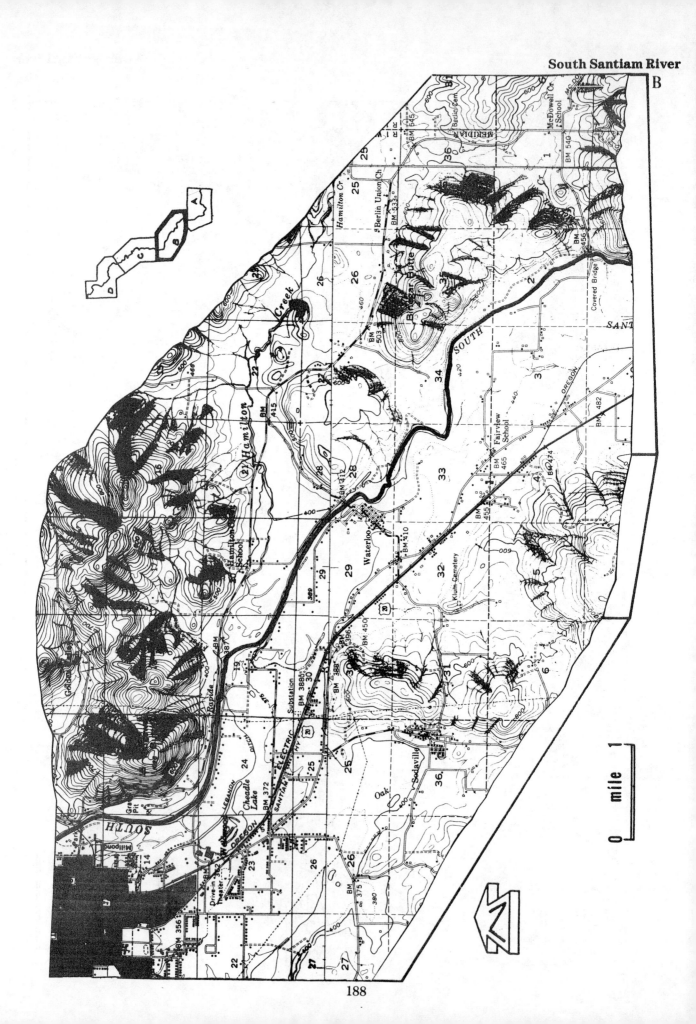

South Santiam River

B

0   mile   1

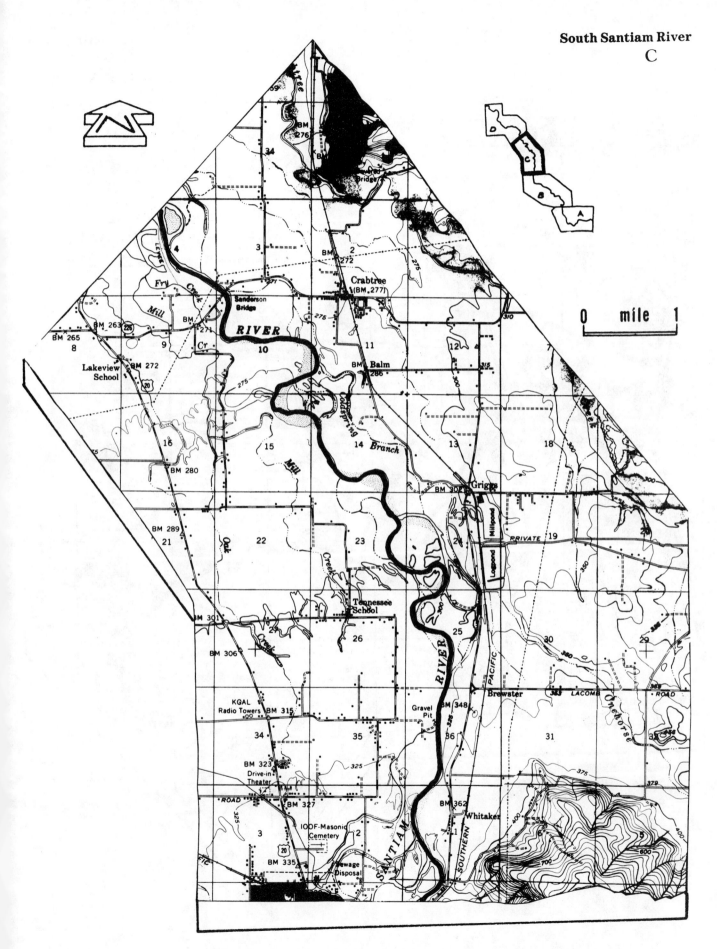

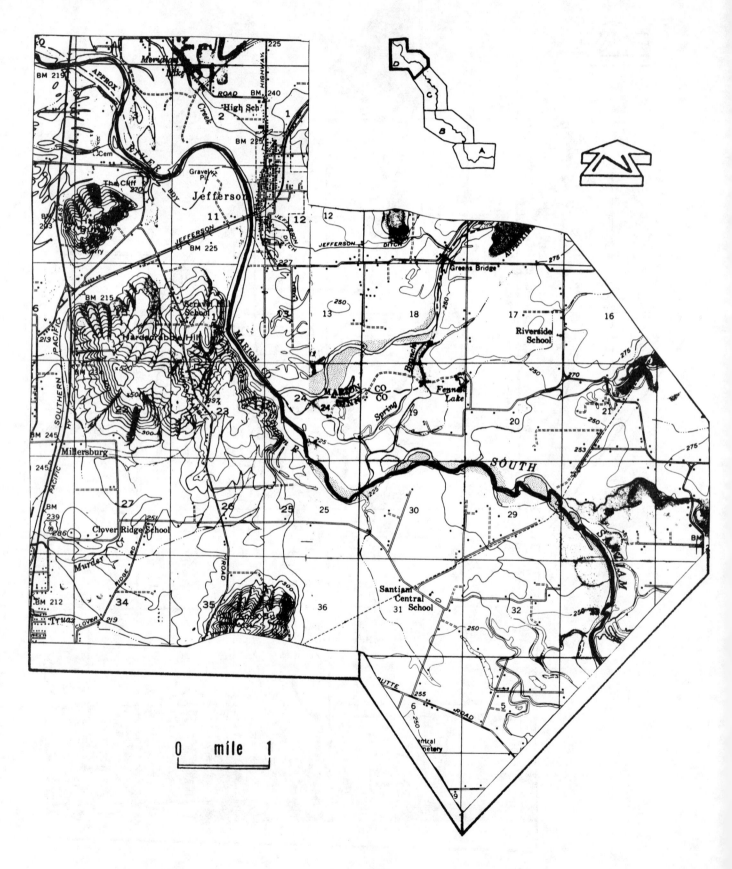

0 mile 1

California Rivers

American River

Placer and El Dorado Counties, California
North Fork (Eucre Bar to Rattlesnake Bar, Folsom Lake)
Length of run: 43 miles
Number of days: 1-2

Middle Fork (Rubicon River to confluence with North Fork)
Length of run: 26 miles
Number of days: 1

South Fork (Chili Bar to Salmon Falls Road, Folsom Lake)
Length of run: 18 miles
Number of days: 1

Three forks of the American River originate in the Sierra Nevada Mountains and flow westward before merging to become the main stream just below Auburn. Although each of the three forks extends for 100 miles into the Sierras, the combined flow below Auburn only stretches for 30 miles before joining the Sacramento River.

The 1848 discovery of gold at Sutter's Fort along the American River was one of the most profound events in western American history. Settlement of the West, which might have otherwise been a slow, orderly process, was brought about overnight by the presence of gold. Thousands of forty-niners from all over the world ended the quiet life Indians had enjoyed for hundreds of years in the forested American River watershed.

Early History of the American River

The American River, flowing west through the San Joaquin Valley, was named Wild River by fur trappers and Rio Ojotska by Spanish settlers. Eventually the river was noted on an 1841 map and named after a ford, El Paso de los Americanos, three miles above Sutter's Fort.

Nisenan Indians, sometimes called southern Maidu, lived along all branches of the American River. Their population was estimated at 9,000 in 1770, 1,100 in 1910, and only 93 in 1930. "Maidu" is a native word meaning "person."

The southern Maidu were much given to wandering throughout their territory, moving camp so frequently that villages were only designated by direction as east or west. Being relaxed travellers, a journey would begin in mid-morning with a stop near water and a food

James Marshall who first found gold along the American River.

break around 11:00 a.m. They rested from 2:00 p.m. to 4:00 p.m. when they resumed walking until evening, thereby avoiding the afternoon heat. Houses were also moved but many times only a few yards. Older people had difficulty keeping up with these nomadic tendencies and were frequently abandoned or killed, reportedly at their own insistance.

During their travels, Nisenan took advantage of nearly all plants and animals for food. Nothing seems to have been overlooked or avoided except dogs, grizzly bear, buzzard, eagles and woodpeckers, the last two being feared rather than not tasty. Salmon and other fish, caught with spears, nets, or weirs, were one of the most important food sources. Salmon was cooked then dried and crushed to a powder. As a delicacy, lampreys were second only to salmon. Fishing as well as hunting was a communal activity, and dogs were prized for their help in hunting.

All classes of insects, grubs, and earthworms were eaten. A fire was built near underground yellowjacket nests and the smoke blown into the opening to stun the insects. The nest was then roasted, the dead larva extracted and eaten with acorn soup. Communal grasshopper drives had everyone beating bushes, forcing the grasshoppers into pits constructed in the ground. Brush was also burned to roast some of

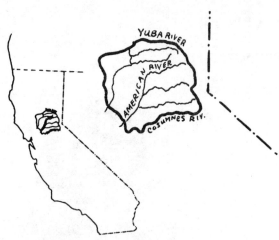

Distribution of the Nisenan Indians in the Sierra Nevadas.

the grasshoppers. Those in the pits were collected, roasted and saved until winter when they were ground up with soup. Grasshoppers were considered very healthful with good medicinal properties.

As with other California Indians, the acorn formed the dietary staple. Black oak acorns were preferred, and water oak acorns were considered inferior. Harvest began in mid-October when an Indian with a long pole climbed a tree and beat off the acorns gathered up by old men and women below. Acorns were pounded into a flour, leached with water, and made into a thick mush. Unused acorns were stored in tall woven

Acorn mush in baskets.

Black oak acorn was ground into mush and flour.

graineries. When making a trip, acorn mush was packed into deep conical baskets and carried along as a food supply. Pine nuts, grass seeds, berries, mushrooms, and a variety of roots and bulbs were eaten as well.

Tule, or bulrush (Scirpus), was used extensively by the Nisenan. Balsa boats were constructed from tule stems lashed together. Tule stems were pounded and used as a "cotton" for articles of clothing — shawls for men in winter and aprons for women. In addition, tule was woven into mats, and the root of tule was boiled or roasted and ground into flour.

At least seventy-three different plants were valued for their medicinal properties. Among them, poison oak leaves were eaten as a preventative cure. For toothache, California buckthorn (Francula) root was heated, placed over the offending tooth, and clamped tightly between the teeth. Mints were used for coughs. Wild tobacco was dried and smoked, and tobacco was occasionally planted and cultivated.

A Nisenan's name was private and personal, hence not revealed, although the Indian's American name was given freely. A woman's name was never spoken on any account, and a husband addressing his wife by her name could be divorced.

California Indians possessed a good sense of humor, and the Nisenan had several slang names for Americans which were said with a great deal of laughter. One name was "road" or "roadmaker", which they felt reflected a humerous penchant for building many unnecessary roads. Americans were also called "red" or "redfaced," and an especially humerous epithet was "whoa-haw," a word early emigrants used in driving oxen teams. A Nisenan, seeing an American, would say, "There comes a whoa-haw," accompanied by swinging arms as if controlling oxen and convulsive laughter.

There was no political leadership. Hereditary village chiefs had no power, and their position carried only whatever influence they could exert on public opinion. Decisions were made at public gatherings with much debate and exchange of gifts.

Murder was punished by relatives seeking revenge. The ultimate in retaliation was to slay the best friend of the murderer, not the murderer himself. Murder feuds could last for generations, and children were raised with the names of men to be killed. Adultery with a foreigner and kidnapping were punished by death. A woman on Dry Creek was killed after being molested by an American even though she was an unwilling participant.

War between groups of organized Indians was rare. Once Indians from a Placerville village

Bulrush or Tule plant.

Tule boat about 8 to 12 feet long.

dug salt from a small lake nearby belonging to another tribe. Indians from Placerville were killed, but warfare didn't ensue. Several hundred people gathered at the conjunction of the North and Middle Fork and engaged in a friendly rock slinging battle until one had a leg broken at which point the "battle" was called to a halt.

Nisenan villages were unique in having a town crier who travelled from place to place, reporting who had died, married, who was sick, and telling where he had seen berries, deer, and other foods. The reporter was always welcomed and listened to without interruption. After he finished, there was much clapping and shouting.

The calendar for these Southern Maidu had descriptive names for summer months while winter months remained nameless. Months were called Birds go north, Smaller month, Leaves on tree, Flower month, Ripe seed month, and Winter.

California was under Spanish domination until 1822 when Mexico, followed by California, declared independence from Spain. Spanish influence was limited to military garrisons and religious missions along the coastal area.

Fur explorers in the 1820's to 1830's crossed the Sierras from the east although the exact location of crossing was not reported. Jedediah Smith, "among that most extraordinary group of mountainmen" was the first to lead a party across the Sierra Nevadas in search of beaver in untapped rivers. Leaving Salt Lake and heading west in 1825, his exact route isn't known, but at one point Smith doubtless camped on the American River near present day Folsom.

A combination of adventurer and soldier, John Fremont of the U.S. Topographical Engineers wandered frenetically in the West, partially under government orders to explore and partly from his own desire for action. In February, 1844, journeying south from The Dalles, Oregon, and into Nevada, he decided to tackle the Sierras directly despite warnings from Indians "rock upon rock; rock upon rock; snow upon snow . . . You will not be able to get down from the mountain."

With Kit Carson leading the way, the trek began. Snowstorms, gale winds, and scant supplies hampered them. Cartographer C. Preuss wrote "we had tonight an extraordinary dinner — pea soup, mule, and dog."

The indefatigable Fremont had attempted to drag a howitzer over the Sierras, but snow and difficult terrain forced them to abandon it at a very steep hill which proved to be the "last and fatal obstacle to our little howitzer." The cannon had been brought all the way from St. Louis. Fremont persevered and arrived approximately one month later, on March 6, at the American River. A Mexican conducted the men to John Sutter's Fort where they rested and were refitted.

The establishment of John Sutter's Fort near present day Coloma on the South Fork was permitted under a grant from the Spanish government. Forsaking his wife and children along with bad business debts in Switzerland, Sutter had made his way to California where he impressed the Mexican government into giving him 50,000 acres. In nine years he built a fort and acquired a total of 146,454 acres. Loving pomp and ceremony, he was happiest when "associating with colonels, judges, or governors" (Jackson, p.42).

Dreaming of producing flour, lumber, and leather on his ranch, he employed Indians, Mexicans and Mormons as a work force. Most of his Indian slaves, as he called them, were Nisenan. Indians were paid in board, food, and two beads

Earth-covered house.

193

Sutter's mill when gold was discovered there along the American River.

a day. One story is that Sutter put his "slaves" out in a cloverfield to eat their fill. This is perhaps not surprising considering clover was one of the plants natives ate normally.

A visitor described mealtime: "The Capt. (Sutter) keeps 600 or 800 Indians in a complete state of slavery and as I had the mortification of seeing them dine, I may give a short description. Ten or 15 troughs 3 or 4 feet long were brought out of the cook room and seated in the broiling sun. All the laborers grate and small ran to the troughs like so many pigs and fed themselves with their hands as long as the troughs contain even a moisture."

Since there was a shortage of cut lumber, Sutter hired James Marshall, a carpenter from New Jersey, to construct a sawmill for him. Marshall chose a valley called by the Indians, Culuma, meaning beautiful vale. The valley widened at a point where the turbulent American River could be dammed. In building the dam, Marshall noticed something shining in the riverbed which proved to be a quarter-ounce gold pebble. Marshall and Sutter gave the pebble every known test for gold before they were convinced that the rock was really gold.

Marshall began mining the site, threatening to shoot anyone who worked in the place he and Sutter claimed. Sutter's farm laborers, many of whom were Mormons, moved downriver to discover one of the richest placers at Mormon Bar.

Work was never completed on Sutter's mill, and the presence of gold brought financial di-

saster to Sutter. He was unable to pay his debts and moved to the East where he lived on a small pension until his death in 1881. Marshall's gold claim came to naught, and he died in 1880 working as a gardener in Coloma.

The presence of gold in California was actually known as early as 1820. Explorer Jed Smith had found quantities of it which he was carrying back with him when his men were attacked and killed in Oregon. This quenched Smith's interest in the California gold scene. Two thousand ounces, worth at the time about $35,000 had been sent from San Francisco to the mint at Philadelphia. Missionaries, who had been aware of gold in the rivers, suppressed the information fearing it would interfere with their plans. Travellers had talked about the abundance of gold, and a number of experienced miners were working in California at the time of Marshall's discovery on January 24, 1848.

News of gold on the American River was received in the outside world with skepticism. Walter Colton, chief administrative officer in Monterey, sent a rider to the American River about 400 miles to the east to confirm the rumor. Perhaps skepticism contributed to the fact it took the first gold seekers months to show up.

In October, Colton journeyed to the gold fields himself. He found miners up to their knees in mud, tearing up a bog to obtain gold underneath. "Not having much relish for the bogs and mud," he used a crowbar and split open some rocks to extract slivers of gold. At Weber

Creek, south of Coloma, Charles Weber mined $20,000 in six weeks time. Rich Dry Diggings, later Auburn, yielded $16,000 from five cart-loads of dirt, and a day's earnings averaged $800 to $15,000. At Placerville, William Daylor and P. McCoon made $17,000 in one week. James Carson, a former sargeant in the U.S. Army, made $50 per day by scraping rocks with his knife. Irregardless of this, most miners could have done better remaining at home. A sense of adventure, a desire to throw off responsibilities or escape from boredom motivated many gold seekers.

Colton, after careful study, concluded the average earnings were less than 1/2 ounce a day or $600 per year. He felt much of a miner's time was spent running hither and thither looking for better prospects. "I never met with one who had the strength of purpose to resist these roving temptations." Miners were kept in a perpetual state of turmoil by rumors of rich discoveries just upriver. "They who get less are discontented, and they who get more are not satisfied."

All in all, it is interesting to look at some figures on gold taken from the Middle Fork, for example, from 1848 to 1849.

Volcano Bar $1,500,000
Greenhorn Slide 1,000,000
Mud Canyon 3,000,000
Willow Bar 600,000
Junction Bar 150,000
From all the hills 300,000

Frenzied gold miners inhabited banks and bars of the American in cities of tents. In 1848 there were 2,000 Americans in California,

53,000 by the end of 1849, and 92,597 by 1850. By August the hills around Mormon Diggings were covered with tents. Two hundred men worked there with pans and cradles. At Michigan Flat, below Coloma, the first store on the South Fork opened in a canvas tent. The settlement included a butchershop, hotel, and boarding house in one establishment where a drink cost .50 cents. In 1850 Leland Stanford and his brother began their fortune by operating several stores here.

Indian Diggings was so named when a company of Americans discovered some Indians panning for gold there in 1850. A town soon grew up after rich gravel deposits were uncovered. Texas Bar, Hoosier Bar, and Louisiana Bar attest to the cosmopolitan nature of the mining inhabitants. Foreign emigrants as well as Americans worked the river. Chili Bar was mined by Chileans who went off by themselves after they were accused of jumping claims elsewhere.

Placerville was first called Old Dry Diggings and for a time Hangtown. Three different stories take credit for the title, Hangtown. The true version, according to a local resident, went that a man gambling in a tent saloon lost his gold to the proprietor. He and his two companions "recovered" the gold, but were arrested the next day and sentenced to be flogged and exiled. The sentence was carried out immediately in the true style of justice in the mines. Instead of leaving camp, two of the robbers remained drinking whiskey and verbally abusing members of the impromptu jury. A second meeting was called, the men arrested again and hung from a leaning oak tree near the saloon.

Placerville,
California,
in 1849.

Indians and whites fighting during the gold rush days.

The second and third stories involve hangings of robbers from a gigantic oak tree which formerly stood on the corner of main street.

As crime increased in the mining camps, so did quick trials and summary hangings. James Page was executed for the murder of an unknown man near Folsom, and C. W. Smith was hanged for murder near Pleasant Valley. A crowd of six to eight thousand witnessed the execution of J. Logan and W. Lipsey for murder at Cold Springs in 1854.

Things became so bad by 1865 that a number of Chinese were robbed near Placerville in broad daylight. The Chinese supplied fresh food to several mining companies on the South Fork. On Oregon Bar, miners convinced a Mr. Walden had stolen $100, put a rope around his neck and swung him up, hoping he would confess. This failed, and he was released, thinking he might lead them to the money. When Walden didn't comply, he was strung up again. Finally, disgusted, he demanded to be hung, "without humbug, harangue or torture." whereupon the chagrined miners released him.

Indians were subjected to many crimes from abuse to murder, but no miner could be tried for killing an Indian. Miners who had come down from Oregon were especially harsh, and their treatment of the Nisenan even brought protests from other miners. In April, 1849, the Oregonians returned to camp on the Middle Fork to find their goods ransacked and five companions missing. A posse was formed, led by the Oregon men who blamed Indians. An unsuspecting village was attacked, four men killed and scalped on the spot. A second posse killed an additional twenty Indian men and brought eighty to Coloma as prisoners. Seven stood trial for the murder of the five missing Oregonians. Marched to the center of town, the Nisenan broke and ran only to be shot or stabbed. One, who reached the river, was stoned to death.

In the absence of laws in the mining camps, public meetings would be held to define a district and devise some regulations to be followed. For example, Smith's Ranch Mining District on the American River adopted the following:

"A claim shall be 150 feet front and run to the center of the hill; a claim must be worked within ten days from the time at which it is taken up, and as often as one day every week afterwards; two or more holding claims may form a company to work any of them without being bound to work each claim."

In El Dorado County, other regulations were soon enacted to control that valuable commodity, water. Perpetual streams of flowing water were essential to mining operations, and water had to be supplied to dry areas. Companies formed to dig canals and ditches. South Fork Canal took water out of the American and supplied Placerville. Coloma Canal cost $42,000. Negro Hill Ditch cost $20,000, and the Pilot Creek Canal running parallel to the Middle Fork carried water twenty-six miles to Spanish Dry Diggings and mines there at a cost of $180,000. Water so delivered was sold and local regulations prevented a second canal from tapping the water source used by the first. Water rights became a reality. Other regulations dealt with property damage in the construction of these canals. The Pilot Creek Canal charged $1.00 per inch for its water, even though that price was higher than that of others who charged .12 to .20 cents per inch. The higher charge was for delivery under pressure.

In retrospect although the gold rush lasted only a few years, the sudden presence of more than 100,000 miners left no room for the Native Americans who had lived along the American River for more than 2,500 years. In the face of muddied streams, oak trees cut down, and game killed, Nisenan tried to adapt as best they could. At first they labored in the mines in exchange for food and clothing, and at one point more than half the workers were Indians. An increased feeling of antagonism led to the formation of vigilantes who attacked and killed off whole Indian villages at the slightest provocation. In the wake of these Indian hunting excursions, disease, and hunger, little remained of the Nisenan.

"Never before in history has a people been swept away with such terrible swiftness, or appalled into utter and unwhispering silence forever and forever, as were the California Indians by those hundred thousand of the best blood of the nation" (Powers, p.404).

Indian Life Along the River

The presence of Nisenan can be felt by the rafter along the American River where villages were abundant on all three forks. Each permanent village had a special name. Barno was southwest and Ekele-pakan west of Placerville. Kolo-ma was on the South Fork, Kulkumish at Colfax, and Moloma at Auburn. Yamaku was located near the junction of the South Fork and main river. Villages varied in size. Kolo-ma, for example, had two to three hundred persons.

Materials left by the Nisenan can be found wherever the villages were located as well as in caves along the river. Hawver Cave, near Auburn on the Middle Fork, was excavated in 1910. Human skeletons as well as artifacts were found. Bones were broken and scattered, and no complete skeleton was remained. Bones on the lower level of the cave indicate corpses were thrown in for quick disposal. Stone and bone artifacts here include arrowheads, flaked stone tools, blades and knives. Fossil mammoth, sloth, and other animal bones were also found in the cave but were earlier than the human material. The cave was thought to have been used as late as 4,000 years ago.

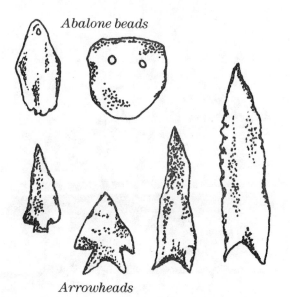

Abalone beads

Arrowheads

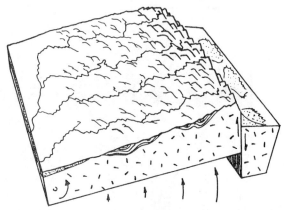

Rotation of the Sierra fault.

Geology of the American River

The American River is divided into three branches, the North, Middle, and South, all of which flow into Folsom Lake before joining the Sacramento River. The North and Middle forks merge just north of Folsom Lake and form a fan pattern from north to northeast. Most of the pathway of the river is controlled by the geology following channels between resistant rocks and along faults and fractures. The American River drops 6,000 feet in fifty miles, and its watershed covers 2,000 square miles between the Bear River to the north and the Cosumnes River to the south.

The Sierra Nevada mountain range is the top of a massive piece of the earth's crust which has been rotated upward on the east side causing it to tilt toward the west. This configuration gives it a steep east face and a more gentle western slope. The area drained by the American River in the Sierras is referred to as the northern block.

Geologically the western Sierras are a series of elongate, smaller blocks oriented north/south and separated by large-scale faults. Across the western slope, the oldest rocks are to be found toward the east, and the younger rocks on the west. Although the oldest rocks in the western Sierras are in excess of 400 million years old (Silurian), much of what is seen along the American River is only 250 to 150 million years old (late Paleozoic to Jurassic age).

The history of the Sierras can be divided into a series of separate geologic episodes. Beginning in the Paleozoic, over 200 years in the past, sediment from lands lying to the east poured into a large basin where the Sierras are today. Thousands of feet of mud, clay and volcanic ash from this eroding continent mixed in with debris from oceanic volcanoes. The collision of the Pacific Plate against North America pushed

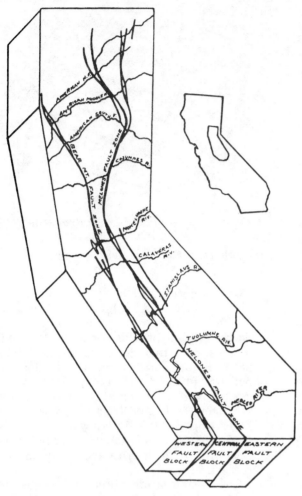

Major fault systems and fault blocks of the western Sierra slope.

posits were eroded and exposed. Gold found its way into the torrential streams draining the Sierras, but its extreme weight — nearly twenty times heavier than water — dictated that it was quickly deposited deep in the stream channels. The gold was gradually dispersed out across the western Sierras into a series of old river channels that roughly parallel the present streams. The weight of gold caused it to accumulate at the bottom of streams under the coarsest gravels.

During the middle Tertiary period or about 30 million years ago, the Sierras were the site of intensive volcanic eruptions. In addition to ash, great volumes of lava were extruded out onto the landscape. This lava, in the form of a rock called basalt, poured down the hills and eventually into the stream valleys themselves. Moving like a runny syrup, lavas flowed down stream channels where they cooled and hardened. The effect was like pouring plaster into a mold, and many of the old channels were thus cast in stone.

The present topography of the Sierras was carved out within the past three million years by the processes of stream runoff and glaciation. In this final episode the Sierras were broken up into blocks by faulting and again uplifted with a pronounced tilt to the west. Many of the large scale faults and fractures that domi-

deep ocean rocks to the surface as well as thrusting them beneath the edge of the continent. Accompanying this process of continental drift, granites from deep within the earth invaded these sediments to cool and crystallize after baking the host rock thoroughly.

As the granites slowly hardened, gold formed in veins and fractures along with quartz near the end of the cooling process. Seventy to 50 million years ago (late Cretaceous to Tertiary), the whole Sierra block was slowly raised to well above sea level. This uplift process triggered rapid erosion by the rivers.

Soils and forests developed atop the eroding block. Fossil leaves from those forests which are 50 million years old (Eocene) can be found in the clay deposits associated with the older gold-bearing gravels. These leaves are typical of a tropical climate. Acids and related chemicals from the plants speeded up the process of rock decomposition during this time.

Along with other rocks, gold and quartz de-

Ancient (Tertiary) gold-bearing stream channels in the western Sierra Nevadas.

nate the courses of the rivers today were formed in this period.

Glaciation, brought about by the Ice Ages, chiseled out the high Sierra crests. In the North American Plains far to the east, continental ice sheets slowly moved southward in several glacial phases punctuated by warm intervals. Although the Sierras were too far south for continental glaciers, the uplift of the earth's crust prior to this period brought the mountain crests well within the realm of temperatures cool enough to develop mountain glaciers. Today glaciers are still one of the most important geologic agents shaping the peaks and ridges of the high Sierras.

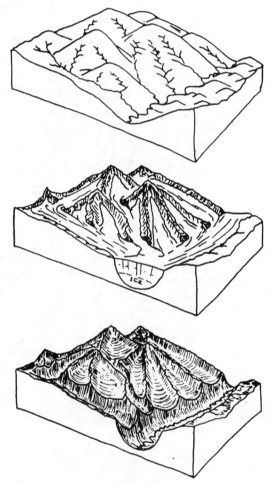

Glacial erosion of a stream valley.

Increasing streamflow and erosion 50 million years ago (Eocene) saw the exposure and removal of much of the Mother Lode gold which had been crystallized with quartz along with the Sierra granites. This gold was in turn quickly reburied with the Eocene gravels in the ancient streambed network across the Sierras. Today streams continue to erode and redeposit placer gold in small amounts as the constant weathering of the Sierra block continues.

The original discovery of gold in 1848, which brought over 100,000 miners to the state by the following year, was made on the American River. At a site on the South Fork, thirty-five miles northeast of Sacramento, James Marshall found the first nuggets to set the great rush in motion. This site, now named Coloma, is near the northern end of a northwest by southeast meandering strip across the western Sierra slopes called the Mother Lode belt.

Geology Along the Route of the North Fork

The North Fork of the American River is run in two intervals from Eucre Bar to Colfax, and from Colfax to Folsom Lake. At Eucre Bar the river exposes a series of 250 million year old Paleozoic oceanic rocks. The rocks locally have been chopped up by a complex series of intersecting faults. These faults and fractures contain mineral deposits, including gold, which find their way into the river as placer deposits.

Only 1.5 miles below the put-in, the river slices through a major fault zone. This fault marks the boundary between the Paleozoic rocks and a series of dark-colored, iron-magnesium rich rocks called ultrabasics. Paleozoic rocks at Eucre Bar are sandstone and shale that have been heated and squeezed by metamorphic processes. During this process new minerals crystalized within the older rocks giving them the greenish color typical of many metamorphic rocks. The high temperature and pressure has destroyed most prehistoric traces of animal remains in the rocks, but occasionally a faint fossil imprint is found.

At the fault zone, the rocks change abruptly from Paleozoic to Mesozoic age. These rocks are rich in iron and magnesium (ultrabasics) and were formed under very high temperature and pressure. After a short interval the river runs back into the Paleozoic Calaveras Formation for over 15 miles.

Most major rock types of the western Sierra slope are separated from each other by large faults. Moving downriver, we are really moving from fault block to fault block.

After the 15 miles of Calaveras rocks, the river begins to intersect small pods of ultrabasic rocks (iron and magnesium-rich) which have invaded the Calaveras. Adjacent to Colfax, the river turns south where it is confined to softer sedimentary layers lying between two hard ridges.

Four miles downstream from the ultrabasics, the river cuts across the Melones fault zone into the Mariposa Formation. Much of the Mariposa is shale that was baked into slate by metamorphism. Only three miles of exposure of the black Mariposa slates give way to more Paleozoic volcanic/metamorphic rocks and sediments which continue through Lake Clementine and past Auburn.

The Melones fault here is represented by a series of fractures formed when the two rock masses of the central and eastern fault blocks slid past each other. Shortly after the faulting, mineral laden fluids from volcanic sources deep in the earth worked their way upward into the cracks to deposit the major Mother Lode to Sierra gold.

Geology Along the Route of the Middle Fork

Because of steep river gradients and sheer canyon walls, rafters on the Middle Fork have only a limited number of places to put in, and several portages are necessary. This part of the American River is commonly run from the Rubicon River to the confluence of the Middle and North Forks at Route 49. Often these segments are divided at Greenwood Bridge.

As with the North Fork, the first several river miles are in Paleozoic marine rocks which include limestones, quartzites, cherts, schists and slates of the Calaveras Formation. This variety of rocks is not typical of the Calavaras, and the local Formation is called the Calaveras Complex.

The Melones fault zone on the Middle Fork marks the boundary between Calaveras slates and schists and Jurassic volcanic rocks. After less than a mile, the river enters exposures of the Jurassic Mariposa Formation. Here the Mariposa is mostly slates, conglomerates, and dirty-grey sandstones (greywackes).

Yet another large fault separates the Mariposa rocks from a complex series of Paleozoic metamorphic rocks. Between this fault and Texas Bar, several limestone lenses in these metamorphics intersect the river. Caves in these limestone pods are natural tombs for the remains of animals that sheltered in them.

Evidence of prehistoric animal life was found in a cave along the Middle Fork about five miles east of Auburn and 700 feet above the present stream level. This cave, Hawver Cave, has unfortunately been destroyed since by construction, but when discovered it had a diverse fossil ice age mammal fauna. In addition to classical ice age species such as sabre tooth cat, large ground sloths, bison and mammoth, the cave had human remains. Other mammals entombed here included dogs, bear, ox, squirrel, horse,

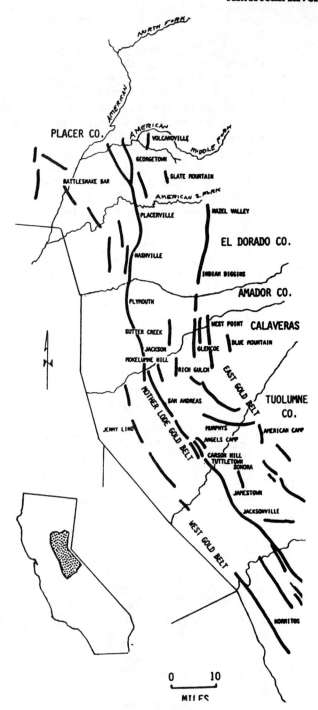

Major lode deposits of gold in the western Sierra Nevada.

rabbit and several types of mice. Hawver Cave animals were interesting because they were a mixture of forest and plains dwelling animals. Curiously, the cave had no animal remains of such modern types as mountain goats living here now.

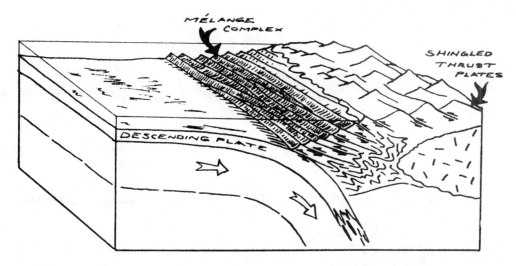

Development of a melange complex and shingled thrust plates.

Geology Along the Route of the South Fork

The South Fork of the American is normally run from Chili Bar to Folsom Lake. Just upstream from the Chili Bar put-in, the Melones fault separates the Calaveras and Mariposa formations. Adjacent to Chili Bar, slates of the Mariposa Formation are visible in the river bank. Marine (oceanic) muds and shales of the Mariposa Formation have been baked hard by contact with the local, molten Cretaceous Sierra granites to create the fine-grained slate. The Mariposa slates have been quarried here for over a hundred years. This high quality, hard platy rock was originally mined and sold for such diverse purposes as countertops, headstones, and coffins. Today it finds most of its use as roofing slate and decorative stone for sidewalks and floors.

Just below Chili Bar, a short section of grey crystalline granite is exposed in the river bed. A brief interval of volcanic rocks altered by heat and pressure (metavolcanics) separates the slates from the granites. After following a two to three mile section of the dark Mariposa slates, the riverbed abruptly cuts into an eight mile long zone of granite. The towns of Coloma and Lotus are in this interval. The former is near the spot where the 1848 gold rush began.

The next section downriver from the granites is a highly faulted sequence called a "melange" which means mixture. This four mile interval includes cherts, lavas, and metamorphic rocks of all descriptions. Within this region "greenstones" are particularly common. Greenstone is a term given to basalt lavas that have been slightly altered by metamorphic activity (heat and pressure).

The melange terrane here marks the zone where two major earth plates came together. Melange rocks are deep sea floor sediments that have been raised to the surface and scraped off as one plate thrusts beneath another. This distinctive melange terrane can be traced north and south for miles along the Sierra front.

Just at the western margin of the melange, about one half mile of faulted exposures are dominated by the rock serpentine. Serpentine is derived from iron-magnesium rich (ultramafic) rocks that have been treated by heat and pressure, and it commonly bears the mineral asbestos. It is a fairly soft rock that breaks and shears (faults) easily. Faulted surfaces in serpentine are typically smooth and polished in appearance.

Downriver from the serpentine exposures, an interval of amphibolite occurs. This rock is rich in elongate black crystals of amphibole and is formed at considerable temperatures and pressures. Another high temperature rock, gabbro, also occurs in this interval. It is usually dark-colored with a coarse, crystalline texture. The Bear Mountain fault here separates the western and central fault blocks.

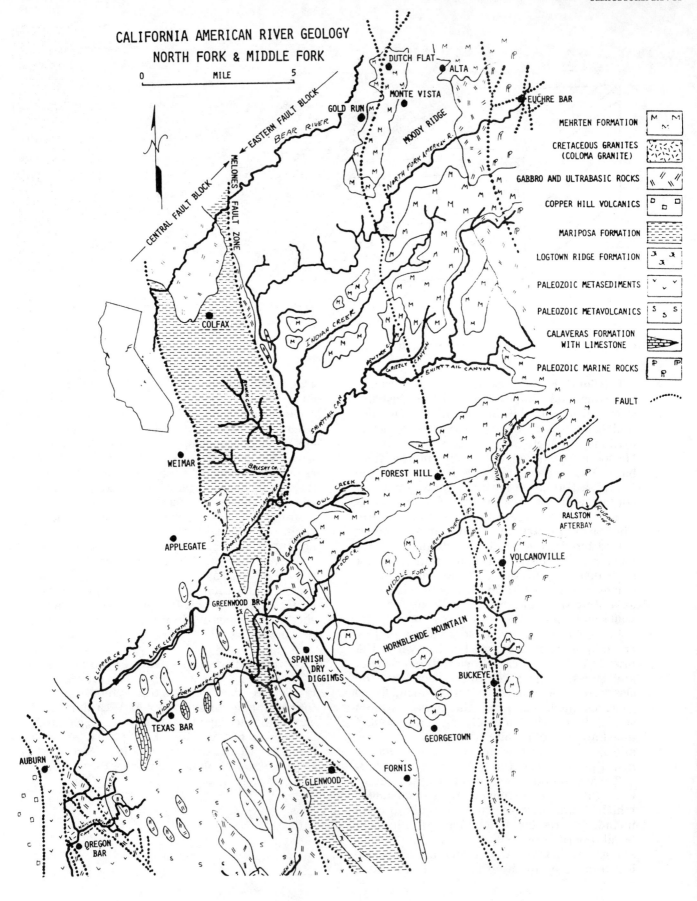

CALIFORNIA AMERICAN RIVER GEOLOGY
NORTH FORK & MIDDLE FORK

0   MILE   5

MEHRTEN FORMATION

CRETACEOUS GRANITES
(COLOMA GRANITE)

GABBRO AND ULTRABASIC ROCKS

COPPER HILL VOLCANICS

MARIPOSA FORMATION

LOGTOWN RIDGE FORMATION

PALEOZOIC METASEDIMENTS

PALEOZOIC METAVOLCANICS

CALAVERAS FORMATION
WITH LIMESTONE

PALEOZOIC MARINE ROCKS

FAULT

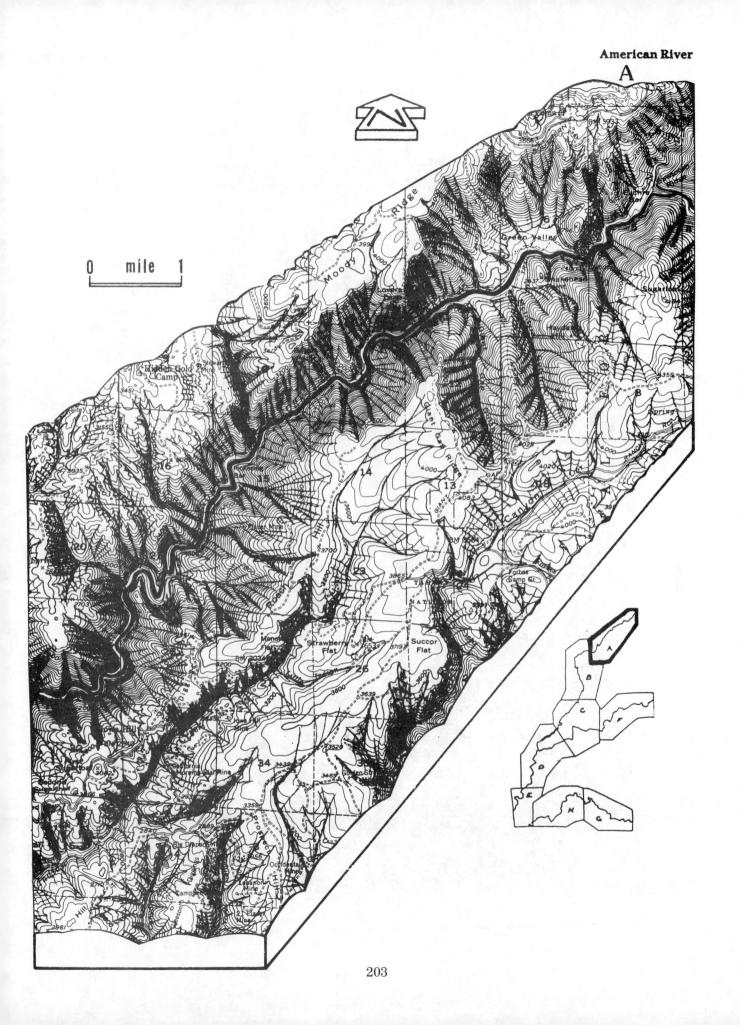

American River

A

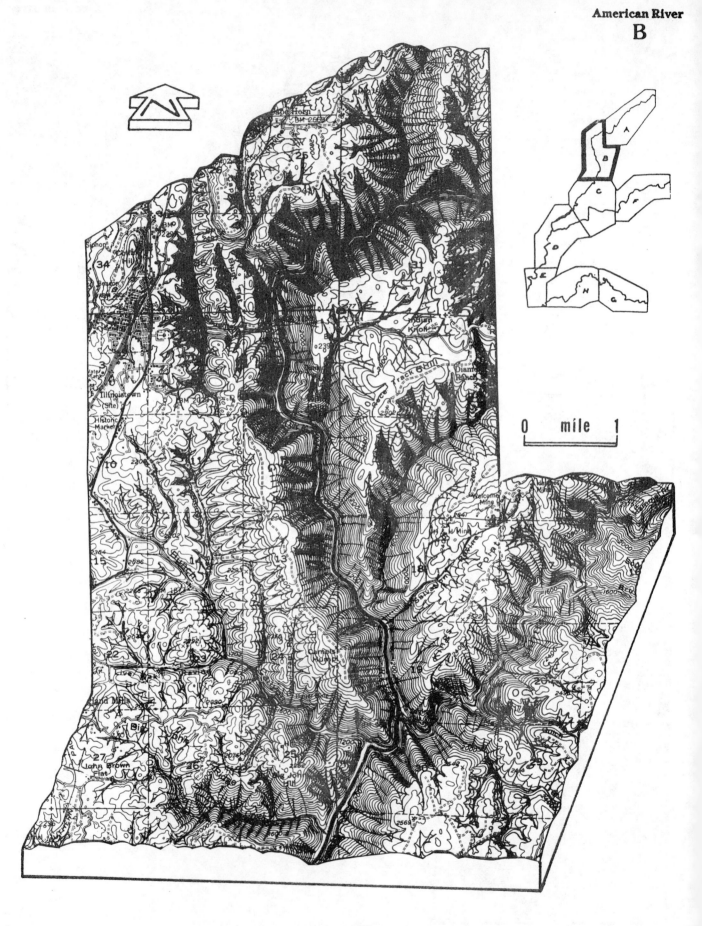

0 mile 1

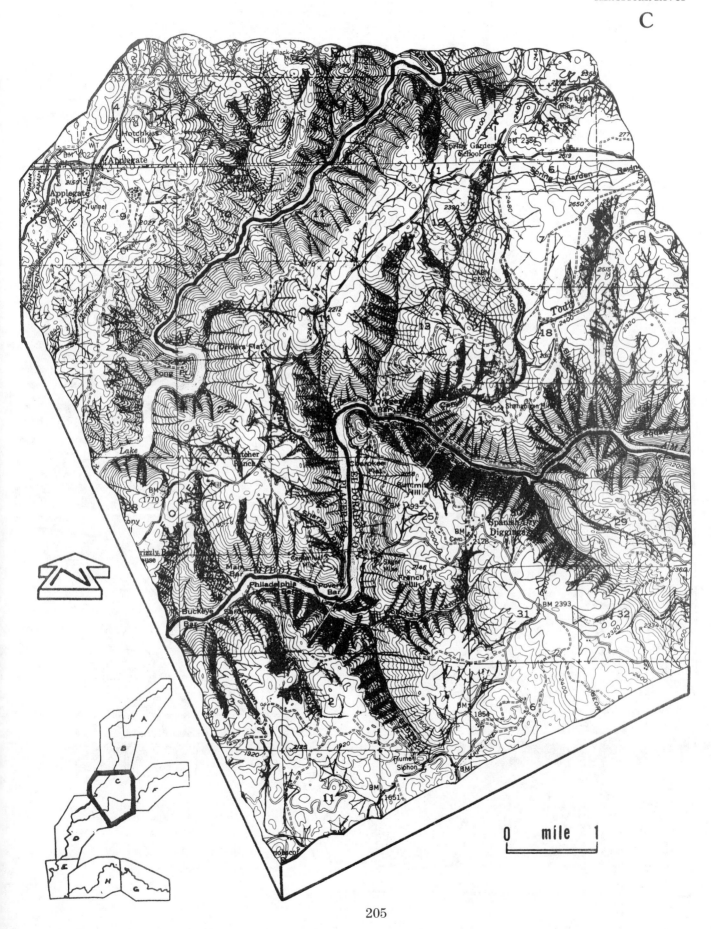

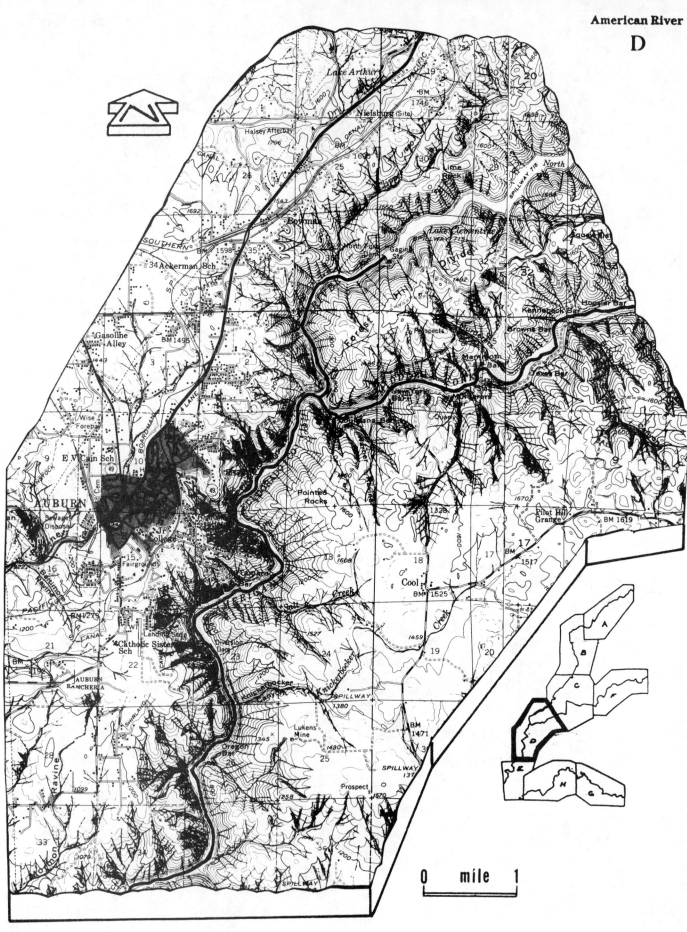

0   mile   1

206

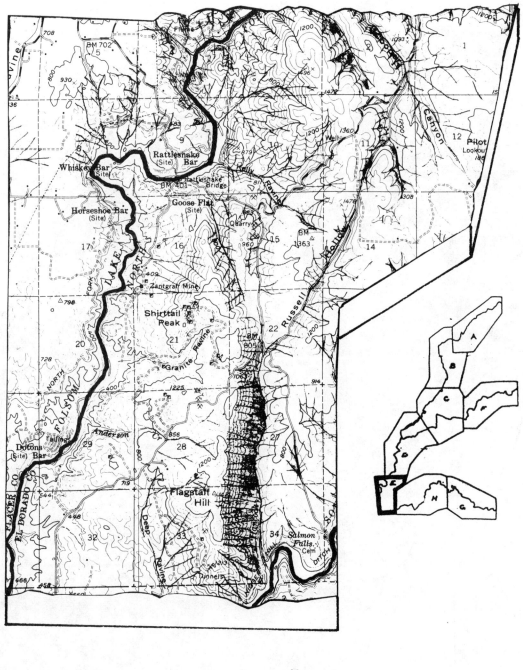

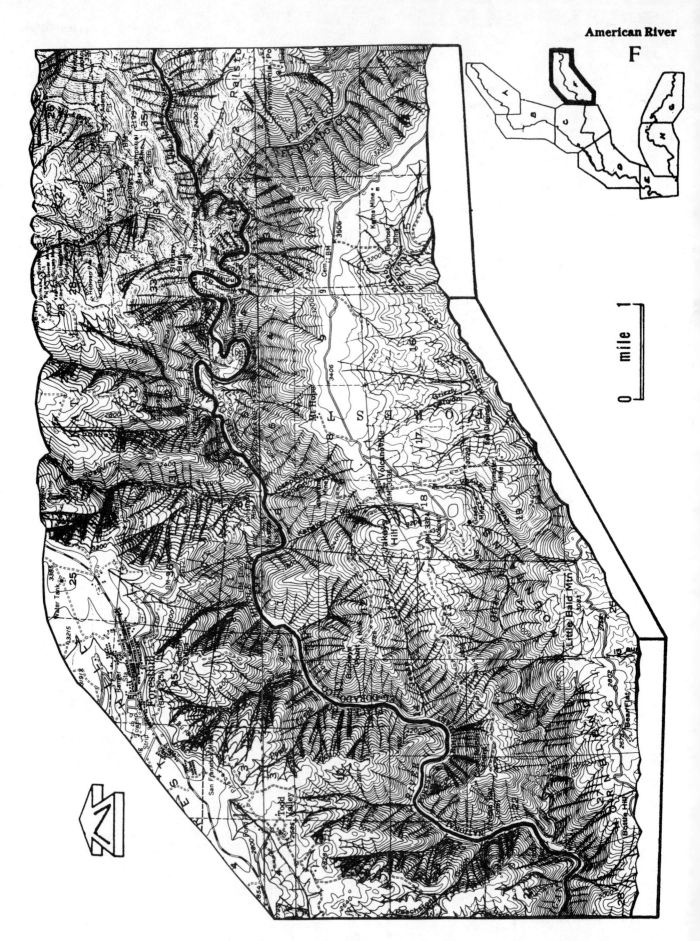

American River

F

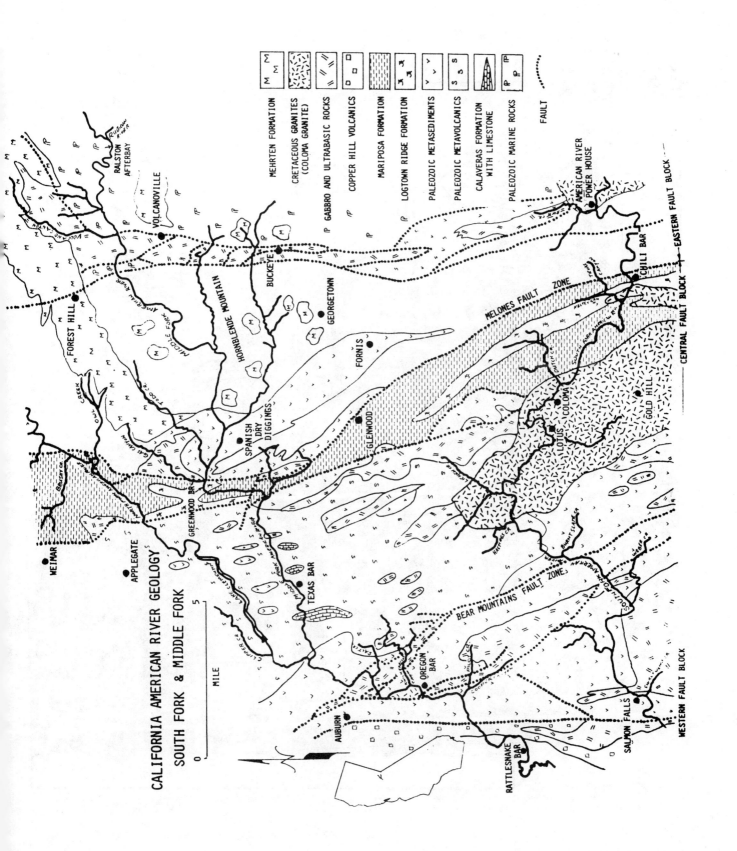

CALIFORNIA AMERICAN RIVER GEOLOGY
SOUTH FORK & MIDDLE FORK

MEHRTEN FORMATION
CRETACEOUS GRANITES (COLOMA GRANITE)
GABBRO AND ULTRABASIC ROCKS
COPPER HILL VOLCANICS
MARIPOSA FORMATION
LOGTOWN RIDGE FORMATION
PALEOZOIC METASEDIMENTS
PALEOZOIC METAVOLCANICS
CALAVERAS FORMATION WITH LIMESTONE
PALEOZOIC MARINE ROCKS
FAULT

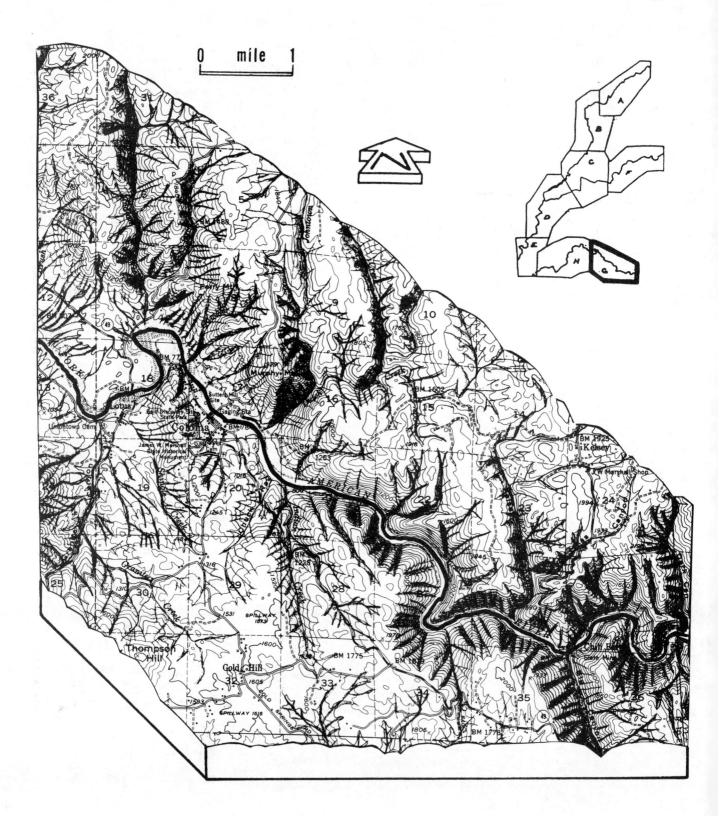

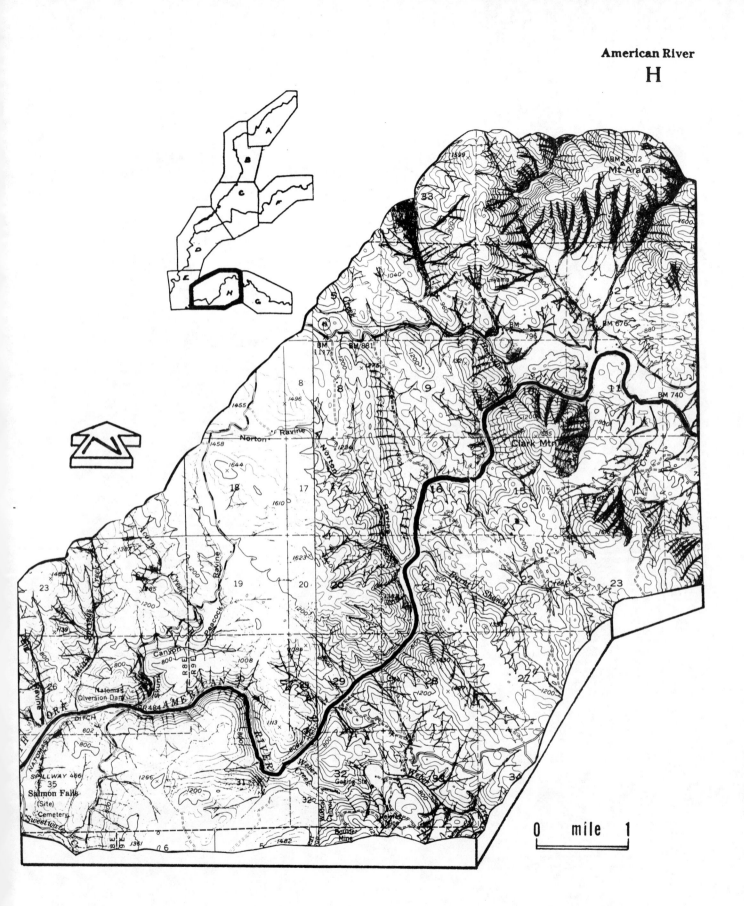

Carson River

East Fork Carson River

(Wolf Creek Campground to Route 95)
Alpine County, California to Douglas County, Nevada
Length of run: 29 miles
Number of days: 1

Beginning as snowmelt in the high Sierras, the East and West forks of the Carson join near Genoa, Nevada, to form the slower moving Carson River. The river then flows northeast for 125 miles into the Lahontan Reservoir. Before the reservoir was built in 1915, the Carson went unhampered all the way to Carson Sink and Carson Lake.

The runnable section of the Carson is not long because the eastern face of the Sierra Nevadas is steep giving the river a high gradient. Unlike gentle west slope rivers, waters of the Carson flow from glacial fed streams to dry desert topography over a very short distance. The change is profound. The flat, dry terrain allows the rafter to look back and see where the trip started earlier in the distant, high Sierras.

Early History of the Carson River

Washo Indians lived in the high Sierra Nevada Mountains around Lake Tahoe and along the watershed of the upper Truckee and Carson Rivers. Their area was described by trapper William Hamilton in 1844 as a "hunters paradise". An abundant, easily accessible supply of food meant a higher population than for any other Great Basin tribe. The Washo averaged two to three persons for each square mile, while other tribes generally had one person for every fifteen square miles. The maximum Washo population was put at 1,500 persons. "Washo" comes from the native word, Washiu, meaning "person".

Washos hunted and fished in the river valleys where pine nuts and rabbits were most commonly found. Acorns, berries, and plant seeds were gathered when ripe. Small animals were hunted and eaten, but the larger animals as antelope and bear were killed for the value of their skins rather than for the meat they provided. Rabbits were caught in a long net, and the position of the rabbit in Washo society was so important, one "rabbit boss" alone was selected and given authority to organize hunts.

Pinon nuts were the basic plant staple. Pine nut areas were marked off in long strips by a se-

Kit Carson, famous scout, trapper, and mountain man who gave his name to the Carson River.

ries of stones about one to one-half mile wide uphill. Since the nuts at the bottom ripened first, gatherers began there and worked their way up to the top with the last nuts collected in mid October. Some of the nuts were stored in pits for use over the winter, although storage facilities were inadequate and frequently the stored food molded or was taken by animals.

Despite the fact that property was owned communally, one band might have more rights than another to a pine nut area or to a fishnet constructed across a stream. Owners of a pine nut picking ground could destroy the baskets or other property of any gatherers invading their pine nut territory. In other words, individual members had the right to hunt or fish in a certain place which "belonged" to them. In effect each individual was actually hunting on a small ground which he knew best and not in Washo territory as a whole.

This concept was not understood by Americans who settled in Washo lands. John Reese in 1851 thought he had purchased land from Gumalange, of the Washos, who took a payment of flour. Gumalange, however, said he thought the gift had been in exchange for using the land temporarily. Only when a person died was their "usage" of this property lost.

Each household was organized around the family, and one, sometimes two, families occupied one house. Summer houses were flimsy structures made of branches and brush thrown around in a circle to form a wall about five feet high as a protection against the wind. A winter slab house was constructed by shingling pieces

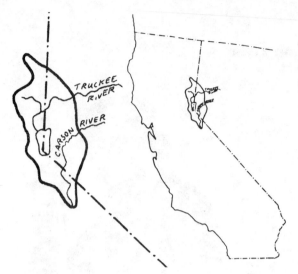

Distribution of Washo Indians in California and Nevada.

of cedar or conifer bark overlapping to give the house a conical shape.

In what can only have been a painful process, the face and body of boys and girls were tattooed when they were about fourteen or fifteen years of age. The designs were usually a series of dots, lines or circles. Soot mixed with acorn juice or grease was used as the dye. Red ochre gathered south of Markleeville was used as a body paint. The ochre was mixed with water, rolled into a stick, and heated to make it hard, after which it could be used for drawing designs.

Gift giving was a Washo social custom for every occasion: births, marriages, social ceremonies, and for greeting strangers. The importance lay not in the value of the gift, but in the act itself, and frequently pine nuts were the item presented. Explorer John Fremont, on his trek across the Sierras, was greeted by a "party of twelve Indians . . . and whenever we met an Indian, his friendly salutation consisted in offering a few nuts to eat and to trade . . . " Fremont went on to relate "when we got near enough they immediately stretched out to us handfuls of pine nuts, which seemed an exercise of hospitality."

On the other hand, the Washo seldom took part in trade with surrounding tribes. Language difficulties and a feeling of dislike for others outside their territory contributed to this. The rugged terrain of the Sierras gave the Washo a physical isolation other tribes didn't have. In historical times, communication increased with the Maidu and Miwok to the west after the building of wagon roads. It is not surprising, then, that a number of trade goods from those tribes found their way among the Washos. Items

most frequently imported were acorns, soaproot combs, and numerous medicinal plants. The Washo, in turn, traded salt, pine nuts, obsidian, skins, and baskets.

Little contact with either Americans or Spanish in western California occured before 1826. Occasional trappers or explorers crossed the Sierras going east or west. American trappers, Jedediah Smith and William T. Hamilton, passed through Washo lands, Hamilton commenting on the miserable condition of the Indians when he was on the Carson River in 1844.

Lt. John Fremont, with his party surveying for the U.S. government, accompanied by the famous scout, Kit Carson, camped near Markleeville on the East Fork of the Carson River before trekking on up over the Sierras in February, 1844. Fremont reported "during the day a few Indians were seen circling around us on snowshoes, skimming along like birds . . . the Indians had only the usual scanty covering and appeared to suffer greatly from the cold . . shivering in the snow which fell upon his (the poor fellow) naked skin."

Washo winter and summer dress was a robe or cape made of animal skin. Moccasins weren't standard dress, even though Fremont commented that his guide's moccasins were about worn out. In cold weather bark was wrapped around the feet inside the moccasins for extra warmth.

Winter house

Summer house

Pinon nut pine was the most common food of the Washo Indians.

Fremont may have introduced the Washo to horses, not a very practical animal to have high in the mountains. He put one on horseback and "enjoyed the unusual sight of an Indian who could not ride. He could not even guide the animal, and appeared to have no knowledge of horses."

Eventual changes came to Washo territory with the construction of permanent wagon roads. These in turn led to the establishment of an outpost at Woodfords on the West Fork of the Carson in 1847 and the beginning of the Genoa trading post in 1849. Roads not only opened up contact with whites, but allowed Washos to visit the adjoining Indian groups. Washos took to begging along the roads, and travellers often commented on their ragged appearance "clad in skins, coarsely sewn . . . filthily clad . . ."

The Genoa post, built by John Reese, called Reeses Station, then Mormon Station, was a combination store, hotel, and stockade. It was designed for and brought a handsome profit to its owner. Emigrants on the last lap of their journey to California paid exorbitant prices for flour, bacon, and other staples.

The discovery of gold along with the Comstock silver at Virginia City, Nevada, in 1858 initiated the "rush to Washo." Twenty thousand people accomplished the task of destroying the streams, diverting rivers, and cutting timber. Game animals retreated. Lake Bigler, or Biglow, as Lake Tahoe was called, was taken over by white settlers, and the Indians were left with no sources of food.

Once a fishing reserve for Indians, Lake Tahoe was opened to commercial fishing, with the Burket Company taking tons of fish out daily using a l/2 mile seine. Sawmills were set up in the Tahoe area to cut and send lumber to the mines. Logs were hauled by rail to the summit nearby and allowed to slide down a V-shaped flume to the Carson River. The Carson River Lumbering Company then sent the logs on downriver, adding some of their own. By 1860, 6 million board feet of logs had floated to Virginia City and gained the Company $75,000. These magnificent forests had vanished by 1870.

More prospectors arrived with the discovery of silver at Koenigsberg, later named Silver Mountain, in Alpine County. Miners scurried up and down canyons around Markleeville, Monitor and Bullion. A three mile long ditch near Markleeville diverted water from the Carson River to the mines where over three hundred claims were being worked in 1863. Tunnels and shafts were dug, blasts rang in the air, and the feeling of wealth was just around the corner for everyone.

The Carson River gained a certain amount of notoriety with the events at Hangmans Bridge and Markleeville. Markleeville was started in 1861 by Jacob Marklee who built a cabin and toll bridge over the river. Marklee was shot and killed a few years afterward in a land squabble over where the town was to be built.

Hangmans bridge, downriver a short distance, became famous as the spot where Ernst Reusch was executed. When Mr. Reusch's new bride left him for one E. H. Errickson, Reusch took his revenge by shooting Errickson. Inspecting the body to make sure he had done a good job, Reusch turned himself over to the authorities. While being transported to the Bridgeport jail on the evening of April 18, 1874, Reusch was intercepted by a party of masked men who proceeded to throw one end of a rope around the bridge and the other around Reusch's neck, pushing him off the bridge. The vigi-

John Fremont, who explored and mapped much of the unknown territory in the West.

215

Old Alpine County Jail, Markleeville, California (Calif. Div. Mines and Geology).

lantes were never identified, but were purported to be residents of Alpine County who were worried about the high cost of a prolonged trial on their already depleted coffers.

One interesting sidelight to Carson River history is the Von Schmidt Line between California and Nevada. Although Nevada was created as a state from the territory of Utah in 1864 by President Lincoln, the exact location of the California-Nevada boundary was in doubt. Earlier surveying contracts came up with different state lines including one done by A. W. Von Schmidt in 1873. Von Schmidt's survey added several million extra acres to Nevada, and the matter of the present day line wasn't settled until the official U.S. Coast and Geodetic survey of 1899. Just across the line in Nevada a toll road and bridge were set up by David Oles who charged travelers .50 cents to cross the Carson.

Indians attempted unsuccessfully to collect rent for use of their lands where they could no longer hunt or gather food. In one incident, the Potato War, hungry Washo harvested three acres of potatoes from a William Morehead near Milford, California. During the severe winter of 1859 Indians were starving to death. "A few days ago a Wa-sho died from actual starvation and exposure in the vicinity of Lake Bigler . . . and another was found dead at the base of those mountains yesterday, from the same cause" (Forbes, p.2).

Both curious and fearful of whites, Washo ultimately developed a reputation for passiveness and merely stood by while their land was overrun. They moved slowly into white settlements dependent on begging and scavenging for survival. By 1900, Indians had been given 64,000 acres of land in the Pine Nut Hills east of the Carson, as well as approximately eighty acres given to thirty individual families in Alpine

County, California. Near Dresslerville, in 1917, William F. Dressler gave the Washo tribe "in trust" with the U.S. government forty acres which constitutes the Colony at Dresslerville. In 1938, the federal government purchased three ranches for the Washo although this land later reverted to white ownership. Finally, the Carson Colony of 160 acres was purchased by the government for Washos in 1840. Most of the tribe presently resides in small houses in Dresslerville, in neighboring towns, or ranches, working as ranch hands or as skilled laborers.

Indian Life Along the River

Very little remains of the former Washo way of life along the Carson. Near Hangmans Bridge a cave yielded animal bones, obsidian flakes, and arrowheads. In the overlying layer, iron nails and glass bottle fragments indicate white

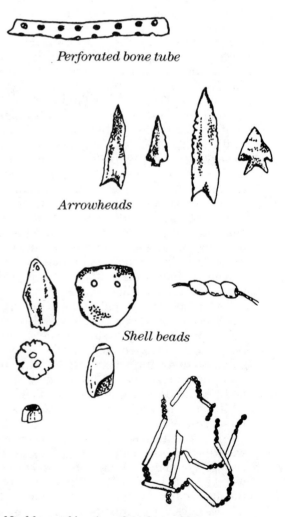

Perforated bone tube

Arrowheads

Shell beads

Necklace of juniper beads and bone.

visitors along the old wagon road stopped here as well.

Near Woodfords store, a campsite was known to have been used historically by Washo, and three miles northwest of Markleeville, near the route of the Old California Trail, abundant obsidian flakes indicate this was probably a "workshop".

Rock art of the Washos can be seen two miles east of Genoa where pecked designs can be found on granite boulders. The Nevada State Museum at Carson City has a boulder with petroglyphs said to have come from somewhere near Gardnerville. The boulder is about two feet long and eighteen inches high with wavy lines, possibly representing snakes, incised into the face.

Rock art of the Washo at Genoa.

Geology of the Carson River

Unlike rivers flowing down the Sierra's western slope, east slope rivers, including the East and West Walker and Carson, tend to have short runnable stretches and very high gradients. The underlying Sierra block tilted only about five million years ago during the Pliocene epoch. If we view this massive block as a gigantic stone trap door that opened slightly in Pliocene time, then the hinge must be hidden somewhere below the San Joaquin Valley to the west. The east side of the Sierras is an enormous fault — so recent that the scarp on the east face is still obvious to even casual inspection.

Beginning over a million years ago, alpine glaciation has intensively dissected this area of the Sierras, but the underlying structural features are still obvious. Throughout the runnable stretch of the East Fork of the Carson, the geology itself changes very little. Quartz-rich volcanic rocks dominate along the route. Locally these rocks contain deposits of gold and silver.

Mixed in with these are volcanic rocks of the type that accompanies particularly violent intermittent eruptions. Although most of the volcanics are relatively young at 5 million years, the episode that generated these rocks came at the end of a long history of eruptions beginning

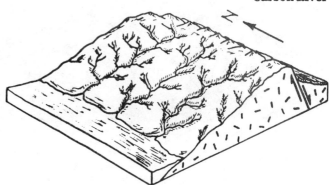

Sierra Nevada fault block

as far back as 40 million years in the past.

Volcanic history of the Sierras may be divided into two parts. The northern Sierras erupted first followed by the southern Sierras. As each episode drew to a close, the stream valleys became choked and clogged with a mixture of wet volcanic ash and lava changing many of their courses. The volcanic mud formed by a mixture of water and ash is called a "lahar." Under wet conditions lahars will flow several miles like syrup on even a slight slope. Volcanic activity along the Sierras continued right into glacial times (Pleistocene), and several of the glaciers show clear evidence of having been cut off and even dammed by lava flows.

Only a few miles to the south in the vicinity of Mono Lake and at Bishop, remnants of a "nuee ardente" can be seen. This type of volcanic deposit, known locally as the Bishop Tuff, represents the most spectacular and destructive of all eruptions. A nuee ardente is the by product of a catastrophic explosion where a charge of glowing, incandescent ash is suddenly shot from a volcano. The weight of the hot cloud exceeds that of air, and it rapidly settles back to earth flowing at enormous velocities down the flanks of the volcano. Hot, air borne particles kill or burn everything in the path of the fiery cloud. Often the fragments are hot enough to anneal themselves together after settling to earth developing a welded tuff.

Many of the Sierra glaciers of this era suffered

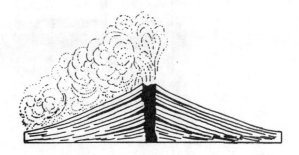

Nuee ardente eruption — "fiery avalanche".

217

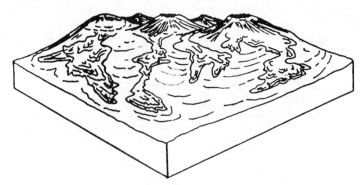

Lahars (mud flows), a volcanic ash/water mixture.

lava flows atop the glacial ice itself. It is fascinating to picture the hot runny fluid pouring out over the ice through a cloud of steam. A chilled layer of stone from the first flow over the ice forms the foundation for successive waves of molten lava. Later caves would develop as ice melted beneath the flows, but eventually the great weight of the cooled lava sheets would collapse leaving a rubble of lava blocks.

Six principal mining areas dot the Carson River drainage in the eastern Sierras. In these districts, silver is often economically as important as gold. Tailings piles are visible in the hills all along the river. Only six miles south of the put-in at Wolf Creek, the rafter comes across the mining district of Silver King. Quartz veins were actively mined here during the 1860's, but the district was never very productive. West

Walker, only five miles to the northeast of Silver King, was not overly rich. Gold occurs in thin quartz veins which were intruded into metamorphic rocks.

Silver Mountain in southcentral Alpine County lies seven miles south of Markleeville. In 1860 both gold and silver were discovered here, but in spite of original expectations total output was never high. Most of the mines in the vicinity were controlled by a British owned stockholders group, the Isabel Mining Company. Silver Mountain City was established by Scandinavian miners and boasted a population of 3,000 in 1863. Gold and silver occur in altered younger volcanic rocks.

Ores from Monitor and Mogul located in central Alpine County came from very young (Tertiary) volcanic rocks. These rocks stand out visually due to their distinctive leached iron stain (yellow, red, tan) appearance. Associated with the gold and silver in these volcanics are a variety of sulfide ores of copper, iron, zinc and lead. Both areas were mined just prior to and after the Civil War, and Monitor was named after the famous Union ironclad warship. Three to five million in gold and silver was recovered from these mines, and mining continued as late as 1930.

Hope Valley is a very small district in northwestern Alpine County ten miles east of Markleeville. Here the principal ores of gold and tungsten were mined well into the 1940's and 1950's. Ores occur in metamorphic roof

Mining town of Monitor, California in 1872 (Calif. Div. Mines and Geology).

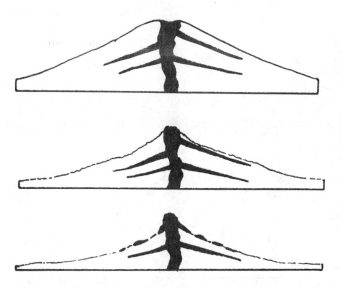

Development of a volcanic erosional plug.

the stream pattern changes completely. Upstream from this point the erosion behavior of the East Carson is notably "dendritic" or tree-like. This pattern reflects the nearly uniform composition of the volcanic rocks cut by the stream.

Below this point at Markleeville the East Carson makes a near right angle turn then proceeds at a straight line toward the Nevada/California border which it crosses at a right angle. Close inspection of the local map shows that the river has followed a well defined fault that trends to the northeast. In this ten mile interval, the stream meanders back and forth across the fault "trace" to stray only a few hundred feet before returning to the main fault trend. This "structural" control of the stream pathway ceases at the state line.

Just inside the Nevada line, the river swings abruptly north to follow the north/south trace

pendants suspended into granites of the Sierra Batholith. In cross section these rocks appear to be draped over the granites of the batholith. Gold occurs in quartz veins, but the tungsten ore develops in garnet rich "tactites." This rock type owes its occurrence to the contact metamorphism (cooking) and chemical alteration of a limestone by nearby volcanics.

Geology Along the Route

The Carson River run is broken into two sections, the upper East Carson and the lower East Carson. The upper East Fork is just over seven miles long between the put-in at Wolf Creek Campground to the take-out at Hangmans Bridge. The lower East Carson run is between Hangmans Bridge two miles east of Markleeville and Ruhenstroth Dam, a distance of a little less than twenty miles.

Five million year old Pliocene volcanic rocks are exposed almost continuously in the first five to eight miles below Wolf Creek. Only a mile north of the creek, an intrusive volcanic plug is apparent on the west side of the river. This plug was one of the many volcanic passageways that fed enormous amounts of Pliocene volcanics to the surface. Upon cooling, the rocks in the volcanic throat eroded more slowly than the adjacent stone, and it now stands out as a hard, resistant nob.

At evenly spaced intervals Wolf Creek, Silver Creek, and Monitor Creek join the Carson East Fork. A score of mines dot the landscape where Mogul Creek enters the East Carson.

At Markleeville, five miles downriver, Markleeville Creek marks the location where

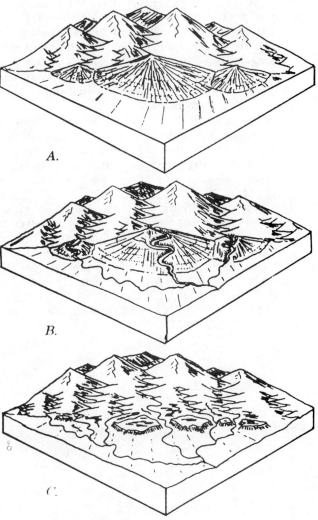

A.

B.

C.

Development of isolated mesas of fanglomerates (C) from alluvial fans (A) after being dissected by streams (B).

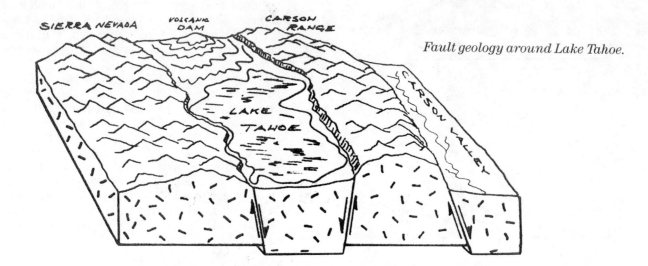

Fault geology around Lake Tahoe.

of yet another local fault. Since the gradient of the river is much less than in California, the river valleys change from a youthful to a more mature pattern. One hallmark of this change in stage of erosion is the increase in the size and frequency of stream meanders. Topographic maps display several meander features as cutoffs and oxbow lakes.

The estimated difficulty for river running also changes at Markleeville from class III (intermediate) to class II-III (beginner to intermediate). Equally profound is the change in vegetation to sagebrush, cottonwood, and pines. To the west, the fault scarp of the east face of the snow-covered Sierras makes a spectacular view.

Complex meanders along this Nevada stretch are in the process of dissecting some of the older "fanglomerates" of the area. These features are typical topography in a dry, desert climate. Coarse sand and conglomerate debris is dumped by streams at the base of steep slopes onto the desert floor forming a fan-shaped de-

posit. These alluvial fans are deposited and later eroded by seasonal torrential floods. Huge alluvial fans being dissected in this way locally stand out as flat top mesas gently tilted away from the mountains. The take out is 2/10 mile south of Ruhenstroth Dam at the BLM boat ramp.

Lake Tahoe, only a few miles to the north of the Carson watershed has a fascinating geologic history. The lake sits in a basin that formed between two fault blocks. At the north end of the lake, a natural dam of volcanic mud forms one side of the basin. The geologic structure here does not augur well for Lake Tahoe. It is estimated that the faults which generated the lake basin were of a type which slowly built up stress for hundreds of years without showing any appreciable creep before faulting. When the fault does occur, it is typically catastrophic, generating the largest of all known earthquakes registering up to 8 or 8.5 on the Richter earthquake scale.

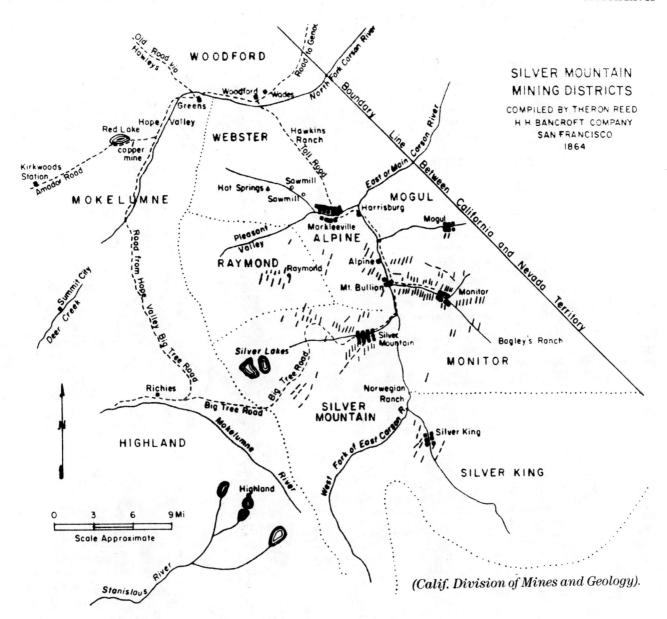

SILVER MOUNTAIN
MINING DISTRICTS
COMPILED BY THERON REED
H H BANCROFT COMPANY
SAN FRANCISCO
1864

(Calif. Division of Mines and Geology).

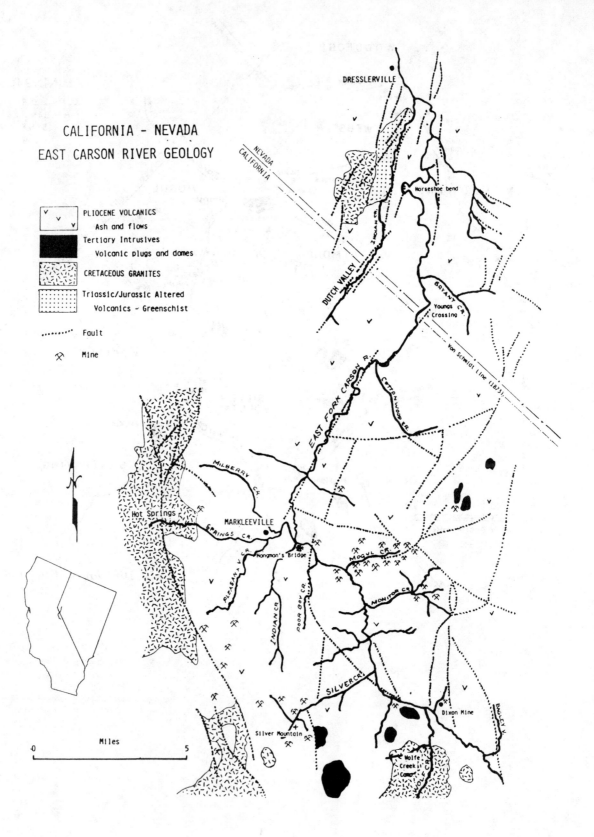

CALIFORNIA - NEVADA
EAST CARSON RIVER GEOLOGY

PLIOCENE VOLCANICS
  Ash and flows
Tertiary Intrusives
  Volcanic plugs and domes
CRETACEOUS GRANITES
Triassic/Jurassic Altered
  Volcanics - Greenschist

.... Fault

⚒ Mine

Miles
0   5

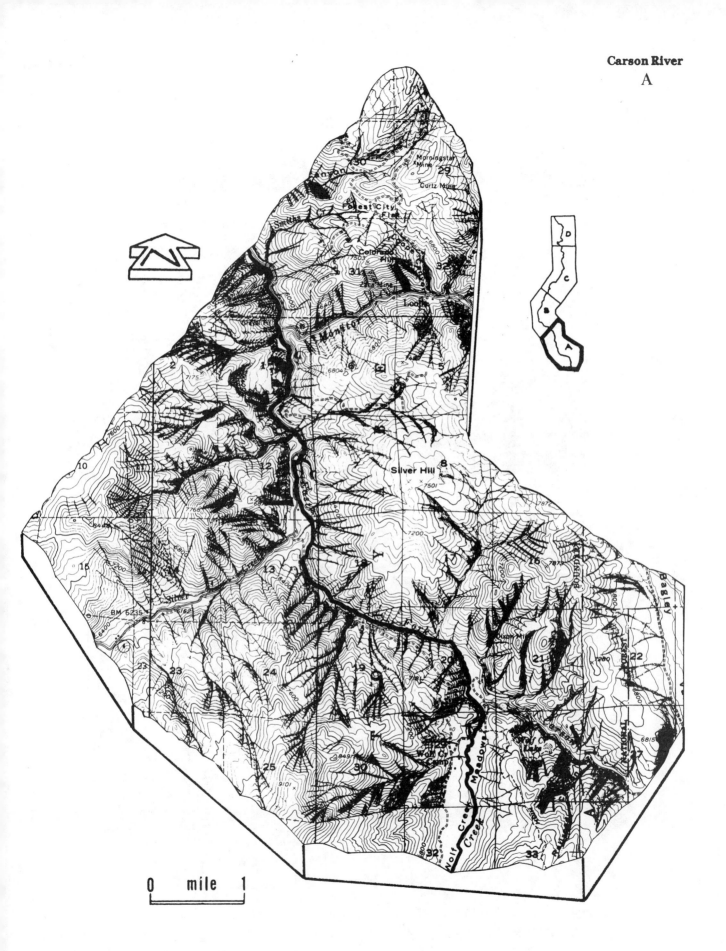

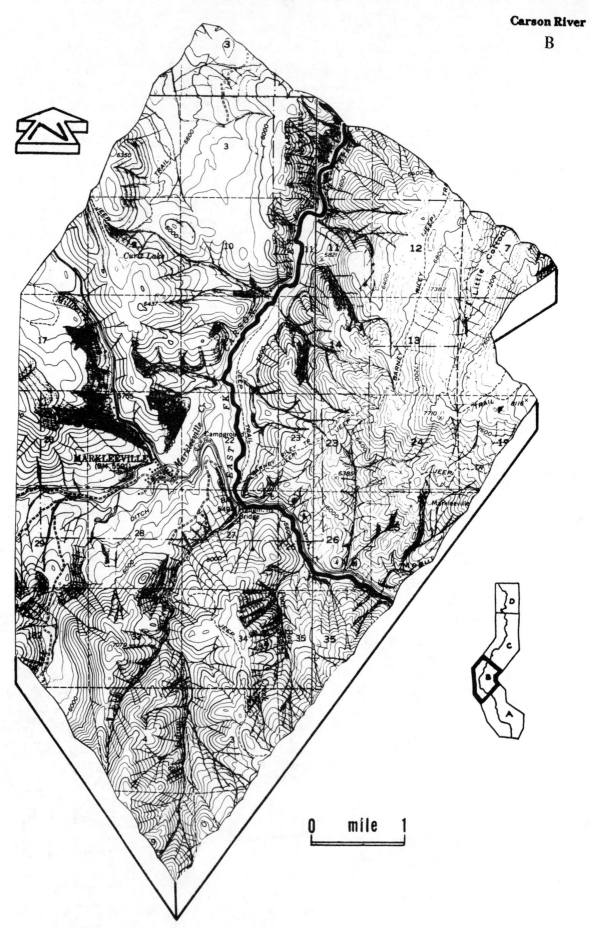

0   mile   1

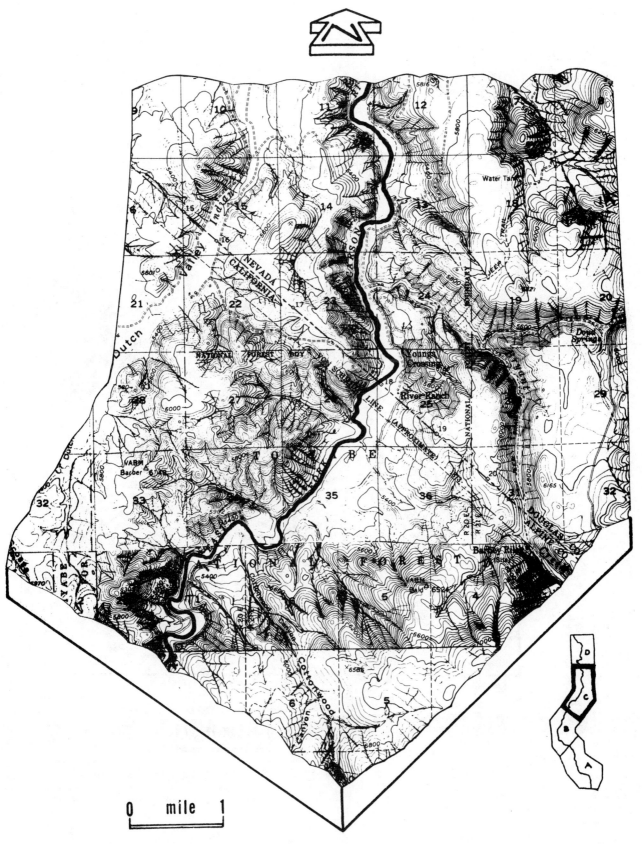

0 mile 1

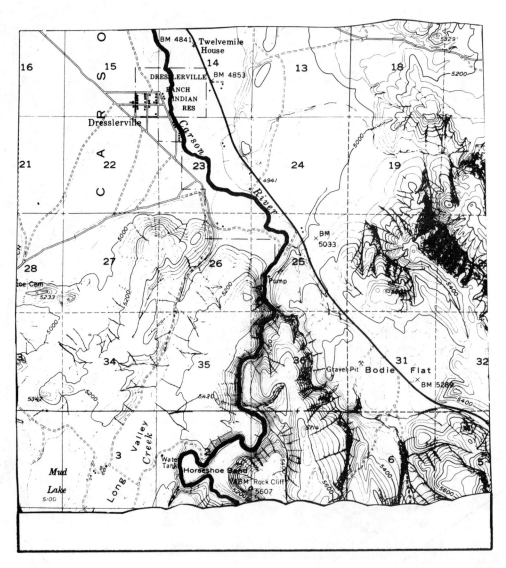

Eel River

(Dos Rios to Alderpoint)
Mendocino, Trinity, Humboldt Counties, California
Length of run: 46 miles
Number of days: 2-3

Draining the Coast Range of northern California, the Eel River flows southeast then west to northwest to enter the Pacific Ocean south of Eureka. Much of this 200 mile run follows the local grain of the geology, meandering between hard and soft ridges of rock. Heavy rainfall and unstable slopes cause landslides all along the banks of the river.

The Eel River Valley was the last frontier of California to be settled, and the conflict between settlers and Indians was cruel and harsh to the very end. Traditionally fierce warriors, Indians along the Eel stubbornly defended their homeland until all men had been exterminated and only a few women and children remained of the tribes. The names Spyrock, where Indians could overlook the Eel and spy on settlers, and Bell Springs where unfortunate Indians, pursued by whites, were betrayed by two cowbells they carried, echo this conflict from the past.

Early History of the Eel River

No less than eight small Indian bands lived along the lengthy Eel River Valley in what is now Mendocino, Humboldt, and Trinity counties. At that time, the main fork of the Eel from Dos Rios to Spyrock was the domain of the Yuki. The Wailiki lived from Spyrock to just south of Alderpoint, and the Lassik occupied only a small area around Alderpoint.

Wailiki woman with tattoos. (Powers, 1877)

Distribution of Indians along the upper Eel River in northern California.

ties. At that time, the main fork of the Eel from Dos Rios to Spyrock was the domain of the Yuki. The Wailiki lived from Spyrock to just south of Alderpoint, and the Lassik occupied only a small area around Alderpoint.

The Wailiki and Lassik, and most of the other small tribes, spoke the same language, Athabascan, whereas, the Yuki spoke a unique language unlike any other in California. This language as well as an unusually short stature and long face have led to speculation that the Yuki were the only remnants of the original native Californians.

Yuki means "stranger" or "enemy". In 1770 there were an estimated 2,000 Yuki, but only 95 in 1910. The Wailiki or "north people" numbered 1,000 in 1770 with 227 in 1910. The Lassik were named after a village chief and, along with two other small tribes, were numbered 2,000 in 1770 and reduced to 100 in 1910.

The Lassik, who lived at Alderpoint only in the winter months, roamed from place to place during the summer. An early observer described them as "a band of gypsies, or rather of thugs, houseless and homeless nomads, whose calling was assassination, and whose subsistance was pillage. Their hand was against every man, and every man's hand against them."

None of these small groups would have been particularly desirable neighbors. Yuki were said to be "fierce and truculent . . . indisputably the worst tribe among the California Indians" while the Wailiki were thought to be "choleric,

Cane whistle

vicious, and quarrelsome'' (Powers, p.125).

These two bands were often involved in conflict. The time and place of battle were arranged beforehand, and during the battle itself both parties stood and shot arrows at each other until they "get enough . . . and go home." Each warrior had approximately 300 arrows to expend. The dead were retrieved and scalps were taken from enemies. Scalps were softened with deer marrow and stretched on a frame. The head of a slain enemy was put up on a tall pole. During battle, Wailiki held up stiff, tough shields which easily repelled arrows and which could protect several men at a time.

Indians along the Eel were not adverse to eating most plants and animals found. Deer, hunted with bows or snares, or run down in relay teams, formed the main source of flesh supplemented by birds, rabbits, rats, and squirrels. Most varieties of birds as robins, doves and blackbirds were trapped. The whole unplucked bird was merely thrown into hot coals, the feathers burned off, and the carcass consumed along with the entrails. Salmon fishing with nets, spears, or poison was a year-round occupation. Wailiki were adept at poisoning calm pools of water in streams with the soaproot plant until the fish became sluggish enough to seize.

After the first frost, acorns were gathered,

Digging stick

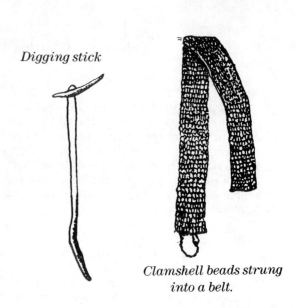

Clamshell beads strung into a belt.

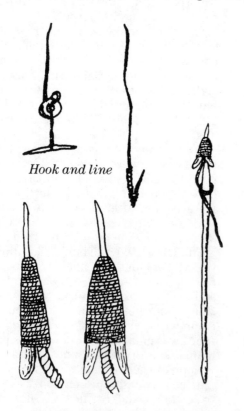

Hook and line

End of a fishing spear showing how its made.

pounded into a flour, leached in water, with the resultant mush made into bread or soup. Clover was eaten raw or stewed. Tubers, "Indian potatoes", berries and nuts were other plant foods harvested.

With the first good rain in Fall bringing out the earthworms, Yuki women, with a traditional stick — a pole about 6 feet long, sharpened and fire hardened at one end — dug and poked into the soil turning up a basketful of earthworms. A "rich and oily soup, aboriginal vermicelli" was the final product.

Clamshells and magnetite beads, dentalia and obsidian were objects of the wealthy. Clam shells, traded from Bodega Bay, were worth $1.00 for a 100. Magnetite beads, known as "Indian gold", were 1 to 2 inches long and 1/2 inch thick. One bead was worth $5 to $20. Dentalia, tooth shells, also obtained in trade, were owned only by the rich, and obsidian chunks were so valuable as to be buried with their owner.

These Indians created few material items, and what they did use was crudely made. Houses, clothing and tools were simple. No agriculture was practiced, not even for tobacco. On the other hand social and ceremonial life among the Yuki was elaborate and filled with rituals. The children's school ritual, the ghost dance, an obsidian ceremony, girls' puberty rites, acorn observances, and an unusual belief in a personal diety who created the world and punished or rewarded humans accordingly all called for special ceremonies.

House made of bark slabs.

Social life revolved around family activities, and one village might consist of as many as 25 individual houses. A village chief, usually an inherited position, wielded considerable power, directing hunts or acorn gathering groups, giving advice and solving problems. There could be any number of chiefs at one time.

Tattooing was common but practiced more by women than men. Facial tattoos were preferred, and patterns varied from tribe to tribe. Soot from burned green plants was rubbed into patterns cut with a sharp bone.

It's not known whether fur traders from Hudson's Bay Company trapped on the Eel River, although they were on the Trinity River to the south in 1830-1835. Gold miners were also on the Trinity by 1849. A party of eight miners, led by Dr. Josiah Gregg, journeyed across the mountains in search of a calm bay to the west along the coast. They reached Humboldt Bay in December, 1849, exhausted by the trip. These men had travelled up the Eel River which they named after watching two Indians catch and eat eel.

The group split up, Gregg and three men going south along the coast, and the other four men following the Eel to California. Facing heavy snows, these four men were forced to eat tree bark and boiled deer skin before reaching Sonoma in February, 1850. They were the first Americans to traverse the length of the Eel.

News of the discovery of Humboldt Bay brought ship traffic to the mouth of the Eel and eventually up the river itself. Eureka was established, developing as a prosperous town, supplying settlers and miners inland. Ft. Humboldt nearby was platted in February, 1853, by Col. R. C. Buchanan whose role was to protect everyone — settlers, miners, and Indians alike — as well as to build roads and open up the interior to whites.

The government appointed three commissioners under Redick McKee to make a study of Indians in northern California including those along the Eel, Salmon, and Klamath rivers. Commissioners were told to "make an impression upon the savage by a display of pomp, by expending an enormous amount of money in the distribution of a few cheap presents, and by making treaties." The sum eventually spent was $800,000. McKee did as he was told. He concluded treaties, passed out gifts, and reported that the Indian population everywhere was overestimated and their intelligence underrated. McKee's treaties and promises were later ignored by the U.S. government.

In a second report Col. Buchanan concluded that "no general war with them (Indians) need ever be apprehended . . . steady encroachments of the white man, from every direction will produce . . . the result of their utter annihilation."

Buchanan's prediction proved to be correct. After 1850, settlers and miners took over the Eel River Valley. While many pioneers were intelligent with noble intentions, others on the frontier were ignorant, vicious, shiftless and treacherous. This element was allowed to carry out depradations of the worst sort with no restraints by the government or military. An 1858 article in the New York Times adds "it is the custom of miners to shoot an Indian just as he would a dog, and it is considered a good joke." When Indians retaliated, newspapers carried reports of Indian outrages.

Kidnapping was turned into a lucrative business by Americans who seized Indians and offered them for sale. Whereas the Spanish and Mexicans used Indian slave labor on ranches and in the missions, American ingenuity turned to selling Indians. Northern California, especially Mendocino County, was the area most heavily exploited for slaves who were transported and sold in Sacramento and San Francisco. Indian children were preferred, and it is estimated that over 4,000 Indians were abducted. In Humboldt and Mendocino counties prices for an Indian ranged from $30 to $150 depending on quality. Children three and four years of age were sold for $50.

Children occasionally rebelled and ran away. Three Yuki children from Round Valley attempting to poison their owner were executed for their effort. One rancher along the Eel solved the problem of a runaway boy by killing the boy's family of six persons.

Humboldt City, 1851.

The first open conflict between Indians and whites occurred in June, 1853, when a party of old men and women digging roots along the Eel were shot by 15 soldiers and volunteers for the supposed murder of two brothers and the theft of some mules.

In 1859 twenty armed soldiers under W. S. Jarboe rode into Round Valley northeast of Dos Rios, the junction of two rivers, to stop Indian attacks on white settlers. Nineteen settlers were said to have been killed. Jarboe and his men killed 283 Indian men, wounded many others, and took 293 prisoners. Four of his men were wounded. Later information brought to light that no white settlers had been killed except one man who had molested an Indian woman. Like many reports of Indian killings, this one was unfounded.

Game was systematically killed off — one J. Longley killed 500 deer one winter in the Eel valley — Indians in a starving condition were forced to steal grain and cattle to stay alive. "Much horse meat and pork (stolen by the Wailiki) was found and destroyed."

"The Indians burned Pardee's ranch (Humboldt Co.). They also dug up the potatoes and threshed out the the oats. In Mendocino county Mr. Woodman lost 109 horses, 74 of which were found dead, upon the bodies of which the Indians were having a good feast" (Cook, p.292).

Through 1864, the gradual attrition predicted by Col. Buchanan occurred. "Lt. Collins' detachment killed 25 savages and apprehended 130 prisoners. Volunteers and soldiers killed nearly 117 hostile Indians with no loss of their own. They had 15 encounters with natives in which 75 Indians were killed. Almost all the males of the Lassik and Wailiki were killed during this round-up of Eel River Indians" (Hoopes, p.130).

The final blow was dealt by Col. H. M. Black and two companies of California volunteers whose orders were to "take to the field and make a vigorous campaign and clean sweep of those rascally Indians." Black harried the Indians who finally made peace in 1864.

The problem arose of what to do with the subdued Indians. Many pioneers advocated slavery. Some felt the Indians should be sent to the Nevada desert, while others liked the idea of extermination. The notion of reservations prevailed, and areas were set aside at Smith River, Klamath River, and eventually at Round Valley north of Dos Rios. These reservations were under Army command.

Once the remnants of these tribes had been brought to the reservations, there was little clothing or food. Mismanagement of funds — "a large amount of money was annually expended in feeding white men and starving Indians . . ." — neglect, disease, and incursion by settlers onto their property continued to be problems (Forbes, p.155). In 1944 the Yuki at Round Valley "live in wretched cabins, containing iron bedsteads, a few chairs . . . running water and electric lights are impossible luxuries" (Foster, p.155).

By the 1870's the Lassik and other tribes in the Eel River Valley had vanished. One Lassik woman, Lucy, born at Alderpoint, was approximately 90 years old in 1942. She told of white settlers and soldiers killing all the men in her tribe, including her father. Lucy escaped from a group of children who were rounded up and "never heard of again," presumably taken south and sold as slaves. She lived with several white men and in 1902 moved back to Lassik territory where she lived with an aunt and cousin. By 1871 only three remaining Lassik could be found.

Indian Life Along the River

Eden, Williams, and Round valleys within the drainage of the Eel River were surveyed in 1950 revealing a large number of depressions where Indians had built their houses. Artifacts found in connection with the house pits showed that

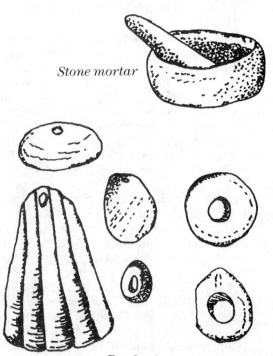

Stone mortar

Beads of stone and shells.

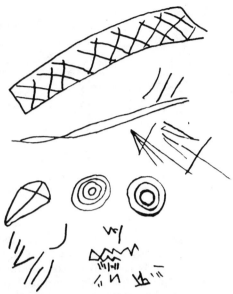

Pictographs at Bell Springs.

Indians lived here up to the time of white settlement. In one area there were nine burials with accompanying mortuary offerings, pestles, mortars, and shell beads. Several cremation sites were located along streams.

Pictographs and petroglyphs are scarce to non-existant along the Eel River. One large boulder near Bell Springs is covered with carved figures. A rock carved with geometric figures and "horse tracks" was reported in the vicinity of Eden Valley, but its exact location is uncertain. The location of petroglyphs on a rock southwest of Spyrock has also been lost.

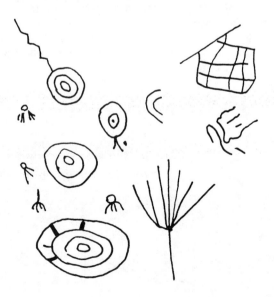

Pictographs at Spyrock.

Geology of the Eel River

Although many runs are possible on the Eel River, the route from Dos Rios to Alderpoint on the main channel is the best for a number of reasons. This interval lacks many of the tedious portages and short runs of the other routes. Equally important is that it provides a good representation of the geology in the Eel watershed.

Stream patterns within the Eel drainage in the vicinity of Mendocino, Lake, Trinity and Humboldt counties show parallel channels running from the southeast to the northwest. Although there are shorter routes to the sea, streams here run in a northwesterly direction for a number of miles before turning left (west) to empty into the ocean.

The geology of this area readily explains this drainage pattern. Rock formations in the northern coastal ranges follow a northwest trend. This trend becomes due north at the Oregon/California border then swings on around to the northeast through the southern Oregon Klamath Mountain region. Streams develop valleys in the soft layers between the harder ridges. The parallel channels created in this way are referred to as trellis drainage.

Much of the rock exposed along the river from Dos Rios to Alderpoint is a formation known as the Franciscan. More properly, the Franciscan is referred to as a terrane, or a group of similar rocks which occur in one area. The Franciscan forms a belt from northern California well down into the southern coast ranges. It is a distinctive group of 150 million year old sediments (Jurassic age) mixed with metamorphic rocks which were deposited here along an ancient volcanic island chain. Hard chert (silica-rich, flint-like rock) as well as fine-grained shales and mudstones dominate the Franciscan. Also present in this terrane is a mixture of several types of volcanic rocks.

Much of the Franciscan was deposited on the ocean floor as deep as 15,000 feet. Transport by way of "sea floor spreading" took these sediments up against the deep sea trench that paralleled the California coast. Portions of the Franciscan rocks were swallowed whole in the trench as one plate slid beneath the mainland. What we see today in the Klamaths was scraped off by the mainland plate as if by a bulldozer blade and piled up along the coast. The prehistoric ocean trench that adsorbed much of the Franciscan Formation was as much as five miles deep and lay parallel to the present coastline just to the east of the Eel River drainage.

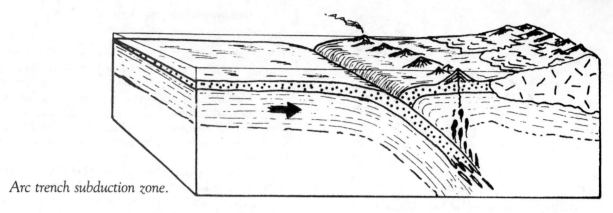

Arc trench subduction zone.

Recently, considerable attention has been focused on the Franciscan because of the role it played in the continental drift story. Obviously such a deep sea environment will not have the fossil molluscs found in shallower water. There are some fossils in these deep sea rocks, however. Microscopic radiolaria, a type of protozoa, have been successfully used assign an age to the Franciscan cherts.

In addition to deep sea sediments, the Franciscan contains mineral deposits including sulfide ores of iron and copper. Small amounts of gold, silver, and platinum are also found associated with these minerals. The origin of sulfide ores is a process that geologists have only begun to understand in light of the phenomena of continental drift. Discoveries of rich deposits of sulfides forming today along the crests of deep ocean ridges in the Pacific Ocean prompted renewed exploration and research into the origin of such ore deposits.

The peculiar geology of the Franciscan Formation itself has a direct effect on the location and nature of the rapids along the Eel River. The Franciscan is largely composed of chert, black shales, and similar deep sea sediments. The sand deposits which do occur in the forma-

tion are "greywackes" or dirty-looking sands reflecting rapid rates of erosion and deposition. These sands as well as the shales decompose easily. In a climate with rainfall like that of the northern Coast Range, soils which develop atop the Franciscan are remarkably thick. Erosion occurs rapidly here, and in many areas the slopes above the river become oversteepened or unstable resulting in large landslides.

Landslides are divided into slumps and debris flows. Of these, debris flows are by far the most common. Very little of the Franciscan Formation holds together well enough to develop a slump block. As landslides develop, they move directly into the main stream channel from the adjacent steep banks. Streams are briefly dammed until the rapid procedure of erosion begins to remove the rocks and soil debris. Large blocks are often left behind in the stream channel while the finer sands and shales are swept downstream. As the debris flow snout is undercut and flushed downstream, more of the landslide slowly moves into the stream channel. Consequently at several places along the Eel landslides are being constantly fed into the river channel.

Debris flow topography in the Franciscan Formation.

Geology Along the Route

At Dos Rios the river takes a northward turn and begins to follow a northwest oriented structural trend. This trend includes a series of faults and valleys developed in easily eroded, soft rock. The stream channel trend continues almost uninterrupted due northwest to the take-out at Alderpoint.

Most of the meanders along this route are caused by debris flows which clog the river. As a landslide moves into the stream channel, the river will meander out around the obstruction while it cuts away at the softer fragments. In this stretch from Dos Rios to Alderpoint, the river channel is in landslide terrain approximately 50 percent of the distance.

More than 25 miles downstream from Dos Rios at Island Mountain, the river meanders around a group of hard volcanic rocks before flowing on northwestward to Alderpoint about 15 miles downriver.

The rock exposures at Island Mountain are of economic importance because metallic sulfide ore deposits of copper were found with the old volcanic rocks. The Island Mountain deposits developed in a soft shale in the Franciscan Formation. Breaks or faults will often occur in shale, and in this area faults often shear or cut through shales at many localities. Along one of the main sheared areas in the Franciscan shale, mineral rich fluids from a nearby volcanic source invaded along the cracks and fractures to deposit the ore minerals. The ore mass is lense-shaped and plunges off to the northeast.

Metallic ores from the mine include iron and copper (pyrite, chalcopyrite, and pyrrhotite) with lesser amounts of lead (galena) and zinc (sphalerite). Gold and silver occur in modest quantities in association with the copper ores. During the period of mining activity at the Island Mountain mine between 1915 and 1930, some 4,000 tons of copper were recovered from 131,000 tons of ore. In 1937 a huge landslide completely buried the mine's shafts and tunnel openings.

Landsliding has obviously had a long history in this particular area of easily erodable Franciscan rocks. In addition to uneven, hummocky topography, landslides also characteristically carry along with them huge boulders that have been weathered free, dislodged, then slowly moved along with the mass in the landslide by gravity.

One such immense boulder just downslope from the mining area on the north bank of the river at Island Mountain was 40 feet in diameter and contained more than 5,000 tons of copper ore. That boulder and another of about 200 tons nearby were broken up and shipped off for processing early in the mining operation. This type of ore is known as float, and approximately 8,000 tons of such ore were recovered from the landslides near the mines. Along the banks of the river here, fragments of ore can still be found.

Proceeding downriver from Island Mountain to the take-out at Alderpoint, it is relatively simple to pick out the landslide areas. The latter have a rumpled, broken up surface appearance rather than the ordinary smooth hills characteristic of normal erosion.

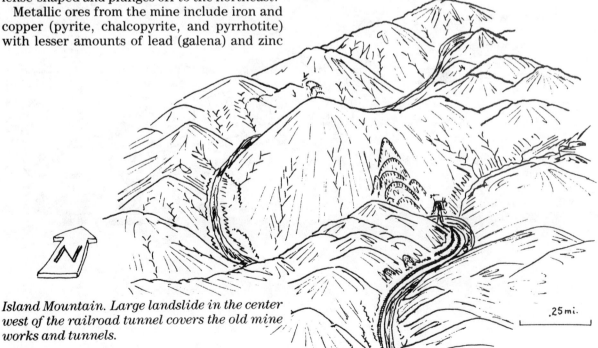

Island Mountain. Large landslide in the center west of the railroad tunnel covers the old mine works and tunnels.

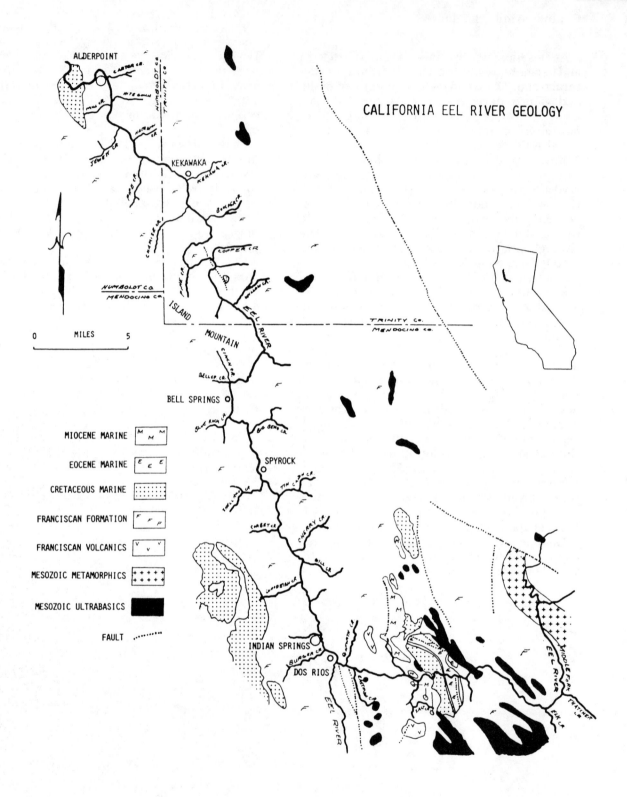

CALIFORNIA EEL RIVER GEOLOGY

MIOCENE MARINE

EOCENE MARINE

CRETACEOUS MARINE

FRANCISCAN FORMATION

FRANCISCAN VOLCANICS

MESOZOIC METAMORPHICS

MESOZOIC ULTRABASICS

FAULT

0   mile   1

0 mile 1

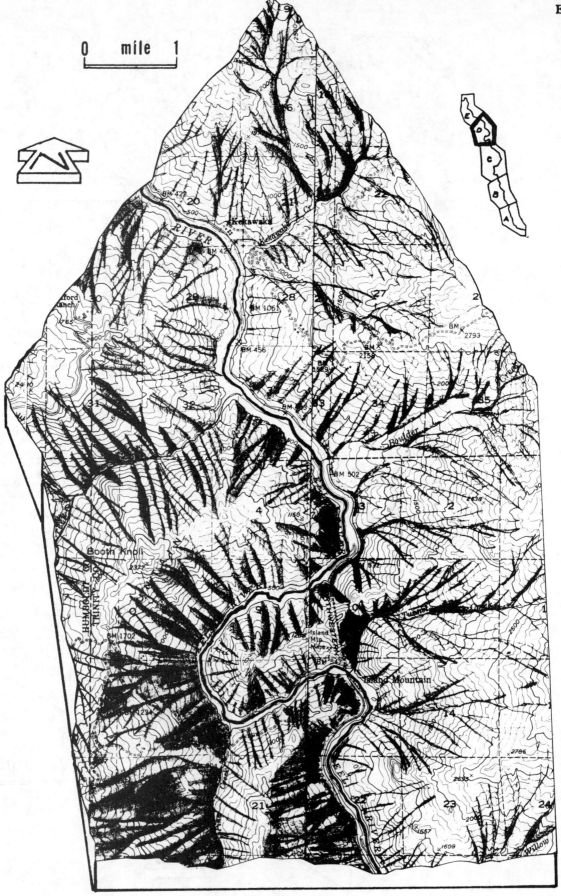

0 mile 1

Klamath-Salmon River

Klamath River
(Happy Camp to Orleans)
Siskiyou to Humboldt Counties, California
Length of run: 47 miles
Number of days: 3

Salmon River
(Butler Creek to the Klamath River)
Siskiyou County, California
Length of run: 50 miles
Number of days: 2 to 3

Remote and unsettled, the Klamath River originates in Lake Ewauna near Klamath Falls, Oregon, crosses into California and continues to gather speed for 200 miles until it enters the Pacific Ocean. From Happy Camp the river flows southward through a narrow, forested canyon less than 500 yards wide. The spectacular geology in the high canyon walls is clearly exposed to the river rafter. Whitewater intervals are caused by a variety of geologic processes such as landslides and narrow channels. Rafters drifting quietly see a number of plants and animals including well over 200 species of birds.

The rapids and steep drop of the Klamath and Salmon rivers in this section are unhampered by modern construction, and there are few changes since the Karok Indians lived along the banks. Although the record is there to be read, the quiet canyons of the these rivers barely reflect this turbulent history of geologic upheaval, Indian wars, and mining for gold. A scene from prehistoric times occurs at Orleans where even today the Karok Indians annually perform their world-renewal ceremony.

Early History of the Klamath and Salmon Rivers

The name "Klamath", with a number of spellings including *Tlamatl* and *Claminitt*, referred to a local tribe of Indians related to the Modocs.

Four Indian groups lived in the Klamath basin. They were the Yurok, meaning "down river", the Karok, or "upriver", the Modoc, or "head of river", and the Shasta, probably named after a Chief Sasti. The Yurok lived along the Klamath River from the mouth up to Orleans. The Karok, or Karuk, lived from Orleans up past Happy Camp and through the long, narrow canyon to Hamburg as well as

Young Karok woman with tattoo marks on her chin. Drawn by George Gibbs, 1851.

along the lower Salmon River where it joins the Klamath. The territory of the Shasta extended east to where the river approaches the Oregon/California border, and the Modoc/Klamaths lived in the northeast corner of California and into Oregon.

Karok population was placed at 1,500 in the 1700's and 800 in 1910, although population estimates varied widely. This would be an average of two to three persons per square mile for 40 miles along the middle Klamath.

Physically the Karoks were tall compared to other California Indians, with a narrower face than those living to the south. Women tattooed their chins with leaf patterns and a blue coloring made by mixing plant juice and soot. The Karoks were highly regarded by their neighbors, as "they had better-made things than anybody, and they did everything fancy" (Holt, p.302).

Basic Karok foods were acorns and salmon. Acorns from tan oak trees were preferred over those from black and white oak, and acorns from certain areas along the river were thought to taste better than others. An oak tree be-

241

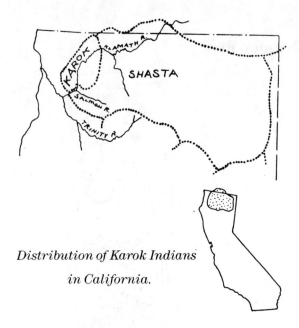

Distribution of Karok Indians

in California.

$2.50 to $5, and dentalium strung to about two and one-half feet was worth $40. When a Karok was killed, relatives of the dead person would be satisfied with the payment of one string of shells. A wife could be purchased for one-half string, but, if talented or of a particularly good family, she could cost up to two strings of shells, or $80.

As with many California groups, Karok were not prone to warfare, so that the village chief served mainly in an advisory role or as an arbitrator in two-party feuds. These feuds were settled in hand-to-hand combat with sharpened stones where opponents effectively mauled each other. In the unusual event of warfare, jagged stones were still preferred to a bow and ar-

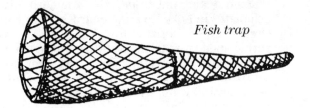

Fish trap

Fish net woven from fine roots or grass.

longed to the family whose house was nearest, giving that family the right to pick there first. Acorns were made into meal for bread by a scorching, pounding, and leaching process.

Acorn bread was eaten with salmon. Weirs, nets stretched across eddies in the river, barbed spears thrown from platforms, or bag nets attached to a triangle of poles were popular ways to fish. Two large weirs were reported on the Klamath with one at Happy Camp. A weir might belong to only one or two men, but anyone would be allowed to take away as many fish as they could carry.

About one-half of the diet of Indians throughout northern California consisted of acorns, one-fourth was fish, and the remainder was small seeds. Approximately 7,000 tons of salmon were caught each year. A Yurok family along the Klamath might have a ton of hanging salmon drying. Salmon was stored in baskets layered with laurel leaves to keep out insects.

The Yurok tribe along the lower Klamath produced canoes which were sold in nothern California. Enormous canoes were constructed from redwood logs by spreading and burning pitch over the area to be dug out. The wood was then worked and polished with stone for up to six months until the sides and ends were smooth and a seat carved out for the oarsman. The resultant craft was sold for the equivalent of $10 to $30 in the 1800's and could carry five tons of goods. Fish caught on the Klamath would be transported by canoe 22 miles down the coast to Crescent City, California.

Dentalia shells from the coast and the red scalps of woodpeckers were used exclusively for money. Woodpecker scalps were worth

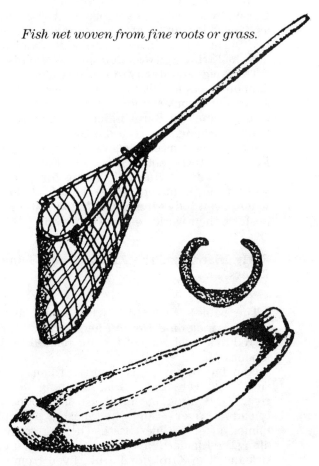

Canoe made from redwood log. Usually 6 to 8 feet wide and 24 to 30 feet long.

row. Slain enemy were decapitated, and their heads put on display.

Two kinds of shamans, or doctors, the barking doctor and the root doctor, were both women. The barking doctor hunched down and barked in front of the patient while the other attempted a cure with plant medicines. The barking doctor was regarded as more important as she could rid the patient of diseases caused by demons inhabiting the body. The Karok felt most diseases were caused by evil spirits. When syphilis first appeared among them, they deliberately infected themselves, feeling it was caused by a demon and could be passed on to their enemy. Of course, the effect on subsequent generations was disasterous.

The Karok performed an annual world-renewal ceremony. The ceremony was to ensure plentiful food for the coming year, freedom from sickness, and, in general, to prevent disasters. People from nearby towns participated in the dance lasting two days and nights while priests carried out secret rituals before the dance began. Three renewal village sites were at Orleans, at the mouth of the Salmon River, and at Clear Creek.

Karok were not significantly troubled by early contact from fur trappers of Hudson's Bay Company and the American Fur Company working the rivers as far south as the Sacramento Valley. Led by American Jedediah Smith, a group of traders in 1827 established a route north from the American River, through the Sacramento River valley, west toward the ocean, and on to Oregon. In following years, this Siskiyou Trail sustained a considerable amount of traffic by emigrants to the Willamette Valley in Oregon as well as by miners to the gold fields.

Initially Indians here were friendly and curious. The U.S. Exploring Expedition of 1841 under Charles Wilkes, accompanied by botanists, geologists and artists, camped along the Klamath during a hot September when the temperature reached 100 degrees. Not finding food, they instead succeeded in capturing two frightened Indians who agreed to trade salmon, as well as bows, arrows, and other goods, for a knife.

Lt. George Emmons described the salmon as of a whitish color, and not "all delicate to the taste; their tails were worn off and the fish otherwise bruised and injured ". He may have been describing a late spawn salmon. The expedition photographer, Mr. Agate, hoped to persuade a group to stand for a picture by showing them a miniature of his mother. Unfortunately the picture had the opposite effect. The Indians decided Mr. Agate wanted to put an enchantment on them.

However, friendly encounters were not the rule, and most Americans dealt callously with the natives. Stephen Meek, described as a "mountain man who had many hair breadth escapes and battles with savage animals and no less savage men", led a party of explorers through to the Willamette Valley. He encountered a village of men, women, and children drying meat from stolen cattle. Meek ordered a charge, capturing the surprised Indians and taking the meat. The prisoners were released, but the Indians retaliated by attack that night. Aggressive acts on both sides set a pattern for miners who appeared on the scene shortly thereafter.

Disruptive miners in groups of 40 to 50 arrived in northern California during the Spring of 1850. Looking for a quick fortune in gold, they crossed over the ridge to the Klamath and Salmon rivers from the mines on the Trinity River. Overnight several hundred men collected at Forks of Salmon working along both

Dentalium shells used in trade.

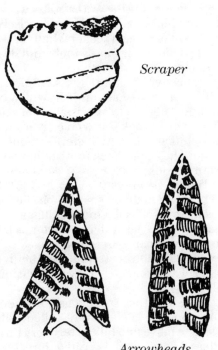

Scraper

Arrowheads

Armored Karok warrior (Siskiyou County Historical Society).

branches. Miners proceeded north to Happy Camp where rumors of the rich sand bars of the Klamath brought in others.

In a very short time the presence of thousands of white miners disrupted Indian life on rivers and streams by building canals and tunnels, muddying the water. Waterways, so necessary to Indian survival, were precisely the areas taken over by gold seekers. Indians were forced away from their food supply and desirable living spots along rivers.

"This river (Trinity) . . . is rated as the best in the country for salmon fish, which constitutes almost the whole subsistance of the Indians. The whites took the whole river and crowded the Indians into the sterile mountains, and when they came back for fish they were usually shot" (Cook, p.281).

Happy Camp was begun in 1851 by miners who constructed a small cabin just downriver from an established Indian village. The name is said to have been derived from a particularly boisterous party. One man stayed at the cabin while all others scattered over the river bars for gold. Everyone met back at the cabin on Sunday.

In a short time trouble began when two miners and one Indian were killed. Miners who tried to move into Indian territory up river were chased away. Indians visited the cabin regularly much to the annoyance of the miners. Deciding to improve the situation, miners gathered together a group of 15 to 20 and attacked the village at dawn killing all the men and some women, thus dispensing of the Indian problem.

"It is needless to say that bows and arrows were no match for the guns of the whites, and that the savages, without exception, were given a free passage and quick dispatch to the Happy Hunting Grounds" (Bledsoe, p.9).

Poor weather conditions and lack of food were other problems in the mining camps. Miners were fearful of wintering over on the remote Salmon, and in December, 1850, only a few men could be found there in cabins well-stocked with food. Unfortunately, the unexpected occurred during January and February when hundreds of miners, thinking the worst of the winter was over, returned from the coast. A heavy snowstorm in March blocked all passes, and supplies couldn't get through until the end of April when help arrived from Orleans on the Klamath. Miners, in a starving state, had eaten everything in camp. One man turned down $16 for two grouse he had killed.

Rich gold finds were discovered on many bars of the Salmon and Klamath rivers. Diggings at Indian Creek yielded as much as $1 to $3 a pan. To the south on Elk Creek, $10 to $40 was made a day. Orleans, presumably named after the city in Louisiana, was settled in 1850. Shops there did a booming business supplying miners coming through on their way up river. More money was made by shopkeepers than by the average miner from the gold fields. A store might take in $1,000 daily.

The ever-present Chinese did much of the rough mining work, transporting goods, and cutting lumber at a wage of $1.50 a day plus food. In June, 1856, a party of Chinese, working a claim they had purchased on the Salmon River, made a gold strike. The Americans, who had sold the claim, took over again, pushing out the Chinese, who sued. The judge ruled in favor of the Chinese, but the Americans refused to move until the Sheriff arrested them, shooting one man in the scuffle.

Regardless, justice in favor of the Chinese was rare. In other cases they were described as "a contaminating influence," living bunched together in "filthy dens . . . surrounded by mud, filth, and garbage." The "evil of Chinese immigration" was deplored (Bledsoe, p.103). On the other hand, mining camp inhabitants were perfectly willing to employ them in the most difficult and menial tasks.

Miners who came in the 1850's were all but gone by 1856. Towns they had built remained and were taken over by permanent settlers and farmers. Displaced Native Americans, no

Tan oak acorns were ground into a flour.

longer able to hunt and fish on the Salmon and Klamath, were sent to reservations. The Klamath Reservation was created in 1854, and arrangements completed in April, 1855. The reservation included the area from the mouth of the Klamath to 20 miles upriver and 2 miles on each side. Indians were supplied with farm equipment and seeds. Approximately 1,500 Indians lived in 150 huts under military supervision. By 1881 there were 82 Indians living on the reservation and 270 others in the county.

Indian Life Along the River

Villages were located every few miles along the Salmon and Klamath rivers. The village at Happy Camp had 20 to 30 houses while there were 91 houses at the mouth of the Salmon in a very large village. Villages in between were small with 12 to 15 houses each. The number of Indian dwellings at Orleans is not known. Smaller villages were located at Ferry Point, Oak Flat, Cottage Grove, Dillon Creek, Ti Bar, and at Elk Creek.

House made of bark slabs.

Geology of the Klamath River

Klamath Mountain geology is remarkably complex and only recently has been deciphered. Here immense continental blocks, carried eastward across the Pacific Ocean by the forces of continental drift, collided with North America then piled one upon the other. As these blocks, or plate fragments, crashed one by one into the West Coast, each new piece was stuffed beneath the earlier arrivals. Collision and underthrusting created a curious series of shingle-like eastward dipping plates. Because these plates, or terranes, as they are called, arrive at North America from the west, they are progressively older in individual age from west to east.

This arrangement of younger rocks thrust beneath older ones in this fashion thwarted geologists' early attempts to unravel the story of the Klamath Mountains. The discovery and identification of very small fossils within the rocks of these plates gave geologists the means to age date the rocks. Once the strata had been properly fit into the geologic time scale, most of the problems of the area were easily solved.

Prior to the Jurassic time period, 150 million years ago, the West Coast of North America was a broad continental shelf that extended well into what is now Nevada. To the west of this shallow seaway a chain or archipelago of volcanic islands lined up about in the present position of the Klamath Mountains. With continental drift, the on-coming plate collided, not with the main bulk of North America, but with this advance guard of offshore volcanic islands. As the plate fragmented and stacked up, the pieces were thrust beneath these islands. The rocks and sediments from the deep sea floor were eventually forced to the surface. Volcanic rocks from the island chain, as well as granite from deep within the earth, added to the complex rock mass that now makes up the Klamaths.

In the watershed of the Klamath River between Happy Camp and Orleans, the underlying geology is a series of these shingle-like plates. The youngest plate of Jurassic age now visible west of the river has been thrust beneath older rocks to the east. Along much of the stretch of the river between Happy Camp and Orleans, the valley is cut as a slot into the plate of upper, older rocks. Much of the river bed lies in this north/south strip that is called the western Jurassic belt.

The pathway of the Klamath River meanders back and forth across the two plates yielding a striking variety of rock types visible in the streambed and banks. Between Happy Camp

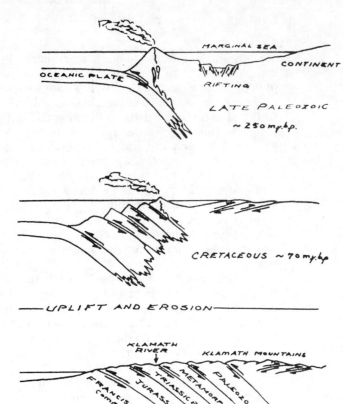

Evolution of the Klamath Mountain thrust fault terrain.

Grove, and Orleans. Nearby Siskon is the site of one of the largest and most recently operative mines. Three million dollars in gold were recovered between 1953 and 1961.

Prior to the late 1800's in California, the most productive technique of mining placer gold was by hydraulics. Using streams of water under high pressure in huge hoses, vast amounts of sand and gravel in streams and older gravels in the terraces above the rivers could be washed out for processing.

Often water was moved great distances by complicated systems of artificial channels to get it to the sites of the gold. Opened in 1851, the famous LaGrande mine west of Weaverville was one of the biggest and most elaborate systems in California. While it was operating, in excess of 100 million cubic yards of gravel was washed out by the system. Water to supply this enormous enterprise was transported by flume and canal 39 miles all the way from the Trinity River. In 1884, after much political pressure, the destructive effect of hydraulic mining on streams was acknowledged, and the operations were shut down by law.

Many of the Klamath Mountain streams have been repeatedly swept end to end by mining operations producing somewhat less gold with each pass. Regardless, gold is continually being bled into the Klamath watershed by the erosion process as time goes by, and with patience one can almost always get a little "color" in a gold pan.

and Orleans, the upper, older plate extends westward in several areas well beyond the Klamath River Valley.

The long geologic history of the Klamaths has included several volcanic episodes as well as the distortion and shattering of rocks by processes of folding and faulting. Often, after the rocks are fractured, precious metals are deposited in the cracks by volcanic fluids. Gold is emplaced in this way, but, because free (metallic) gold does not easily decompose, it also occurs in secondary deposits or placers where it has been washed out of weathered rock upstream.

The presence of rich mineral deposits brought whites into the Klamath region. Gold was first found in the Clear Lake area in 1848 shortly after the historic discovery at Sutter's Mill in California. During the middle to late 1800's, $250 million in gold was extracted from the Klamath Mountains. By today's dollars, this amount would be in excess of two billion.

The bulk of the Klamath Mountain gold was located in areas to the south and east of our Klamath River stretch, but locally gold was mined at Happy Camp, Dillon Creek, Cottage

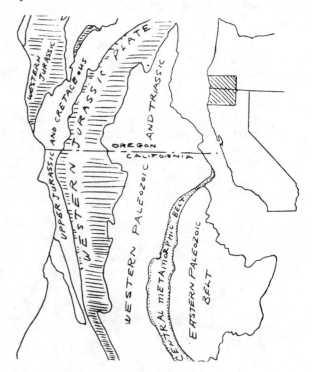

Klamath Mountain thrust fault terrain.

As the price of gold continues to slowly rise, the costs of mining it have proportionally increased. To date, these two curves have kept pace, and the great gold rush predicted when gold broke $500 per ounce in the 1970's has failed to materialize. In the vicinity of Happy Camp, the most economically important mineral has been copper, not gold.

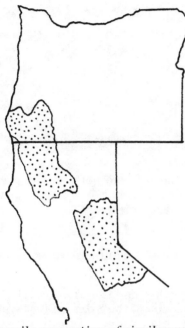

Fifty mile separation of similar rock types in the Klamath and Sierra mountains.

Geology Along the Route of the Klamath River

Rocks exposed in and around Happy Camp are the 150 million year old Galice Formation named for strata near Galice, Oregon. The Galice includes a variety of marine (oceanic) and volcanic rocks originally deposited thousands of feet below the surface at the bottom of the ocean. Today these rocks are being eroded by the Klamath River at elevations of 2,000 to 3,000 feet above sea level. The Galice Formation has endured several invasions by molten rocks accompanied by extreme conditions of heat and pressure (metamorphism). These events have left unmistakable evidence in the form of precious metals and distinctive minerals that develop only under these conditions.

At Happy Camp the Galice Formation is composed of sedimentary rocks which have been subject to this process called metamorphism. Often metamorphism will even destroy the original bedding of rocks, but the layering is still visible here. Locally the Galice is mineralized or impregnated with iron pyrite or "fools gold".

Kamakazie Curve Rapids, about 2 miles downstream from the put-in, marks the contact between the Galice Formation and a very hard dense, dark-colored rock called peridotite. Peridotite formed under conditions of extreme heat and pressure. In the Kimberly area of South Africa and elsewhere peridotites are mined for diamonds. The Galice Formation and peridotite at Kamakazie Curve Rapids are separated by a fault.

Within the next mile downstream, several dikes are exposed in the banks of the river. Dikes are sheet-like masses of rock where molten material has flowed along vertical fractures and cracks. These dikes are composed of granite and are readily identified by their white and grey, coarse crystalline texture. Elsewhere in the Klamaths, granite dikes have quartz veins with gold.

Rattlesnake Rapids is the site of a large granitic mass now exposed by erosion. Several visible dikes radiate outward from the mass. Poseidon Rapids further on marks the contact between the granites and volcanic rocks of the Galice Formation. Wingate Bar, about seven

Mine sluices (Siskiyou County Historical Society).

Hydraulic mining (Siskiyou County Historical Society).

miles downstream from Happy Camp, displays rocks of the Galice Formation that have been intensely heated and squeezed. Garnet bearing schists here are characteristic of these extreme temperatures.

The gravel bars at Ferry Point were, at one time, the scene of intensive placer mining. Debris or spoils from these operations are visible as piles of coarse, rounded gravel scattered about the vicinity.

Rock Garden Rapids is developed in a dense field of large boulders. These boulders, composed of the rock serpentine, act as a trap for smaller rocks being carried downstream. This natural dam produces rapids as the stream flattens out and increases velocity over the obstacles. Serpentine, a dull, dark-greenish rock created by conditions of low heat and pressure, is relatively weak and often broken by cracks or faults. Slippage and movement along these fractures give serpentine a pale green polished appearance. The mineral asbestos characteristically is found crystallized in the fractures of serpentine. Blue Nose Mine, on the west side of the river here, was operated into the late 1800's.

Ti Bar is a popular take-out for rafters owing to its sandy beaches and easy access. In the early 1900's, the entire bar was alive with the activity of gold placer mine operations. Like Ferry Point upstream, evidence of this mining is still obvious as irregular mounds of gravel litter the bar.

Between Ti Bar and Orleans a large projecting sheet of the older plate extends westward out over the younger western Jurassic terrane. Near Orleans, placer gold was mined as late as the 1930's. Although several mine tunnels have

been dug here, most of the mining was of placer deposits. In addition to gold, platinum was mined near Orleans.

Geology of the Salmon River

Normally run in short two to three day trips, this section of the Salmon River winds up back on the Klamath near Orleans after a run over more than fifty rapids. While the main Klamath River flows north/south parallel to the regional geology, the Salmon between Butler Creek and Somes Creek travels west directly across the geologic "grain" of the area. Faulted slabs are stacked up from west to east like fallen dominos becoming progressively younger downriver.

Although gold has been mined along this stretch of the river, areas to the south drained by the Salmon are much more important economically. One of these districts at Cecilville included both placer and lode gold. After the discovery of gold in 1849, three to five hundred Chinese worked the river gravels intensively with both small flumes and wing dams.

Geology Along the Route of the Salmon River

At the put-in near the mouth of Butler Creek where it enters the Salmon, the local rock is about 200 million years old (Permian/Triassic). Much of the rock here is flint-like chert and shales which were baked by heat and pressure. In addition to limestones and greenstones, some highly fractured rocks (breccias) also oc-

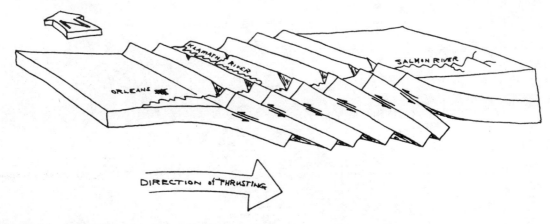

Shingled fault blocks of northern California being cut by the Klamath and Salmon rivers.

cur. Just downstream from the put-in, a rock contact may be seen where the river has exposed younger granitic rocks of Jurassic Age (150 million years old). At other locations in the Klamaths, gold has been found in association with quartz veins in this granite.

The river remains in the lighter gray granitic rocks for five miles before moving back into older, metamorphic rocks. Rocks of this older terrane dominate the river banks down to Somesbar where the river cuts into an exposure of Jurassic ultramafics. Ultramafics are a coarse, crystalline rock, rich in dark minerals which formed at very high temperature and pressure.

Typically this rock type marks the base of a thrust sheet.

At the confluence of the Salmon and Klamath rivers, rafters turn left with the current to follow the Klamath down to Orleans only eight miles distant. Over this interval the river meanders back and forth across granites for four miles before intersecting marine rocks of the Galice Formation which represent deep ocean environments. The contact between these two rock types marks a major fault through the area.

The community of Orleans is situated on Galice Formation rocks.

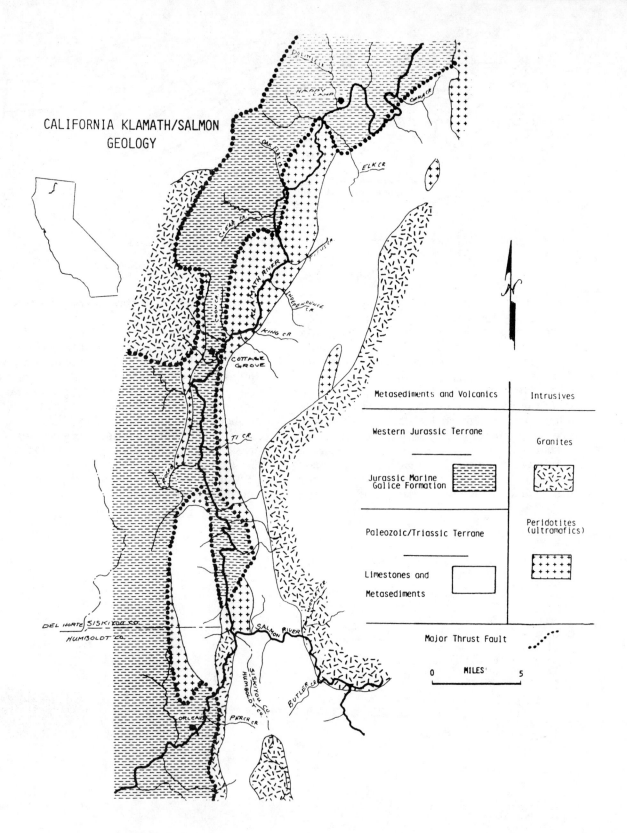

CALIFORNIA KLAMATH/SALMON
GEOLOGY

Metasediments and Volcanics | Intrusives

Western Jurassic Terrane

Jurassic Marine
Galice Formation

Granites

Paleozoic/Triassic Terrane

Limestones and
Metasediments

Peridotites
(ultramafics)

Major Thrust Fault

0   MILES   5

0 mile 1

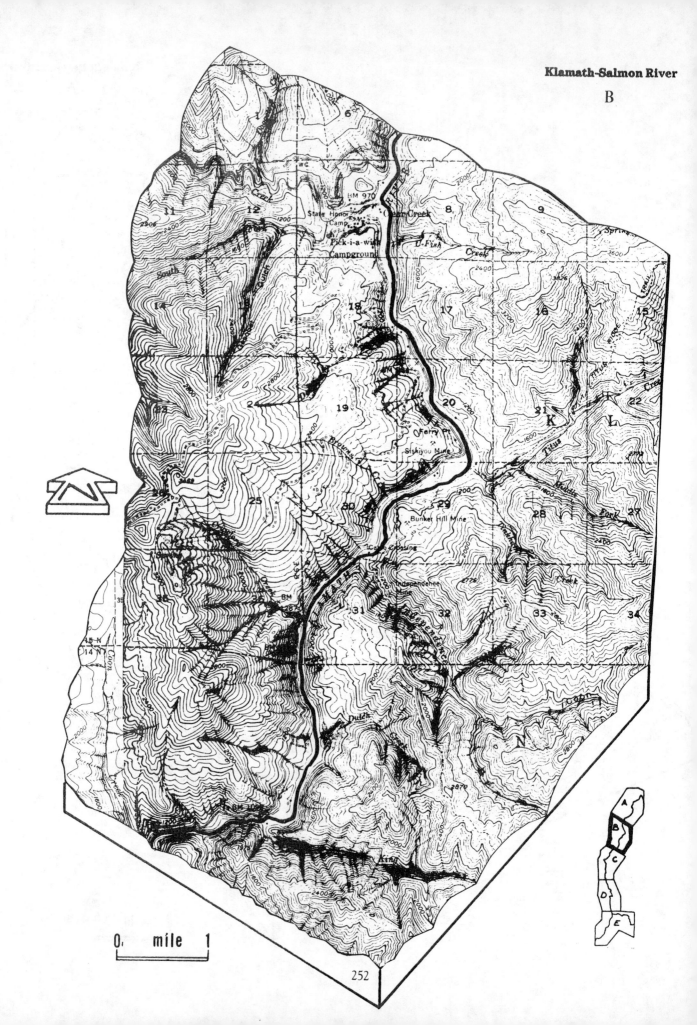

0   mile   1

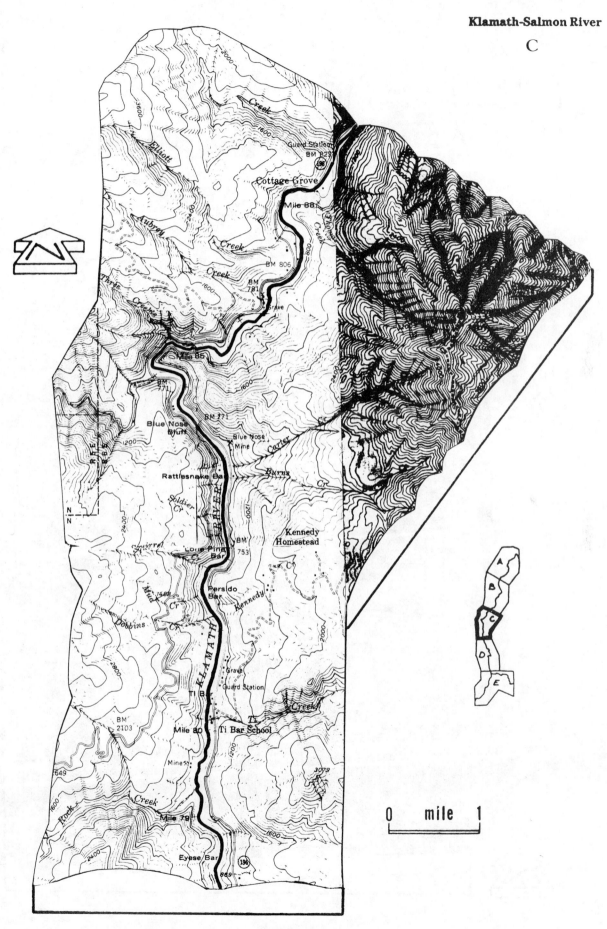

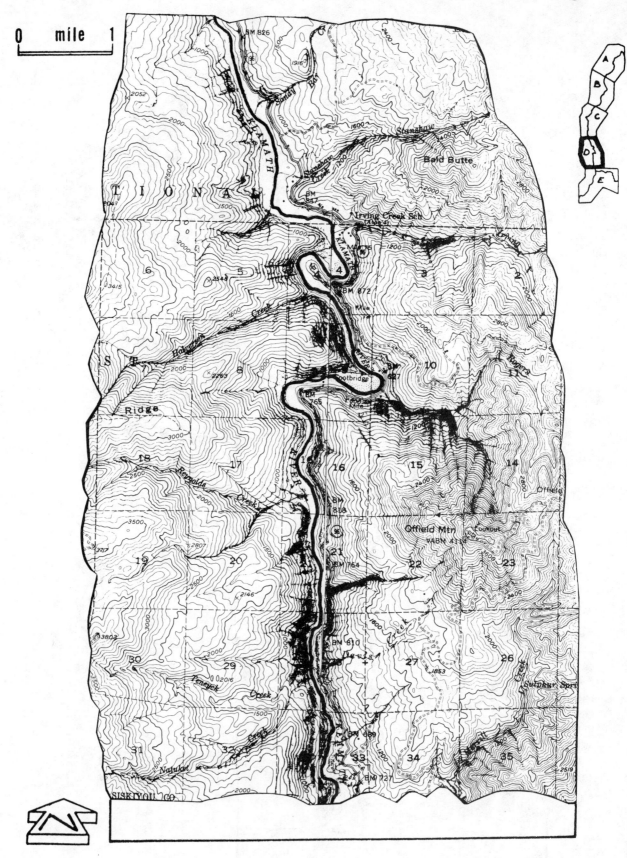

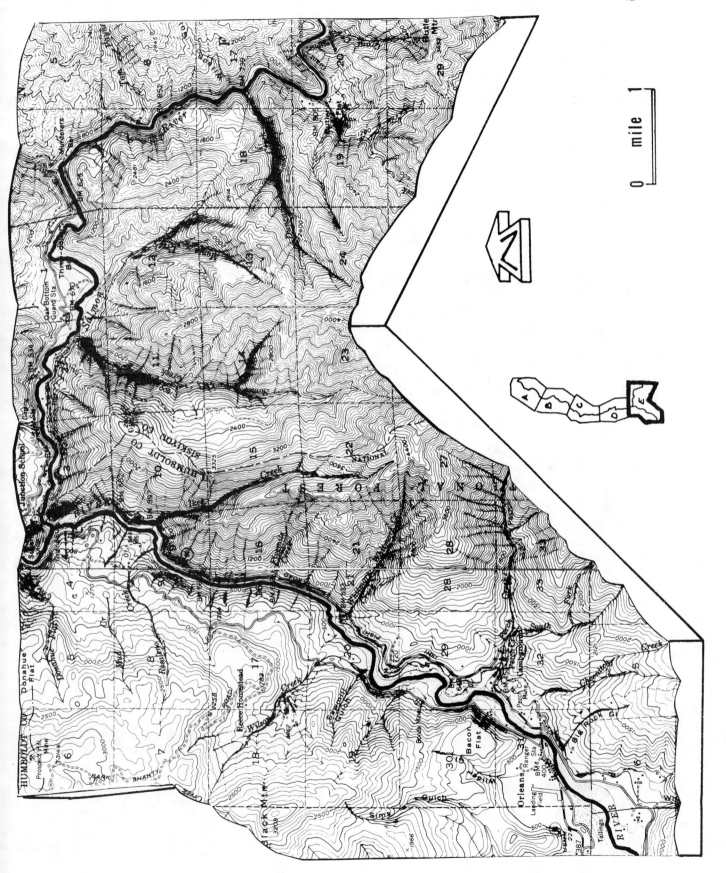

Miwok Indians

In both population and territory the Miwok, or Moquelumnan family, were the largest Native American group in California. They occupied an extensive area along the western slope of the Sierra Nevadas, from the Cosumnes River south to the Fresno River and west to the San Joaquin Valley. The coast Miwok lived in the San Francisco Bay region, and the lake Miwok dwelt near Clear Lake in Lake County.

The Sierra Miwok inhabited the basins of eight major rivers including the Stanislaus, Toulumne, Mokelumne, and Merced. Population for the lake Miwok in 1770 was 500 persons, for the coast Miwok it was set at 1,500, and for the Sierra Miwok it was put at 9,000. The 1910 census recorded 640 Sierra Miwok, and the 1930 census had a total of 491 persons. Miwok is an Indian word signifying "people".

The rich alluvial soil of the Stanislaus, Tuolumne, and Merced produced enough food to support many people, and numerous villages lined both sides of the rivers. Villages were named by general direction, as Village to the East, or Village to the West, as well as by a specific village name if the village was large or permanent. "Lumni" meant "people of," so that Moke-lumni meant "people of the village of Mokel," and Otceha-mni signified "people of Otce".

Miwok woman with tattooing on her face (Drawn after watercolor by L. Choris, 1822).

As with most other California Indians, the principal food of the Miwoks was the acorn. Seven species of oak were eaten, but black oak acorns were preferred. One large tree is said to have yielded 500 to 1,000 pounds of acorns. Tanin, which gave acorns a bitter taste, was removed by grinding the acorns into powder and leaching warm water through the meal. The ground meal was baked into bread or eaten as mush.

Portable as well as bedrock mortars were used for grinding, although the Miwok preferred bedrock mortars. Here women in a village would congregate and socialize. One mortar might contain as many as six to seven holes in a large, well established village. Holes were no longer used when they became five or more inches deep.

Generally enough acorns were collected during a season to last a family for two years in case of crop failure. In a good harvest year, excess acorns were stored in a woven grainery hung out of reach of animals.

Acorns were combined with pine cones collected when tender and green in the Spring, roasted, and eaten. Cones taste sweet and syrupy. Nuts, berries, bulbs, tubers, grass seeds, and clover from among the variety of plants available were eaten as well. Manzanita berries were the most common but least desirable plant.

Land animals consumed by the Miwok included snakes, lizards, numerous birds, and

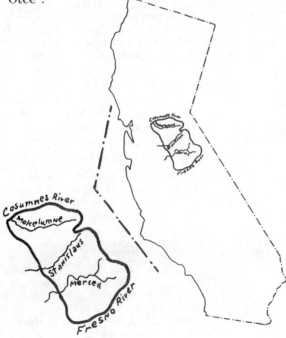

Distribution of Miwok in the western Sierra Nevada.

Bedrock mortars where Indian women gathered to grind acorns.

Black oak acorns were a common food of the Miwok.

larger game animals as elk, antelope, bear, and deer. Grasshoppers and other insects were dried, along with animal meat, and stored. Deer meat was the favorite flesh, followed by that of the California Gray squirrel. Salmon and other fish were caught in nets or speared in most rivers and creeks.

Salt was collected from locals on the Stanislaus River, and from two holes near Coulterville, Mariposa County. Salt was also obtained as a trade item. Along the south bank of the Mokelumne River about six miles south of Silver Lake, an elevated bench above the river had nearly 100 holes which were filled with salt water dripping down from a shallow pool. It is thought the Indians pounded out the holes to obtain the salt.

It has been said of the Miwok diet they ate "all the creatures that swim in the waters, all that fly through the air, and all that creep, crawl, or walk upon the earth, with a dozen or so exceptions" (Powers, p.348).

Miwok lived in a variety of dwellings. Houses of poles with tule mat doors and roofs were covered with dirt in the wintertime. They moved to the higher mountains in Summer where temperatures were cooler and trees more abundant. Summer houses were quickly built, rough brush affairs. Other houses were for grinding, the hunting blind, the sweathouse, and a large house for assemblies. The Kuksu cult, a world renewal ceremony practiced to guarantee plentiful food, was performed in a specially built roundhouse which was 50 to 60 feet in diameter. Only one family occupied each single dwelling.

Summer camps in the high Sierras were the scene of much trading activity. Most trade occurred between bordering tribes, and travel through the territory of another band was rare. One account of long distance travel was the trading expedition of Indians from northeast Oregon, southeast Washington, and western Idaho. This group rode horseback up the Deschutes River in Oregon, down the Pit and Sacramento Rivers in California to Sutter's Fort on the American River, and onto the cinebar deposits south of San Francisco.

The Miwok traded, among other items, baskets, food, including salt, tobacco, beads, and worked shells. Artifacts and trade goods spread great distances from group to group. Marine shells from the Pacific Coast found their way into Indian culture in the southwest, and woven blankets from the Pueblo groups were used by Santa Barbara Channel Indians. Dentalium and olivella shells, used as money, were traded south, east, and north from Vancouver Island. Clam shells were also traded, the shells pierced in the center and strung. A yard long string was worth $5.

Goods were traded along established routes, many of which were subsequently converted to military and public toll roads under American settlers. Indian trails followed streams in wooded country but ridgelines in open country. Miwok trails were famous for running in a direct line up hills and down valleys, not zigzaging. A trail ran along the North Fork of the Tuolumne and along the north canyon wall, connecting towns along the river. Most trails were merely narrow footpaths and are no longer observable except in the desert.

Rock quarries were important in providing the right material to be used in making a tool. If

Miwok roundhouse

necessary deep pits were dug, but Indians preferred rock outcrops with easy access. One flint mine on Table Mountain had been dug out enough for a person to stand inside. Soapstone, steatite, was quarried and easily worked into pans, pipes, and bowls to be used or traded. It could resist high temperatures without cracking and a broken object would be worked into yet a smaller item.

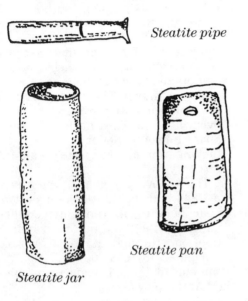

Steatite pipe

Steatite jar

Steatite pan

Although the Miwok language was fairly homogeneous through the length of their territory, there was no unifying social structure. Each village had an hereditary chief who mainly presided over meetings and dances.

There were several instances of village leaders with charisma and the ability to unify the surrounding Indian villages. One such person, Tai-pok-si, described as 6 feet tall and striking looking, resided along the Merced River. In a stentorian voice, he urged his people each morning to go off to work in the river for gold. All day, until 3:00 p.m., men dove into the river bringing up gravel which women and children washed. The gold was used to purchase goods for the group. When Tai-pok-si died in 1857, his burial was attended by 1,200 people.

Miwok were peaceful, seeming to enjoy life. A social gathering or dance was held for all occasions. For example, in the event of the acorn dance, adjoining villages were called together and acorns, roots, and meat prepared for the feast. A clearing in the forest was selected ahead. Here everyone spent the week gambling, eating, and sleeping during the daylight hours, dancing riotously at night.

There was no death dance, but an annual mourning period where all concerned villagers gathered with loud demonstrative cries and tearing of hair. On the Merced, small spots of pitch were stuck over a widow's ears and the hair sometimes singed off. Burial and cremation were both practiced by the Miwok, although burial was more common. Property belonging to the deceased was burned. Along the Stanislaus, near Robinsons Ferry, numerous skeletons were washed out by the river, none of which showed signs of having been burned. On the other hand, Indians near Chinese Camp, north of Jacksonville on the Tuolumne, practiced cremation.

Miwok experienced contact with their first Europeans when Spanish missionaries, located along the California coast raided inland in search of workers. Escapees from missions fled to the Sierras bringing with them European influences and disease.

While missioning as a way to help more "unfortunate" beings was looked upon as a beneficial act, its effect on native inhabitants was less than salubrious. "The Spanish crown in the eighteenth century considered it a religious duty to reduce heathenism and bring to as many native peoples as possible the virtues of Catholicism. Only in this way could non-Christians . . . be made truly human. The means used to carry out this lofty aim amounted to near catastrophe for the native people" (Heizer and Elasser, p.225).

The first California mission was founded at San Diego in 1769 and the last in 1823. Early re-

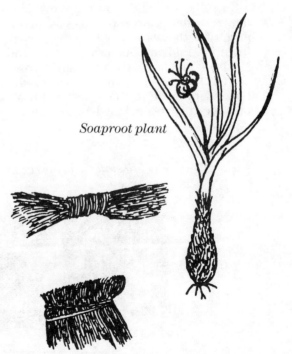

Soaproot plant

Brushes made from the coarse fibers of a soaproot plant.

cruits from among native inhabitants were friendly, and baptized neophytes, it was hoped, would adopt "civilized" ways and go into society as useful citizens. Initially Indians came freely to the missions, lured by curiosity and gifts. However, unpleasant aspects of mission life, unhealthy living conditions, forced labor, loss of freedom and death of family and friends, stopped this flow. Indians ran away and word of harsh treatment spread. Fugitives were captured and beaten.

Armed priests dragged new converts from villages in an attempt to maintain an adequate number of mission Indians to produce food and supplies for the Spanish population. Regular excursions were made into villages for this purpose. These campaigns went further into valleys to the north and east into Miwok territory. Indians who had first welcomed visitors in a friendly manner, resisted. Villages were either deserted at the approach of the Spanish, or recruiters were greeted with arrows and aggression.

The dream of the mission's purpose, during its 65 years of operation, never came to pass. Statistics are impressive. Eighty-one thousand Indians were baptized. Sixty thousand deaths are recorded. When missions were abandoned, the 15,000 remaining neophytes were turned over to Mexican ranches to work as peons or returned to the interior. The mission experiment was disasterous to the natives, and the effect on their culture permanent.

Once the Miwok discovered the value of the horse as a food source, then as a means of transportation, they carried out a series of costly and irritating raids on Spanish missions and later Mexican ranchers, stealing cattle from their corrals. By 1833 complaints of serious depredations were made to the government. Mexican ranchers had been killed by Indians, and in 1840 one thousand head of stock was run off.

As a defense, dozens of soldiers were sent out, but they met with little success in the rugged mountains. Some Indian villages were destroyed and their inhabitants slain, but the raids continued. A permanent border police patrol was begun, and the construction of a stockade in certain passes was suggested, but the plan was abandoned as impractical. The Indian offensive was pushing the border back westward, and the Spanish were proving incapable of overturning the situation.

It is interesting to speculate as to whether the Indians would have succeeded against these intruders, but the outcome of this struggle will never be known. The arrival of American miners from the east proved so cataclysmic as to overwhelm all existing Indian, Mexican, and Spanish events. Sheer numbers of miners, attacking from the rear, so to speak, filled Miwok territory.

The discovery of gold on the American River in 1848 and six months later on Woods Creek near Jamestown, Tuolumne County, brought miners by the thousands to examine gravels in river bars throughout most of Miwok lands. By the end of 1849, 53,000 "wealth-crazed American misfits" had claims on Indian lands (Moratto, p.64).

Miners and white settlers, who might have admired the dashing tall, Plains Indians, found the short, dark California natives whom they characterized as brutish and beastly, not to their liking. "Diggers" was the term applied to the Indians, either because they dug up roots and bulbs to eat, or because of their alleged bur-

Mission buildings at San Juan.

Digger Indians (Drawn by J. R. Browne, 1864).

rowing under the ground. John Audubon, son of the naturalist, noted in his journal that "the Miwoks along the Stanislaus River were not as handsome as the Indians of the east." Charles L. Brace, in the mining camps, considered the Miwoks to be "all disgustingly dirty." California Indians were so dark skinned that one miner commented the term "redskin" was inappropriate. "Their true color is close to a chocolate-brown." Their appearance excluded them from being classified as noble savages, a concept which so enthralled European and eastern Americans earlier in the century (Rawls, p.196).

The gap between the ideal image of the tall, aggressive, light-skinned Indian and the short, dark, "lazy" California Indian was one not readily bridged by white settlers who followed the miners. It was easy for emigrants to disregard such "animals" and dispossess the Miwok of their land.

In 1851 the Fresno Reservation was set aside between the Merced and Tuolumne Rivers, and 100 Indians lived there by 1856. In 1908 a small government reservation was established near Jackson, Amador County, although those people occupying the reservation received little or no aid from the government. Never more than 12 or so persons at a time were on the Amador Reservation. The remaider of Miwok lived on small farms which they owned or occupied houses on ranches owned by American landowners.

Merced River

(El Portal to Bagby)
Mariposa County, California
Length of run: 30 miles
Number of days: 1-2

The unusually clear waters of the Merced River run through Yosemite Valley in the Sierra Nevadas before plunging into steep canyons to enter the San Joaquin River south of Stockton. Walls of the canyon display a sequence of rocks from youngest to oldest so the rafter is drifting backward in time millions of years after leaving El Portal.

The Merced is known for its numerous rapids interspersed with pools of quiet water. Early trappers recorded their experiences in these whitewaters. Joseph Walker's party of explorers in 1833, following the Merced, described the steep walls and turbulent waters which "tossed to such a degree that no Indian has the courage to attempt to navigate it in a canoe."

Early view of Mt. Diablo and San Joaquin River.

Early History of the Merced River

The Merced was unknown territory before 1849. Except for Spanish/Mexican raids to obtain slave labor from the native Miwok Indians, European influence didn't extend beyond the western slopes of Mt. Diablo in Contra Costa County. Early historical accounts don't mention the San Joaquin Valley, and the Sierra Mountains don't appear on maps produced then. The Sierra Nevadas, perhaps first viewed by Fray Pedro Font as he stood on a hill near the junction of the San Joaquin and Sacramento Rivers, were given the Spanish name meaning snow white mountains.

In June of 1806, Lt. Moraga of the Mexican Army led a company of 15 soldiers into the Valley in pursuit of Indians who had stolen mission horses. Moraga crossed the Tuolumne close to where it enters the San Joaquin and in very hot weather marched forty miles to the Merced. Drinking out of the river, they named it El Rio de la Merced, the River of Mercy. Continuing onward, Moraga and his men encountered hundreds of beautiful varigated butterflies along a small stream which they named El Arroyo de las Mariposas, presently the name of the county through which the Merced flows. As a result of Moraga's explorations, land grants in the valley and Sierra foothills were made by the governor of California.

The ubiquitous fur traders, from both American and British companies, had trapped in valleys of the Tulare, San Joaquin, and Sacramento as early as 1820. Because these men didn't like to reveal profitable hunting areas, the San Joaquin remained relatively unknown. Joseph Walker and his men made the first east to west crossing of the Sierras in October to November, 1833, arriving to see at a distance the "lofty precipices" of the Yosemite just before heavy winter snows began. Anxious to get to the San Joaquin Valley, the expedition didn't stop to explore Yosemite but followed the Merced River along Indian trails straight up one canyon and down another until reaching the lower Merced where they "commenced" setting traps for beaver.

In 1837, Dr. John Marsh purchased land near Mt. Diablo, thereby becoming the first white settler in the interior. His Ranchos los Meganos was a popular stopping place for emigrants and travellers. Dr. Marsh was murdered in September, 1856, by two Mexicans as he was going home one evening, but his death wasn't realized until the next morning when his horse and buggy were found. The Mexicans were arrested, and, after a trial, one released while the other was sentenced to life in the state prison.

John Fremont, explorer for the government, who spent many years in the West.

As late 1844, the indomitable explorer, Capt. John Fremont, entered California on a journey for the U.S. government. He travelled down from Oregon and reached Sutter's Fort on March 8. Departing again on March 24, the party traveled southeast, crossing the Cosumnes, Mokelumne, and finally what Fremont called the Merced River on April 1st. He described the area as crowded with animals, with bands of elk and wild horses frequently seen along the river bank. However, the Great Pathfinder was mistaken geographically, and not for the first time. The river he described was probably the Tuolumne, not the Merced. Fremont continued south. He later applied for and received a land grant which included the Merced River area near the present town of Bagby. Fremont originally called the town Benton Mills in honor of his father-in-law, Senator Benton.

Gold was discovered in 1848 on Mariposa Creek which flows into the Merced near Briceburg. News of the strike brought in such adventurers as James Savage. Savage's checkered career included working in the California Batallion organized by Capt. Fremont, as well as leading a band of approximately 60 to 80 horse thieves in the valley. Once on the Merced, he employed Indians to help look for gold.

Primarily a businessman, Savage owned two stores, one on Mariposa Creek and one on the Fresno River. Both served as the meeting place for neighboring Miwok Indians as well as supplying miners with standard equipment and food. Savage himself, living with two Indian women, Eekino and Homut, kept in close touch with problems which might arise.

Savage and the two women, accompanied by an influential neighboring Miwok Indian leader,

Jose Jerez, travelled to San Francisco. Savage's perported purpose was to convince Jerez the whites were too numerous and powerful to be exterminated. In San Francisco, Savage purchased goods for his stores while Jerez, the story goes, became intoxicated. Upon returning to his hotel, Jerez was locked in his room by Savage, who struck "Jerez after a great deal of provocation."

Once the return trip to the Merced had been made, the angry Jerez used his influence to drive out and kill white settlers. Accordingly, in November, Savage's store and home were attacked his two wives taken by their own people. Determined to teach the offenders a lesson, Savage raised 75 volunteers and fell upon Jerez's village. Jerez and 23 of his men were killed. The volunteers didn't stop there, but destroyed other Indian villages as well.

Not satisfied with this victory, Savage applied to the governor for milita, wildly exaggerating the threat of an Indian uprising. Permission was granted, and Savage was commissioned a major. Together with 150 volunteer militia forming the Mariposa Batallion, they followed the Merced to the Yosemite area where they succeeded in capturing the leader, Ten-ie-ya, and 72 of his tribe. Later, guarded only by a single officer, the Indians escaped back to the mountains.

A second military expedition in 1851 managed to capture five Indians, at the same time killing Ten-ie-ya's youngest son. Ten-ie-ya surrendered several days later, overwhelmed with grief. His people stayed behind, fearing they would be killed by the volunteers.

Ten-ie-ya's period of captivity was particularly difficult for the old leader. On one occasion, discouraged with American ways and food, he requested to be tethered in a field of clover to eat. Trying once to escape, he was recaptured when about to swim the river. His angry speech to the soldiers was "Kill me, if you like, but if you do, my voice shall be heard at night, calling upon my people to revenge me . . . ." The last of Ten-ie-ya's people was captured by the military, and Ten-ie-ya was allowed to return to Yosemite where he lived until killed by Mono Indians in 1853. He was stoned to death by the attacking Indians who were seeking their horses stolen by young men of Ten-ie-ya's tribe.

Savage settled back into civilian life supported in style by his prosperous store. Indians brought in gold exchanging it for goods, while white miners paid his prices rather than travel south to Mariposa. He is rumored to have taken in so much gold he had to store it in pork barrels.

Valleys and ravines were soon filled with a

*Joaquin Murrieta, notorious bandit who oper-
ated throughout the gold camps.*

mixture of miners arriving from gold areas on
rivers to the north. Exciting stories of gold, including one of an ll pound nugget, sent gold
seekers scurrying up the canyon, only to find
the nugget was really a baby. The baby proved
to be such a rarity it was put on regular display.
The first woman to appear in Mariposa was
greeted by a band and hundreds of miners on
hand from miles around to welcome her. She
stayed on to open a bakery selling pies for $5.00
each. Her pies, reportedly, left much to be de-
sired.

Miners traversed the Merced at McSwains
Ferry, south of present day Lake McClure, operated by Silas McSwain. McSwain came to an unfortunate demise fatally shot in an argument
over some neighbor's hogs which had been tres-
passing.

Mail delivery to the mines was by Everett's
Express, which went "like lightening over hills,
ravines, gulches, bars, valleys, and everything
else that come in the way." The Express was
started by John Everett who carried the mail on
his back, leaving Knights Ferry on the Stanislaus, reaching as far south as Snelling on the
Merced. Everett failed to return one day and
was found drowned in a flooded mine shaft. His
mail bag wasn't disturbed, and his death was
never satisfactorily explained.

One of the most notorious, successful, and
possibly fictional, mining camp bandits was
Joaquin Murrieta. Murrieta was born in Sonora,
Mexico, and, with his wife, came to San Francisco to meet his brother. Murrieta's dealings
with Americans seem to have been unfortunate
for the most part, as he arrived in time to see
hundreds of miners "running through the
streets shouting noisily "Hang them. Hang
them. Put a rope on their necks and let these
Mexicans be tried in eternity, the devilish
thieves." One of the Mexicans was his brother,
and Murrieta witnessed the hanging. He left for
Sacramento, "bearing in his heart the desire for
vengeance" (Paz, p.6).

Murrieta went to the Stanislaus mines, only
to be attacked by a gang of "mad people" who
told him to leave his claim. When he refused, he
was knocked to the ground and beaten, his wife
killed. Resolving to take his revenge, Murrieta's
first victim, reportedly from this gang, was
stabbed with a sword, crying for pity, while one
colorful account has Murrieta, whose dagger
dripped with blood, looking on.

Murrieta operated from 1851 to 1853 with a
band of highwaymen, stopping caravans, stealing gold, horses, killing and destroying property. One of his psychopathic gang members
was Jack Three Fingers who tortured miners,
"cut their tongues out and plucked out their
eyes and then burned their quivering bodies," a
treatment they presumably deserved being
Americans.

Marauding from the Feather River in the
north to the Merced in the south, his activities
had the whole state in turmoil. Headquarters
for the gang was first on the Stanislaus, then
south on the Merced and the remote reaches of
Tulare Lake after their first camp was invaded.

The bandits came to a predictable ending. The

Jack Three Fingers.

governor of California gave Texas soldier, Henry Love, and his men $150 each and three months to "arrest, drive out of the country, or exterminate the numerous gangs . . . " Crossing the Merced, Love approached Murrieta's camp at dawn. Gang members were shot, Murrieta killed with three or four bullets.

It is not clear if Murrieta was, in reality, one or more persons, but it is probable that Love actually had killed a small group of ranchers or cowboys. A head claimed to be Murrieta's was later displayed for a minimal fee in the California mining camps. Ultimately, some thirsty miner is supposed to have drunk the alcohol in which it was preserved, thus bringing an end to the exhibition.

Amenities of civilization came to the Merced with the construction of roads and the Yosemite Valley Railroad from Merced to the terminus at El Portal. El Portal derived its name from being the "doorway" to the Yosemite Park. From 1907, the Railroad carried passengers for 70 miles along the banks of the Merced to visit the scenic natural park.

While Miwoks had been able to cope with missions, explorers, epidemics, and the occasional settler, the overwhelming influx of Americans accelerated their destruction. Americans felt little sympathy toward these creatures they considered inferior and hostile. Probably feelings of isolation and fear contributed to American-Indian relations. This era was charcterized as "one of the last human hunts of civilization, and the basest and most brutal of them all" (Forbes, p.59).

Indians along the Merced were exterminated as well. On January, 1879, near Mariposa, seven

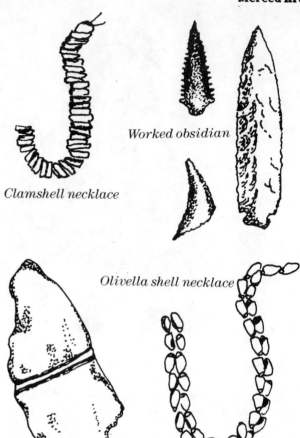

Worked obsidian

Clamshell necklace

Olivella shell necklace

Steatite arrow shaft straightener

white men attacked an Indian encampment, hanging, beating, and shooting all of the Indians. None of the attackers, whose names were known, were brought to trial. After 1855, Indians had virtually disappeared. By 1880 the Indian population of Merced County was seven, and in Mariposa County there were 174.

Indian Life Along the River

The area from El Portal and south along the foothills of the Merced River provided a varied and rich food supply of fruits, nuts, acorns, birds, and large game for the Indians. Villages weren't located down in the Merced canyon itself. Near El Portal, three small villages, four house sites, and three camping areas have been investigated by anthropologists.

Large villages had been built at the location of present day Sonora, Mariposa, Yosemite, and San Andreas. Names of these villages were Alaula-chi, Hikena, Kakahula, and Wilito, but no meanings have been recorded from the past.

Henry Love, who collected a reward for killing Murrieta.

Geology of the Merced River

The Merced River gives the rafter an excellent look at the various types of rocks of the Calaveras Formation. While other Sierra rivers cross the Calaveras, almost all of the runnable Merced is cut through exposures of this formation.

A raft trip on the Merced begins at El Portal and ends at Bagby in the reservoir waters of Lake McClure. Like other rivers along the Sierra's western face, the Merced carves a curious box-shaped meander on its north bank just at the Melones fault. These box meanders are almost invariably chosen as the site of large scale reservoirs. The reservoirs follow a meander then terminate against a dam. Flooded stream valleys at the dam give the reservoir a hammerhead shape. The position of stream channels here is largely is due to geologic structures. Sierra reservoirs are invariably situated in either the western or central fault blocks. The runnable stretch of the Merced is almost entirely within the eastern fault block.

The Merced watershed has a rich mining history, but much of the activity came quite late compared to other districts in California. Examining some of the more important districts from east to west takes us from gold associated with granite rocks of the Sierra Batholith, to rocks intruded into the Calaveras, to the Mother Lode in the Melones fault zone, and finally into the central and western fault blocks.

Geology along the Merced can be summarized as an area where granites of the Sierra Nevada Batholith were forced into rocks of the overlying Calaveras Formation. The batholith was a massive chamber deep within the earth filled

with hot, liquid rock. Today the Calaveras mudstones, shales, and limestones appear to be draped over granites of the cooled batholith. Heat from these cooling granites thoroughly cooked the Calaveras rocks converting the shales to slates and the limestone to marble.

One of the eastern most gold districts along the Merced is Clearinghouse. Named after the principal mine of the area, Clearinghouse is near the center of Mariposa County. Much of the district's gold was not discovered until 1860, and mining continued well into the 1930's. Nearby, along the South Fork, the gold district of Hite Cove was named after John R. Hite. Unlike many miners, Hite owned the mine for 17 years and subsequently became very wealthy. Granite dikes here broke into the above rocks of the Calaveras Formation to precipite gold in quartz veins.

Cat Town is a famous mining camp in northwestern Mariposa County. Like Clearinghouse, Cat Town was not established until the 1880's with active mining proceeding into the 1930's. As elsewhere in the vicinity, gold is mined from dikes containing gold which intruded into rocks of the Calaveras.

Bagby, at the end of the runnable section of the river, is in western Mariposa County. The district straddles the Mother Lode belt in the area of Bagby and Bear Valley. John C. Fremont was given a Spanish land grant here. His original name for the town, Benton Mills, was later changed to Bagby for an early local hotel owner.

The Mother Lode belt in the Bagby district is over a mile wide where the Melones fault cuts through rocks of the Mariposa Formation. The local rocks, dated at 150 million years, are

Geology of the Merced River.

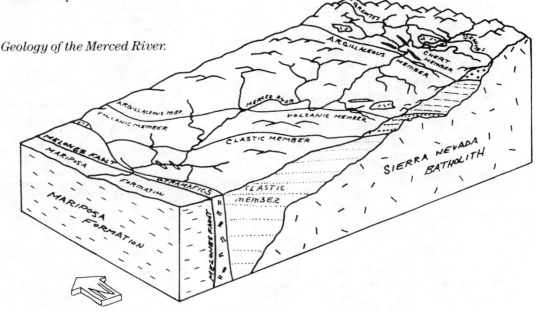

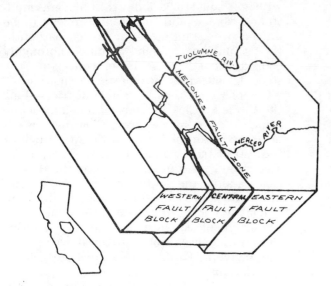

Major fault systems and fault blocks of the western Sierra slope.

greenstones, serpentines, and green schists. These developed where the Mariposa was baked by underlying molten granites. Veins of quartz and sulfide-rich rocks, containing gold, cut through the Mariposa Formation. Several famous large masses of free gold came from this area where many of the veins are as thick as 50 feet. A good portion of the district as well as the site of the original town of Bagby was covered by the Exchequer Reservoir in 1967.

Dog Town was also known as the district of Kinsley. Several smaller districts make up this area in northcentral Mariposa County. As with other districts, principal geology is gold bearing quartz veins intruded into altered Calaveras rocks.

The Whitlock, or Colorado district, in west-central Mariposa County, includes the areas of Sherlock Creek and Whiskey Flat. Altered by heat and pressure, rocks are part of the Mother Lode and belong to the Mariposa Formation. Several enormously large gold specimens came from this district. A curious phenomena here is the recorded tendency for the gold bearing veins to display a roll or bend. The discovery of such a bend caused considerable excitment after it was realized that gold could be found in rich pockets just at the bend or roll.

Geology Along the Route

At the El Portal put-in, the Sierra Batholith has broken into the Calaveras from below. The Calaveras Formation is subdivided into four separate parts which will be seen as the trip moves downriver. These four members from youngest to oldest are the sandy (clastic) member, a volcanic member, a shale (hard claystone) member, and a chert (hard, flint-like, silica-rich) member. Lenses of limestones occur both at the bottom and top of the Calaveras. At El Portal, the uppermost chert member overlies granites of the Sierra Batholith.

Downriver for two to three miles, exposures in the river bank alternate between dark cherts of the Calaveras and gray crystalline granites of the Sierra Batholith. Only five miles below El Portal, the river swings into an odd block-shaped meander. This meander is controlled by local geology, and the river maneuvers back and forth between the hard chert and soft limestone layers. The river follows along one of the soft north/south oriented limestone lenses for nearly 3/4 mile before meandering back to the west. Because the local rocks plunge eastward on the Merced, we are going backwards in time as we go downriver to the west.

The next Calaveras member encountered is a five mile long stretch of the shale unit. These are mostly black slates and baked, altered siltstones. A short, one mile stretch of the volcanic member follows where the river meanders in a northwesterly direction for several miles.

A mile below Quartz Mountain, the river begins cutting the oldest or lowermost unit of the Calaveras, the sandy (clastic) member. Good exposures of this member may be seen along the last six miles of the river above Bagby. The clastic member includes both fine siltstones and coarse sandstones. Evenly spaced along this stretch are three small exposures of limestone.

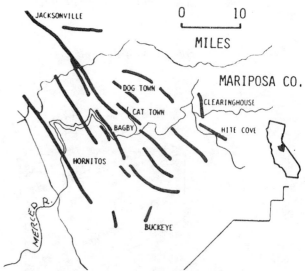

Major mining districts in the Mother Lode and adjacent gold belts of the western Sierra Nevada.

Finely divided carbon gives the limestone in these lenses a dark black color.

At many locations, fossil sea lily columnals can be found. These fossils look like buttons and are relatives of a starfish-like animal which lived in the shallow water of the sea floor over 200 million years ago.

At Bagby, the northwest trending Melones fault slices through as a very wide zone of broken, shattered rock. The Melones fault is the dividing line between the clastic, sandy rocks of the Calaveras Formation in the eastern fault block and the slates of the Mariposa Formation in the central fault block.

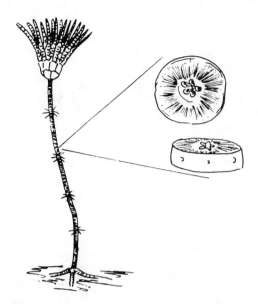

Fossil sea lily from the Calaveras Formation. Button-like columnal elements of the flexible stalk of this extinct form are scattered through the limestones.

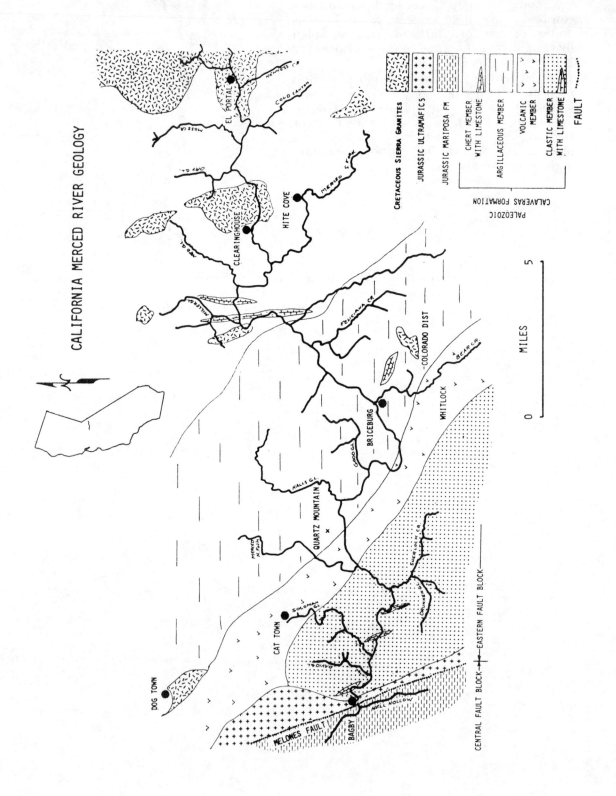

CALIFORNIA MERCED RIVER GEOLOGY

Cretaceous Sierra Granites
Jurassic Ultramafics
Jurassic Mariposa FM
Chert Member with Limestone
Argillaceous Member
Volcanic Member
Clastic Member with Limestone
FAULT

PALEOZOIC CALAVERAS FORMATION

MILES

0   5

CENTRAL FAULT BLOCK — EASTERN FAULT BLOCK

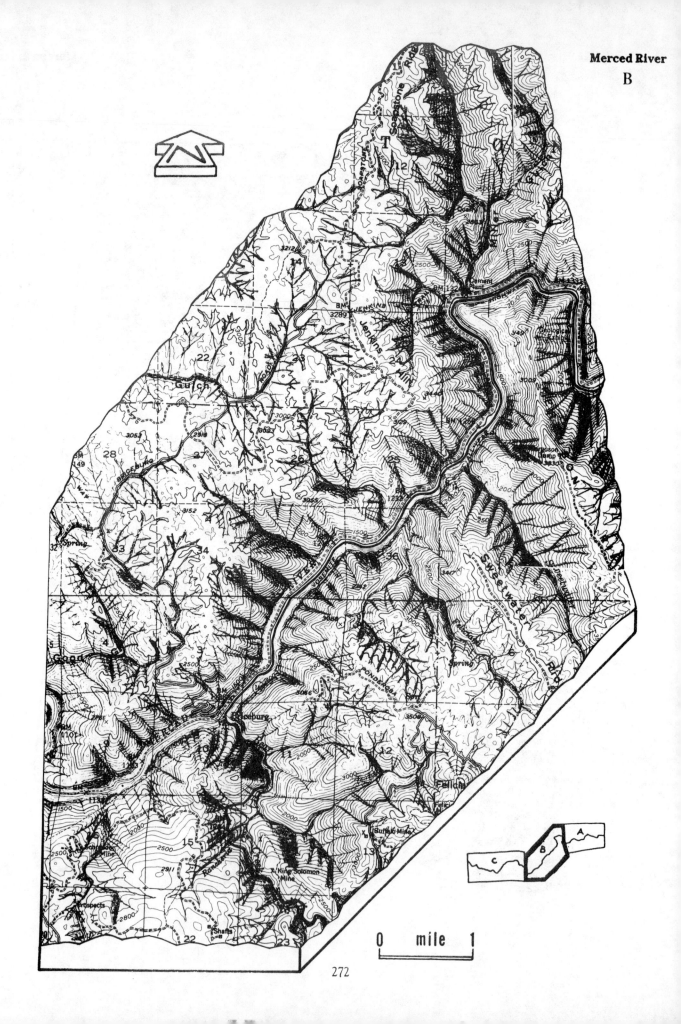

Merced River

B

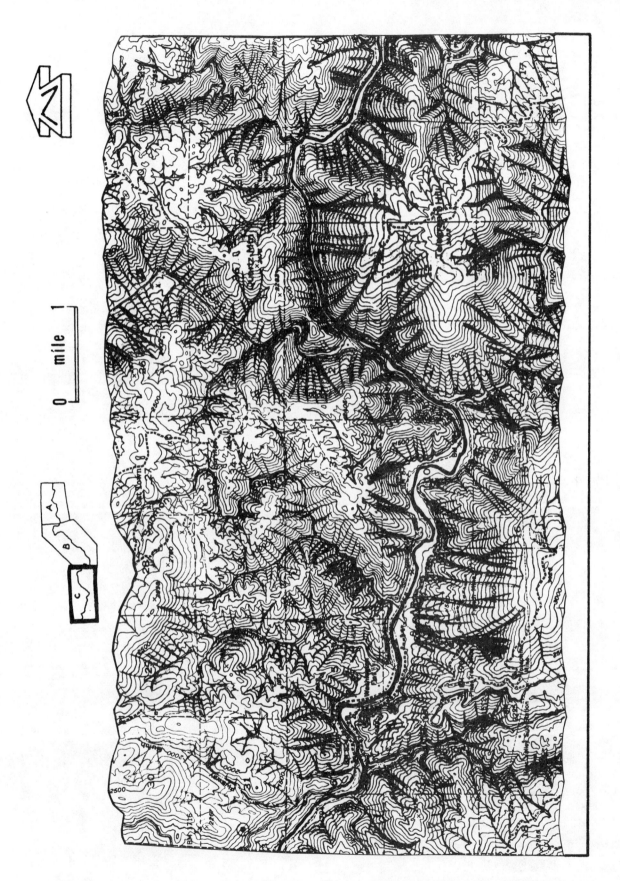

Mokelumne River

Boundary between Amador and Calaveras Counties, California
Bear River to Tiger Creek Reservoir
Length of run: 12 miles
Number of days: 1

Ponderosa Way to Pardee Reservoir
Length of run: 8 miles
Number of days: 1

Three branches of the Mokelumne River rise at the base of the Sierra Nevada Mountains in California before merging to enter the San Joaquin River after a run of 200 miles. Throughout its length the Mokelumne alternates between granites of the Sierras and the somewhat softer rocks of the Calaveras Formation. This river was considered to be the dividing line between the northern and southern gold mining districts. Most of the gold found along the Mokelumne was preserved in prehistoric abandoned river channels cut into the irregular, ancient landscape.

Early History of the Mokelumne River

Lined by tall pine forests, the boulder strewn bed of the Rio Mokellemos was mapped in 1841 by the U.S. Exploring Expedition under Charles Wilkes. Wilkes named the river for a local tribe of Indians, the Moquelumnes.

Fur trappers based in Vancouver's Hudson's Bay Company operated in California during the 1830's. Headquarters was established at French Camp in southwest Amador County with a trail from here running up to Vancouver. Joseph R. Walker, "one of the bravest and most skillful of the mountain men," was the first American trapper in 1833 who had the distinction of exploring the Sierras from east to west. Trapper Jedediah Smith, of course, had crossed earlier going in the opposite direction. As with Smith, Walker's exact route isn't known, although George Nidever, a member of the expedition says that the route lay "through a valley between the Merced and Tuolumi Rivers."

Spanish and Mexicans were ever on the lookout for slave labor for missions along the coast and frequently conducted raids into the Central Valley for this purpose. This same Joseph Walker in 1833 aided in one of these excursions where hundreds of Miwok Indians were slaughtered. Contact between the two groups escalated when mission converts escaped back to

Indians worked for the miners to pan for gold (From Hutchins' Illustrated California Magazine, 1859).

their families and tribes in the Sierras bringing not only European ideas and influences, but disease as well.

An epidemic in 1833, perhaps maleria, spread from Oregon and proved to be "extremely lethal" on the Central Valley population. The "Miramontes epidemic" of 1837 was probably smallpox. Syphilis, brought in by the Spanish, was all pervasive, and, along with later epidemics of fever, dirrhea and vomiting, reduced the overall population from 83,000 persons in 1800 to 19,000 in 1851.

In 1848, hearing of the discovery of gold on the American River, Charles M. Weber, a rancher in the San Joaquin Valley, organized a gold prospecting party consisting of settlers and Indians living on his land grant property. The party explored the rivers east of Stockton, beginning with the Stanislaus, where their frantic activities eventually resulted in the discovery of gold.

This was the beginning of what was called the "southern gold mines" which ultimately expanded gold country from the Trinity River south to the Merced. Towns deserted in the

Joseph Walker, mountain man and explorer.

northern district were in direct proportion to those which sprang up in the southern Mother Lode.

The Mokelumne River was soon cluttered with such daring persons as Lt. Hollingsworth, who "with his covey of senoritas in Los Angeles, had broken free in mid-September and galloped to the Mokelumne." Tent towns appeared at Mokelumne Hill, Volcano, Jackson, and Ione.

Mokelumne Hill, originally called Big Bar, Mok Hill, or The Hill, developed a well-deserved reputation for gambling and murder typical of mining towns. According to the local doctor, for seventeen successive weeks a man was killed every Saturday night, and five men dispatched in one week. A vigilante committee was formed only to be disbanded when one member confessed, on the eve of his hanging, to eight murders around Mokelumne Hill and areas south.

Gulches around The Hill proved to be extremely rich in gold, and the zealous miners eventually drilled into the hill itself. A party of Frenchmen, tunneling into their claim, ate, slept, and cooked in their fifty foot square hole to prevent anyone else from entering. Unfortunately for them, an American nearby shouted out that the Frenchmen had raised the French flag, whereupon hundreds of Americans stormed the hole, carrying away dirt which yielded $50 to $100 per sack.

A Mr. Cowden, one of the thousands of miners, was financed by his father with $500. His letters back home revealed he was working from 5:00 a.m. until 1:00 p.m. with breakfast at 8:00 a.m. It was too hot for work in the afternoon. He and his companions dug a hole ten feet deep and fifteen feet square, using a rocker

to extract the gold. They reached bedrock at which point they used a pick and crowbar. Their hole had yielded six ounces in one week of work. Cowden commented that "digging wells or cellars at home would be play alongside of digging into these hills and rocks."

Prospecting a few yards from Cowden was a man named E. G. Waite who felt the necessities were pick, pan, shovel, rocker, tent, blanket, bread, bacon and beans. The luxuries were coffee, tea, dried peaches and women. Waite averaged nearly $150 a day.

The thrill of the first discoveries faded with the coming of rains, poor food, and unhealthy living conditions. Each year seemed "particularly rainy." One miner complained "it had been raining for about six weeks and our claim had been four feet under water for a month." These miners, so desperate for food, but unable to afford high prices, crossed the swollen creeks at night to steal pork and bread from the store. That fare provided a feast as meals were usually quick affairs prepared by one unfortunate person chosen on a rotating basis.

Anything was thrown together. Hard bread, dried meat, beans, potatoes and onions were common. Fruit and fresh vegetables were rare. When they had no money, boiled barley might suffice. Walter Colton found the food, especially the jerked beef, appalling "when moistened and toasted it will do something toward sustaining life; so also will the sole of your shoe."

In wintertime miners suffered from fever, swelling, blackening of the legs, and bleeding

Headed toward California (Drawn by W. Colton, 1850).

276

Returning from the California mines.

easy gold sources dried up. Untold hours of labor were spent in the construction of lengthy ditches and canals to divert water for mining and eventually for domestic and agricultural purposes.

A party of eight dammed the Mokelumne River near Rich Gulch only to discover the amount of gold in the now dry bed wasn't worth their efforts, and the dam was abandoned. Undaunted, N. W. Spaulding and Co. cut 13,000 feet of lumber to dam and flume the Mokelumne, exposing the channel in a second spot, to find only $160 worth of gold.

The Amador-Sutter ditch of 1853 supplied water from Sutter Creek below Volcano to the towns of Amador and Sutter at a cost of $22,000. The Amador ditch was to take water to Ione thirteen miles away at a cost of $20,000. The Amador Canal, begun in 1870 by the Amador Canal and Mining Company, controlled water on the Main Fork of the Mokelumne. The canal ran a distance of thirty miles, supplying Jackson, Amador, and Drytown. The canal cost $400,000, and water users on the canal paid .20 cents an inch.

The town of Lancha Plana once stood in the area now covered by Pardee Reservoir. In 1850 the town's population was scattered along the river at Poverty Bar, Winters Bar and at other prospecting spots. A dam across the river was built here by putting up a series of pens held in place with rocks. As with other dams on the Mokelumne, gold returns were minimal once the effort had been made to expose the riverbed.

The Spanish name, Lancha Plana, referred to the flat bottom boats used to ferry miners across the river. Later boats were made of casks lashed together. Crossing was made at a charge of .50 cents a person. A somewhat rickety bridge, built in 1852, lasted until the first rains. Another story has the bridge collapsing with the first horse and wagon over it. A second bridge was built with green lumber, but this one fell when the long, hot summer weather shrank the wood.

A dugout ferry operated at Big Bar, just up river from Pardee Reservoir, cost $1.00 per crossing. The ferry was sold and improved with horses swimming alongside to pull the boat. Finally, a bridge was put here in 1853 to replace the ferry.

Jackson, near Big Bar, was the stopping place for travellers on the road north and south. A spring here, where they left a quantity of bottles, gave the town its first name of Bottilleas. This was later changed to honor Col. Alden M. Jackson, a lawyer who settled there and opened an office. Jackson was set up as a tent

gums, which turned out to be scurvy. Edward Buffum was cured by eating bean sprouts growing wild, spruce bark and eventually fresh potatoes. It was not yet known that onions and potatoes would have helped as a cure.

During the season of 1849-1850 a bumper potato crop was grown by small independent farmers near San Fancisco all of whom had come up with the same idea. It is unfortunate that the over abundance of potatoes didn't help the miners a short distance away. One farmer did raise a crop of onions on two acres which he sold for $8,000.

Miners had to resort to more drastic measures of dams and river water diversion by 1849 as

city in 1848, and in 1850 there were seven permanent buildings some of which were empty as miners were already leaving for better gold workings.

Mining was hazardous at the Eureka Mine (After Calif. Div. Mines and Geology).

A somewhat later, but interesting sidelight, is the log from Eureka Mine on Sutter Creek near Volcano which shows the hazards of mining and living in the 1860's to 1870's.

1865: July 1 — Two men injured, one fatally, in the Eureka mine, by the falling of a timber. Hayne bled to death in a few minutes.

October 27 — Martin Collins killed in the Eureka mine.

1867: October 22 — D. R. Whitman crushed to death in the Eureka mine at Sutter Creek.

December 10 — Mrs. Foster, a widow lady at Sutter Creek, was killed by a man called Eureka John by a blow from his fist.

1868: June 25 — A man by the name of Williams fell into the Eureka shaft and was instantly killed.

July 12 — Ed Burns falls four hundred feet in the Eureka mine . . . and is instantly killed.

August 25 — Workmen killed in Eureka shaft at Sutter Creek.

1870: Amador mine (Eureka) took fire on the seven hundred foot level, men all escaping.

1872: March 1 — Austrian killed in the Eureka mine, by the falling of a stick of timber.

June 1 — Amador mine (Eureka) took fire. Tom Frakes seriously injured during the efforts to control it. Loss, one hundred thousand dollars.

November 15 — Richard Jackson thrown into the Amador shaft (Eureka) by the swinging of the hoisting tub and precipitated to the bottom, thirteen hundred feet, tearing his body to pieces.

1873: June 7 — John Everest killed in the Eureka (Amador) mine by falling down the shaft.

Indian Life Along the River

Evidence collected by anthropologists has shown that Indians lived in the vicinity of Camanche Reservoir as far back as 4,000 years ago. Seventy-seven house locations were studied along with forty-four burials, pottery, and worked pieces of bone and stone. In conjunction with these, five cremation burials were uncovered. These burials took place around the year 300 A.D.

North of the reservoir a dry, undisturbed cave was examined. The interior revealed many grasses, seeds, acorns, wood and other plant fragments which were perfectly preserved in these unusual dry conditions. Grass and straw were presumably brought in by the inhabitants for bedding. Wood tools for digging and arrow shafts were among the preserved items. Several large flat stones covered with a thin layer of acorn mush, mammal bones, as well as mussels from the Mokelumne showed what made up the diet of these cave dwellers.

Along the Mokelumne in Amador County there are a number of pictograph and petroglyph locations for the rafter to see. In Bamert

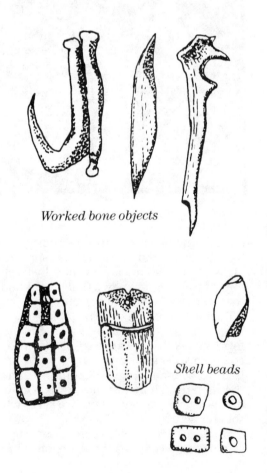

Worked bone objects

Shell beads

Cave north of Camanche Reservoir, petroglyphs have been cut into the back cave wall. Just before reaching the reservoir several caves on the Mokelumne have pictographs painted using red, black and white colors. Linear designs are most common, although an unusual cross-like element occurs in one of the caves. A jumble of petroglyphs in these caves includes curved lines and possible small figures. The viewer will have to decide what many of the designs can be.

South of Volcano near the Mokelumne, petroglyphs have been pecked into flat limestone rock surfaces. The designs here appear to be more "sophisticated" or, perhaps it would be better said, more recognizable to the viewer. A sun, hub of a wheel, and ladder or grid can be seen in association with bedrock mortars.

Rock art at caves along the Mokelumne.

Rock art south of Volcano.

Rock art north of Camanche Reservoir.

Geology of the Mokelumne River

The Mokelumne has two major runs. The Electra run between Ponderosa Way and Pardee Reservoir near the town of Mokelumne is a very popular route, whereas, the upper stretch between Bear River and Tiger Creek Reservoir is more difficult. In addition to many interesting features associated with ancient volcanic activity, the Mokelumne displays an excellent variety of the major rock types characteristic of the eastern Sierra block. Both runs are confined to the eastern block, and the lower run stops just short of the eastern/central block boundary at the Melones fault.

From the beginning of the gold rush, the Mokelumne watershed was an area of tremendous activity. At Mokelumne Hill, originally called Big Bar, gold was mined from both lode and placer deposits. Placers were richer and more productive because of eight separate ancient river channels located near the hill. The lode gold of this vicinity lies in the eastern gold belt, or slightly to the east of the Mother Lode along the Melones fault. The adjacent communities of West Point, Volcano, and Jackson all began as mining towns. Gold mined in these areas included both placer and lode.

At Camanche-Lancha Plana dredging operations yielded very fine-grade gold producing as low as .10 cents per cubic yard of material worked. Dredging continued as late as 1951. Gold here is from 50 million year old, Eocene gravels which have been carefully mapped to locate deposits up to fifty feet thick in places. Gold occurs in both Eocene gravels and quartz veins at Campo Seco and Valley Springs. Along with the gold, deposits of iron pyrite (fools gold) are common.

Mining in the Paloma district was mostly in the Gwin Mine which followed quartz veins in the Mariposa Formation. Rich Gulch in northcentral Calaveras County produced large quantities of gold in quartz veins intruded into the Calaveras Formation. Despite the richness of the initial strike, these quartz veins quickly played out.

The Volcano district was mistakenly named from the local notion that the limestone caves here are somehow related to volcanic activity. Placer gravels deposited on the irregular landscape of the Calaveras Formation as much as 50 million years ago are responsible for much of the gold. West Point, originally known as Indian Gulch, was one of the most productive areas of the Sierra eastern gold belt. Most of the gold is related to the underlying body of Sierra granite which intruded the slates, quartzites and

mon around Mokelumne Hill. Tuff is a cohesive mass of volcanic rock debris including ash, dust and pumice. It is cemented together into a light-colored, mottled rock that can be easily cut and worked. Freshly exposed, damp volcanic tuff may be easily sawed like wood. Once the sawed blocks are permitted to dry, they become as hard as concrete. Building stones of sawed tuffaceous volcanic rock were used to construct almost all of the older buildings in the community of Mokelumne Hill.

Geology Along the Route

Geology along the runnable stretches of the Mokelumne provides a good look at the eastern block as well as granitic rocks of the Sierra core. Granites are part of an immense structure called a "batholith." This was once a huge chamber of molten rock deep in the earth. The hot fluid cooked the overlying rocks as it slowly cooled. Along the upper run, these granites, typical of the core of the Sierras, occur at the put-in near the confluence of Bear River and the North Fork of the Mokelumne River. Along this stretch of the river "xenoliths" appear as dark rocks imbedded in the granites. These xenoliths are older rocks frozen in the process of being adsorbed into the granite just as crystallization took place. There are no significant gold mining areas along this upper part of the river. Most of the mining activity was confined to the lower stretches close to the older river channels and faults of the Mother Lode.

Less than five miles east of the put-in at Bear River, some of the granite rocks in the valleys show glacial scratches or "striations". During the Pleistocene ice ages, the high Sierras were repeatedly scoured by glaciers advancing down the valleys. Much of the evidence of glacial activity at this low elevation has long since been destroyed by the rapidly eroding streams of the Sierras.

Some of the best evidence for glaciation is deposits of glacial debris or till and the finer glacial flour. Unfortunately these soft sediments are among the first to be eroded and carried off by streams. Glacial striations are not made by ice, but rather by rocks imbedded in the glacial ice moving over the rock surface.

Only five miles downstream from Bear Creek is a profound contact between the Sierra granites and the older rocks of the Calaveras Formation. Rocks of this formation are 250 million years old and are schists for the most part. Schists are recognized because the rock splits readily into flakes, and the mineral grains are large enough to be seen. Mica flakes in schist

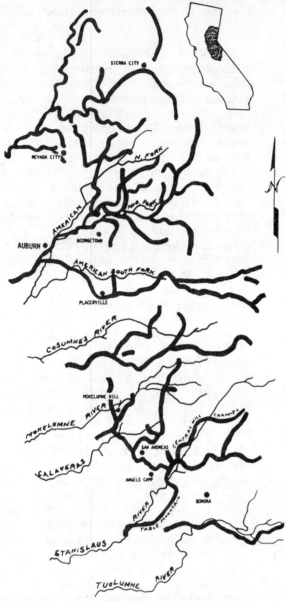

Ancient river channel deposits containing placer gold in the western Sierra Nevada.

schists of the Calaveras Formation. In addition to gold, the area is rich in galena, or lead ore, which almost always occurs with very high grade gold ores. Over 500 mine separate mine tunnels have been dug in this district.

Close to Mokelumne Hill an interesting geologic feature is the volcanic domes. These domes represent eruptive sequences where the lava is of such low temperature and high viscosity that it will not flow but instead makes a distinct bulge or dome. These domes are of a variety of compositions but most fall in to the lighter-colored, quartz-rich rock types.

Lightweight, ash-rich, tuffaceous rocks from 50 million year old (Tertiary) volcanos are com-

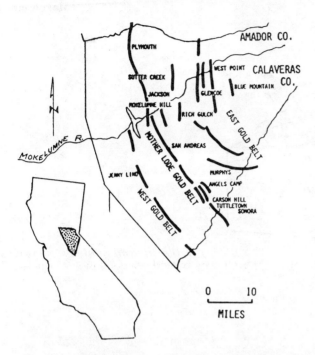

Major lode deposits of gold in the western Sierra Nevada.

After about five river miles of steeply inclined schists, the river resumes its course back into Sierra granites near Tiger Creek Reservoir at the take-out. Another striking feature of the Sierra granites in this upper river stretch is fracturing. After a molten phase, granite cools and hardens under considerable pressure deep in the earth. As erosion slowly exposes the rock, the release in pressure causes a fracture or break. Fracturing may be nearly parallel to the horizon creating domes and rounded nobs or it may be at right angles to the horizon creating sharp vertical joints.

In the lower Mokelumne stretch, the Electra run begins at Ponderosa Way and ends at Big Bar. This is only a six mile run, but it includes many nice rapids that may be attempted by the intermediate rafter.

At Ponderosa Way the river cuts across schists of the Calaveras Formation followed by an exposure of a thin lense of marble. This was Calaveras limestone which has been cooked by the granite intrusives to recrystallize as marble. After only a few hundred feet of Calaveras Formation the river exposes Sierra granites which continue for a mile or two. The granites are part of an underlying structure called a "stock" or irregular bulge atop the main granite dome-like mass of the batholith.

Alternating Calaveras schists and granites are exposed for two to three miles before the river cuts across a small outcrop of Calaveras marble.

are all parallel which gives the rocks a highly crystalline sheen as sunlight hits it. The contact between the granites and schists is worth examining. Pieces of partially melted Calaveras schist can be seen frozen in the process of being melted into the granite.

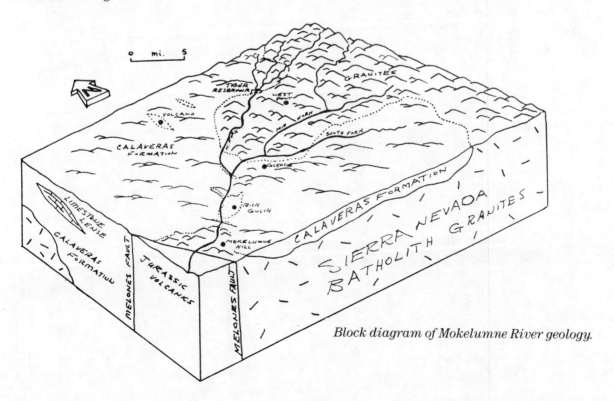

Block diagram of Mokelumne River geology.

281

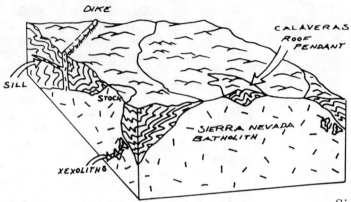

Sierra Batholith and related structures.

This exposure of marble is a "roof pendant" or a piece of the Calaveras that hangs down into the batholith from above. Exposures of granites continue for three miles before intersecting the Melones fault. Within the fault zone itself, volcanic rocks have been sheared and chopped up by the earth movement to form a soft clay-rich "gouge."

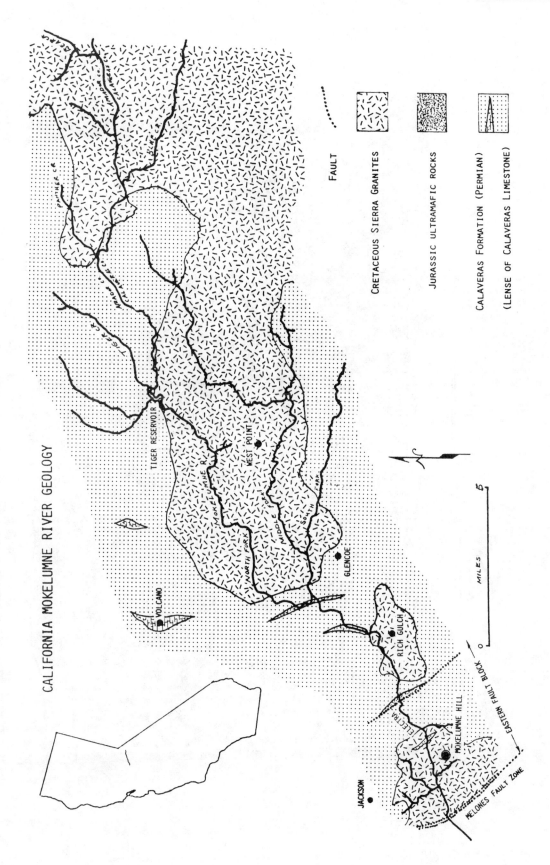

CALIFORNIA MOKELUMNE RIVER GEOLOGY

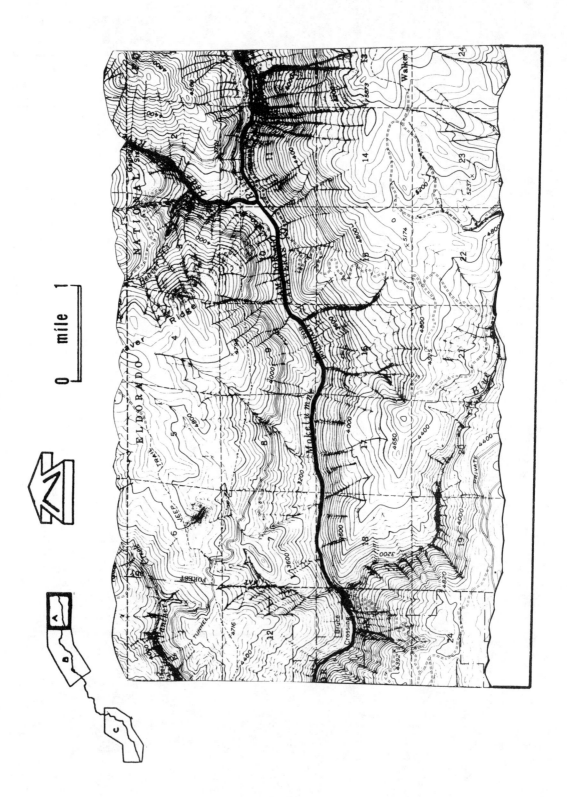

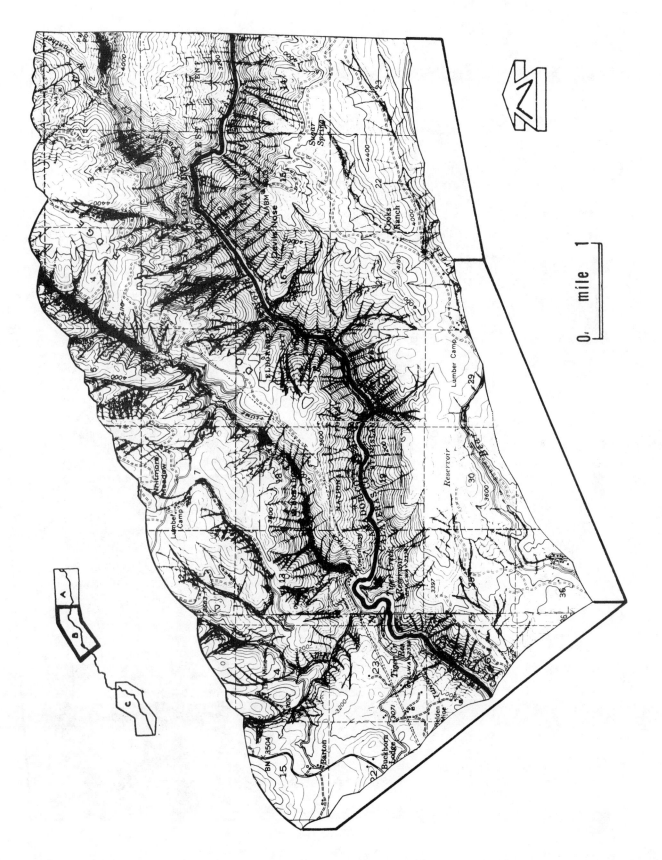

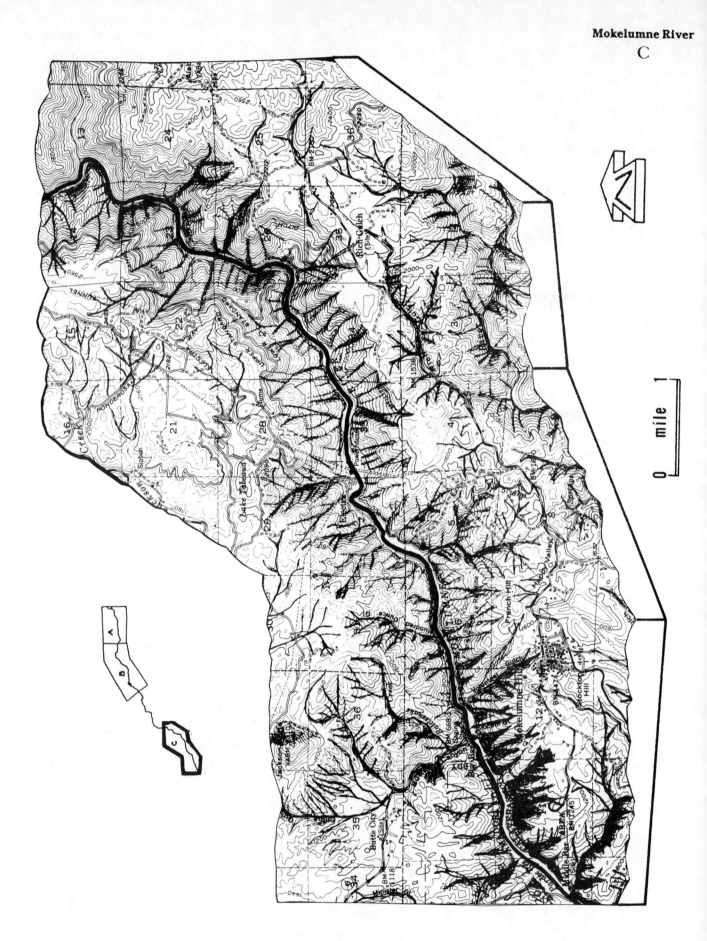

Stanislaus River

Boundary between Calaveras and Tuolumne Counties, California
Camp Nine to Melones
Length of run: 14 miles
Number of days: 1
Goodwin Dam to Knights Ferry
Length of run: 4 miles
Number of days: 1

Formed from the North, Middle, and South Forks which begin as melting snow in the Sierra Nevada Mountains, the Stanislaus River flows into the San Joaquin River after a run of more than 100 miles. In its upper reaches, the Stanislaus cuts through limestones which have been dissolved out to form an unusual number of caves. Elsewhere along the river, contact with molten granites in the geologic past has altered the limestone to marble.

Gold along the Stanislaus formed as as rich deposits in potholes of an ancient riverbed. Unique gold nuggets discovered at Melones, flattened by stream pounding and shaped like watermellon seeds, gave the town its name.

The Stanislaus offers something for all classes of rafters. There are quiet stretches for drifting, rapid waters cutting through narrow channels, and rolling valleys where rafters should keep an eye on the long rapids. Doing the Stanislaus at the right time can provide the rafter with an enjoyable trip through a still quiet wilderness.

Early History of the Stanislaus River

The Stanislaus River may have first been seen by Spanish explorers led by Ensign Gabriel Moraga in 1802. Fray Pedro Munoz who accompanied them described the scenery as "running in a much deeper canyon (than other rivers) . . . bordered by an abundance of wild grapes, a bit of torote and many ash trees." They named the river Our Lady of Guadelupe. Camping for four days near present day Knights Ferry, Moraga and his men found Miwok Indians living above the river in caves only accessible by ladder. As might be expected, the natives were unwilling to come down to be converted by missionaries with the expedition.

Later Spanish-Mexican incursions, contact with fur trappers, and refugees from missions along the coast brought permanent changes to the Stanislaus Valley. The British Hudson's Bay Company employed about 100 fur trappers in California and along rivers of the Sierras until

California Indians.

they left in 1845. They removed thousand of pelts during their 15 years of operation.

Peripetatic American explorer and trapper, Jedediah Smith, in 1827, on one of his trips eastward may have followed the Merced or Stanislaus up to its source, but the exact location of his Sierra crossing is unknown. American Ewing Young, trapping in the valley from 1825 to 1830, was surprised when he "discovered unmistakable signs of another trapping party" which proved to belong to the Hudson's Bay Company commanded by Peter Ogden.

Miwok Indian refugees who had fled from the harsh slavery conditions of mission life banded together on the rivers of the western Sierras. In 1827 an Indian neophyte with the christianized name of Polish St. Stanislas, Estanislao, led a series of attacks against the missions successfully enticing other Indians to revolt.

In May, an expedition of 40 Mexican soldiers equipped with a swivel gun sought out the Miwok who were entrenched near the "River of the Laquisimes," probably the Stanislaus. After two days of futile effort and a number of Mexican casualties, the soldiers retreated. A second effort by 100 soldiers later in the month was equally undistinguished. Yet a third outing proved more "successful" in scattering the Indian band by brushfires set by the soldiers. Mexican commander Martinez was elated. "I rejoyce exceedingly that this scum has been chastized . . . " Little is known of the later life of Estanislao who was reported still fighting as late as 1839.

While few people were involved in this early contact, it brought disease among the Indians in epidemic proportions. By the 1840's no tribes were entirely without sickness. American trapper Ewing Young's trip in the Fall of 1832 found Indian villages "numerous, many of them containing 50 to 100 dwellings . . . on the Tuolumne, Stanislaus and Calaveras rivers. . ." On their river trip in the Summer of the following year, "we found the valleys depopulated."

The first pandemic had occurred. However, what the disease was hasn't been determined. Symptoms ruled out smallpox or maleria and pointed to such illnesses as cholera or typhus. Twelve thousand died in the San Joaquin Valley and 8,000 in the Sacramento Valley. Whole tribes were exterminated, empty villages and large numbers of skeletons were seen everywhere.

Young went on to report that "from the head of the Sacramento to the great bend and slough of the San Joaquin we did not see more than six or eight live Indians, while large numbers of their skulls and dead bodies were to be seen under almost every shade tree near water, where the uninhabited and deserted villages had been converted into graveyards." The odor of decay was overwhelming.

A second epidemic of smallpox in 1837 killed a minimum of 2,000 persons. From that time and for the succeeding ten years, outbreaks of disease accounted for an estimated 5,000 deaths. These diseases emptied the countryside of inhabitants prior to the arrival of gold seekers.

Before the 1848 gold rush, reports of a salubrious climate, fertile soil, and abundant fruit in California, accompanied by a considerable amount of publicity encouraged emigrants from the East. One easterner reflected "that the superb region of California is adequate to the sustaining of twenty millions of people; has for several hundred years been in the possession of an indolent and limited population, incapable from their character of appreciating its resources . . ."

One such emigrant party, influenced by the publicity, was the Bidwell group, organized in Missouri in May, 1841, to make the long trip to California. The party knew only that "California lay west, and that was the extent of our knowledge" (Clelend, p.10l). No experienced guides were along, and the 69 or so settlers were led by John Bidwell, a young man in search of healthier climates, and John Bartleson, of Missouri. At the last moment, trapper Thomas Fitzpatrick agreed to lend his assistance.

Mexican ranch worker in early California.

In Idaho the party separated into two groups, one heading north, the remainder, without Fitzpatrick, turned south across Utah and Nevada. Wagons had to be abandoned when the settlers, fearful of making such slow progress and crossing the Sierras late, took to horseback. Going on ahead, Bartleson and eight men deserted the group. Passing by the Carson River in Nevada, the settlers began their ascent of the Sierras. At this point they were rejoined by Bartleson's men who had become sick after eating diseased fish and pinon nuts.

Wandering without knowing where they were and short of food, the party eventually came to a small stream which proved to be the headwaters of the Stanislaus. Cattle and horses had to be left in the deep river canyon as they pressed on down the south side of the Stanislaus and through Tuolumne County. Crows, coyote and other available game were eaten. Finally, worn out and starving, the party reached Mt. Diablo in the San Joaquin Valley on Nov. 4, having travelled six months.

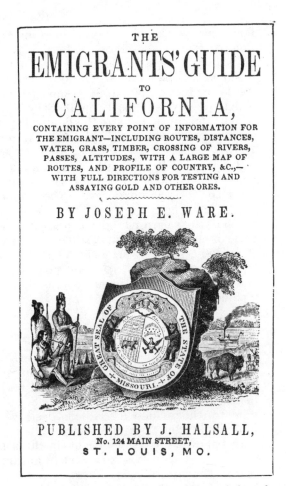

THE
EMIGRANTS' GUIDE
TO
CALIFORNIA,

CONTAINING EVERY POINT OF INFORMATION FOR
THE EMIGRANT—INCLUDING ROUTES, DISTANCES,
WATER, GRASS, TIMBER, CROSSING OF RIVERS,
PASSES, ALTITUDES, WITH A LARGE MAP OF
ROUTES, AND PROFILE OF COUNTRY, &C.,—
WITH FULL DIRECTIONS FOR TESTING AND
ASSAYING GOLD AND OTHER ORES.

BY JOSEPH E. WARE.

PUBLISHED BY J. HALSALL,
No. 124 MAIN STREET,
ST. LOUIS, MO.

Smart travellers to California followed the advice from Ware's pamphlet.

Captain Charles M. Weber was one of the Bidwell party. Weber was born in Bavaria and came to America at age 22, landing at New Orleans before going west. Upon reaching California, Weber went into partnership with Guillermo Gulnac. Using his partner's name, Weber applied for and received a land grant of 48,000 acres in the Central Valley. He moved to his land grant property where he established a store which eventually was surrounded by the town of Stockton.

The store proved to be a booming enterprise after gold was discovered in 1848. A consumate businessman, Weber charged outrageous prices. Indians paid even higher prices at mining stores than those charged to whites. A two ounce weight was substituted for the one ounce when trading with Indians. This came to be called the "digger ounce."

Weber send a group of Miwok Indians to the Stanislaus promising goods from his store in return for gold. The Indians promptly complied in August, 1848, with enough gold to send frenetic miners south to what became known as the southern diggings.

Upon hearing of this new discovery, Antonio Francisco Coronel, a Mexican schoolteacher from Los Angeles on his way to find gold on the American River immediately changed direction. Once on the Stanislaus, he sold several blankets to Indians in exchange for gold nuggets following them upriver to their source near Melones. On the first day 45 ounces of gold was collected and 38 ounces on the second.

In another gulley on the Stanislaus, $2,000 a day was supposed to have been picked up off the ground by a group of Frenchmen. Sargeant James Carson, formerly of the U.S. Army at Monterey, California, wandered up a tributary of the Stanislaus to where he found 180 ounces in ten days on what is now called Carson Creek. Carson Hill nearby was discovered to be the source of gold, and each gulley up the hill was soon exploited by the gold seekers. Upriver less gold was found.

Walter Colton, from Monterey, was working on the American River on Thursday, Oct. 19, when he reported "All the gold-diggers through the entire encampment were shaken out of their slumbers this morning by a report that a solid pocket of gold had been discovered in a bend of the Stanislaus. In half an hour a motley multitude, covered with crowbars, pickaxe, spades, rifles, and washbowls, went streaming over the hills . . . " Unable to resist, on Friday Colton "threw myself into my saddle" and started for the Stanislaus where he and others scoured the hills and creeks, one lucky miner digging out "the largest lump of gold found in California" weighing 23 pounds.

Tents popped up along the banks of the river. Camps and the towns which followed were named after the first miners to arrive. George

Charles Weber who started a store where Stockton stands today.

289

Angel "founded" Angels Camp, John Murphy, who had emigrated in 1844, set up Murphys Camp.

In these camps water had to be imported for some distance. At Murphys a series of ditches, flumes and trestles were built from the Stanislaus fifteen miles away. Lumber was needed and was brought further and further distances as wood supplies close by were depleted.

The Columbia and Stanislaus Water Company was formed in the Fall of 1854 to bring water to the placers. The ditch, carrying water to Columbia from the Stanislaus, eventually reached a distance of 60 miles in length. Of the last 21 miles, 12 were flumed and 9 were excavated at an estimated cost of $350,000. The ditch was completed on Nov. 29, 1858, and opened to a Grand Water Celebration. The event was attended by thousands of persons dressed in their best, who were entertained with fireworks and grand processions.

Ferries on the Stanislaus conveyed men and supplies back and forth from the southern Sonora area to mines north of the Calaveras River. Abbeys Ferry on the Sonora road was run by George Abbey. The San Francisco Herald of July 3, 1860, reported that the trick elephant, Victoria, from the Dan Rice Travelling Show, died three days after swimming across the Stanislaus at this point because the river was too high for the ferry to operate. Victoria was subsequently stuffed, and the Dan Rice Show went on to play in Placerville without her. Downstream a ferry operated by Thomas Parrott was started up in 1860 in competition with Abbeys. Parrotts Ferry ran until 1903 when a bridge was constructed there.

Gold mining traffic in the Melones area was served by both Robinsons and McLeans ferries.

Robinsons was on the main road from Angels Camp to Sonora and reportedly earned its owners $10,000 in a six week period when mining business was booming around Carson Hill. On the lower section of the Stanislaus, service at Knights Ferry was begun in 1848 by William Knight and continued after his death. The ferry was entirely destroyed in the flood waters of 1861.

By late 1849, mining and miners had changed. The majority of miners came and left after a brief try at what proved to be a laborious job. A few weeks were enough for most, with much of their earnings spent in Stockton gambling or on supplies. Mining still continued, but the gold became less accessible. "There is gold still in those banks," said one newspaper in 1851, "but they will never yield as they have yielded."

Indian Life Along the River

Remnants of Indian villages have been found along the river at favorite food gathering spots. These sites could have oak trees and perhaps bedrock mortars nearby. Along the Stanislaus much of this evidence has been destroyed by later mining and farming activity.

Just above Murphys, north of the Stanislaus, a limestone cavern was excavated by anthropologists. The site, Winslow Cave, was used as a burial chamber beginning 4,000 years ago. Burials continued until the year 300 A.D. A large number of human skeletal remains were scattered over the cave floor representing at least sixteen persons. The total amount of bone material excavated would indicate as many as one hundred individuals might have been thrown into the cave as a means of quick burial. A small

Indian corpse being prepared for burning or burial (From Hutchings' Illustrated California Magazine, 1859).

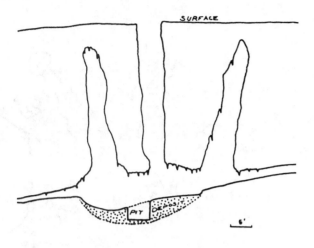

Winslow Cave where a number of Indian burials took place.

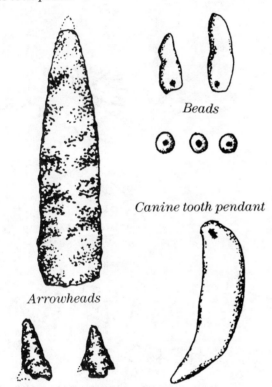

Beads

Canine tooth pendant

Arrowheads

Artifacts from Winslow Cave.

tention from miners, but in 1922 a steel staircase made the entrance accessible to tourists. Excavation showed the cave floor to be covered with human bones, artifacts, mammal remains and pieces of charcoal. Many of the bones were broken and crushed, and no complete skulls or skeletons were collected. Nine adults and two children were counted from the skeletal remains. Indications are that the cave was not inhabited for any great length of time. Miwok tradition relates that a stone giant lived here coming forth at night in search of victims to be carried back to the cavern and eaten.

The cave was probably not used as a place for burial because it would mean the body would first have to be lowered 150 feet into the cave floor from the ledge above before a grave was dug. It is more likely corpses were merely lowered or thrown down into the cave for disposal.

A cave near Abbots Ferry, discovered over 80 years ago, was used for a similar purpose. Here skulls were lying on the surface unburied. Bows, arrows, and charcoal indicate torches were carried into the cave with the body. Shells, obsidian points, pipes, and unknown ornaments found here were 3,000 years old.

Human bones and artifacts turn up in many gold mining operations, but they should be viewed skeptically. A human skull reported from a depth of 250 feet in the diggings near Columbia, 300 stone mortars 180 feet below the surface lava on Table Mountain, and the Calaveras Skull from gravels 50 million years old

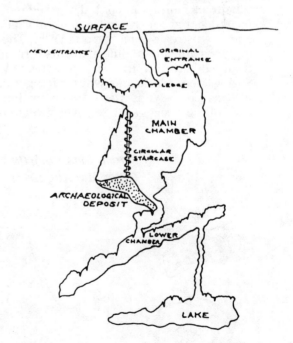

Moaning Cave near Vallecito showing burial locations.

amount of charcoal probably came from torches and not from campfires.

A second cave two miles south of Vallecito in Calaveras County is on a steep hillside overlooking Coyote Creek. This large cave was discovered by early gold seekers who had filed a claim on the spot. The cave didn't receive much at-

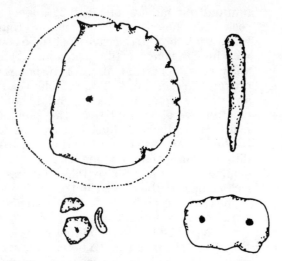

Shell ornaments and beads from Moaning Cave.

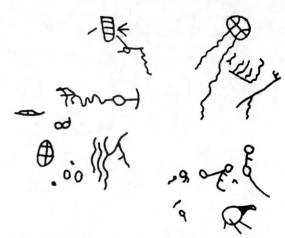

Rock art along the Stanislaus near Parrotts Ferry.

near Angels Camp are examples of impossible finds. Originally these discoveries led to speculation that humans had occupied the Sierras during an extremely early period. However, geologists could not establish any connection between the artifacts collected by miners and the spot where they were supposed to have been found. Sometimes the materials had been carried around for a while before they were reported. In other cases, artifacts were discovered in reworked river gravels. Miners once found a penknife under twenty feet of river gravel which they were forced to conclude had been worked earlier.

Bedrock mortars and pitted, worked rock surfaces occur about five miles south of Columbia and downstream from Parrotts Ferry. The petroglyphs are dot and line elements forming intriguing patterns. In two caves near Valley Springs red, black and white pictographs as well as petroglyphs were drawn by Indians. Most of the designs are a series of straight lines.

Mining remnants along the river.
(After Moratto, 1971).

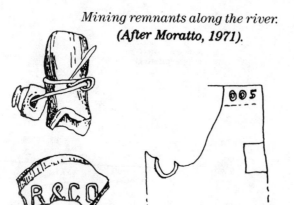

Rock art along the Stanislaus near Columbia.

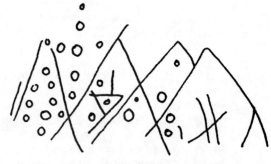

Rock art near Knights Ferry.

Geology of the Stanislaus River

Dam construction dictates that the Stanislaus River is run in two different sections. The upper portion between Camp Nine at the communities of Stanislaus and Melones at Melones Reservoir is itself divided into an "upper Stanislaus" from Camp Nine to Parrotts Ferry bridge and a "lower Stanislaus" from Parrotts Ferry bridge to Melones. Downriver there is a short four mile run between the Goodwin Irrigation Dam and Knights Ferry.

There is at present a proposal to further dam waters of the Stanislaus which would flood the upper section from Camp Nine to Melones. Rationals for the increased dam size at Melones vary from "flood control" to holding ponds for transfer of water to the Central Valley for use in irrigation. Both the Bureau of Reclamation and the Army Corps of Engineers are involved in the project. In view of the fact that the usage of water for power and irrigation goes back much further historically than recreational river running, the outlook for saving this section of the Stanislaus is indeed bleak.

Although the Stanislaus River parallels the course of other major western Sierra rivers, a difference in the geology along its length can be seen in the excellent exposures of the very old Calaveras Formation. Much of the 250 million year old Calaveras limestones have been changed to marble due to heat and pressure during later periods of time. Regardless of whether it's classified as marble or limestone, the mineral constituent is still calcite. The latter mineral is relatively soft and soluble in water, thus caves occur commonly on the upper reaches of the Stanislaus.

The most important economic aspect of the Calaveras is the use of its limestone for cement-making processes. The Columbia-Sonora limestone belt occurs over a twenty square mile area and is the largest single limestone deposit in the entire Sierra Nevada. The northern end of this massive deposit is quarried near the Stanislaus River at Cataract Quarry less than three miles from the put-in. At the quarry, the stone is crushed then mixed with water before being pumped through a seventeen mile long, seven inch diameter pipe to holding ponds at the cement plant at San Andreas. There the limestone is processed by heat to make cement. The lapsed time for the limestone slurry to make the seventeen mile run to the cement plant in the pipe is almost four hours. By using this efficient slurry method of transfer, the company avoids the trouble of trucking the stone or shipping by rail. In addition, the problem of dust is avoided.

Ancient Paleozoic shorelines which developed here over 200 milion years ago can be reconstructed to show the old ocean strand running northeast to southwest. This is almost at right angles to the current northwest/southeast trend of the Sierra Mountains.

Gold has been mined at several localities in

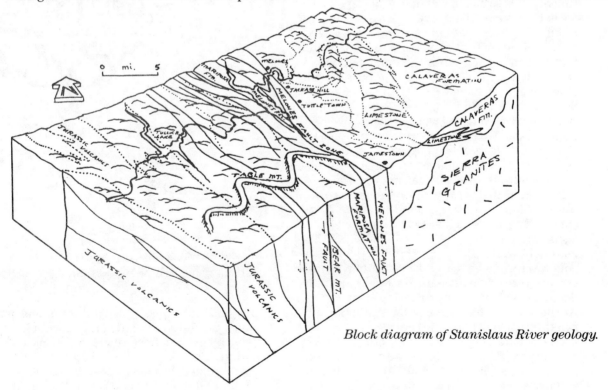

Block diagram of Stanislaus River geology.

the Stanislaus River Valley as well as from
mines within a few miles of the river. Murphys is
one of the better preserved mining camps in the
Sierra Nevadas. Here gold occurs in rose quartz
veins that have intruded the altered rocks of
the Calaveras Formation. In addition to lode
gold, placer gold has been removed from gravels
locally. Although quartz veins in this district are
disappointingly thin, they are very rich with
free (visible) gold.

Vallecito, just five miles south of Murphys,
was predominantly a placer mining area. This
district marks the site of a prehistoric river
channel designated the Central Hill Channel.
That old, main channel and several small tributaries make this a natural place for placer gold
to accumulate in the bottom of prehistoric
stream beds. Gold from this district was relatively coarse and many large nuggets are recorded.

Angels Camp, to the west of the Stanislaus
Valley, lies in an important part of the Mother
Lode belt. In the late 1880's this was one of the
richest districts in the state. Almost all the mining here was from lode gold taken from mines
tunnelled into veins along old faults. The gold
occurs in massive quartz veins that trend northwest by southeast.

A very rich and famous placer mining area
was Columbia, lying only two miles south of the
river and a few miles from the put-in. During
the 1850's and 1860's, Columbia, with a population of 2,000 to 3,000, was one of the largest cities in California. The district is on a prehistoric
river valley that was cut into the ancient erosional surface of the Calaveras limestone. Water
had eroded potholes into the limestone which
later filled with gold bearing gravels as the
older river system was buried and abandoned.
Some of these pothole "pockets" are incredibly
rich in gold. Several nuggets weighing over 300
ounces came from here as well as one monster
that was in excess of 50 pounds.

Originally the Columbia workings were discovered by a group of Mexican miners who were
driven from their claims at Sonora just to the
south. When word got around that the Mexicans had struck it rich at Columbia, they were
again driven out by American miners. Ultimately between $87 to $150 million dollars in
gold was recovered in this district. Gold bearing
gravels were originally hauled straight up out of
deep shafts drilled down to the buried potholes.
At the surface the gravels were systematically
run through long toms and sluices.

The Sonora district was the site of pocket
mining as well. Named for its discoverers from
the Mexican State of Sonora, the district
yielded some very large chunks of lode gold

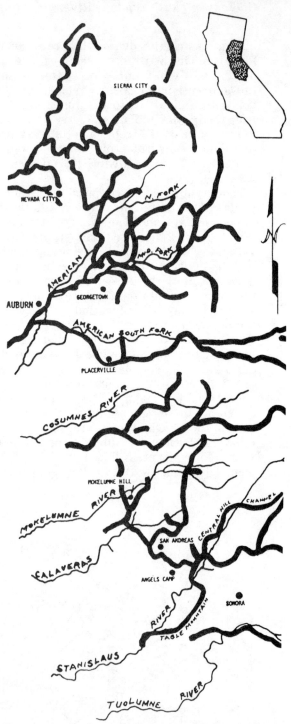

*Ancient river channel deposits containing placer
gold in the western Sierra Nevada.*

that had developed in rich pockets along quartz
veins. In addition to lode mining, gold here has
been recovered from ancient potholes in the
underlying Calaveras limestone.

Although it has never been considered a major gold mining district, Carson Hill should not

be overlooked. The Carson Hill mine sits atop a flattened peak adjacent to Melones. The latter name refers to the configuration of gold nuggets found locally in the streams which had been flattened by rocks to watermellon-seed shapes. Discoveries in 1850 at this site grew more and more fantastic attracting miners from good claims elsewhere. One vein nearly 2,000 feet long was described at the time as the "largest mass of vein gold in the world."

At the height of Carson Hill's fame the famous Calaveras nugget was discovered. Believed to be the largest nugget ever discovered in the United States, it weighed 214 pounds. Gold at this time was valued at $17 per ounce, and the nugget was worth $58,000. At today's prices, with the price of gold around $500 per ounce, the same nugget, measuring six inches wide, four inches thick, and fifteen inches long, would be worth about one and three quarters of a million dollars. For all the publicity and fever, the workings of Carson Hill played out remarkably fast. By 1858 it was a ghost town of crumbling remains.

Smaller mining operations were scattered north and south of the Stanislaus Valley. Collierville, only five miles east of Murphys, lies in the east gold belt. Here quartz veins from underlying granite rocks intruded into the local slate and schist rocks. These veins bear iron pyrite (fools gold) and free gold in small but rich pockets. Tuttletown is near the Mother Lode belt between Carson Hill and Jamestown. The area was mined up to World War II and includes both pockets of lode gold as well as rich surface placers.

Knights Ferry was an important supply center for mines and placers upriver on the Stanislaus. After the initial gold rush in the middle 1800's, the area was reworked by Chinese placer miners. Gold is found in small patches of 50 million year old (Eocene) gravels deposited in terraces as well as in buried gravels under the volcanics of Table Mountain.

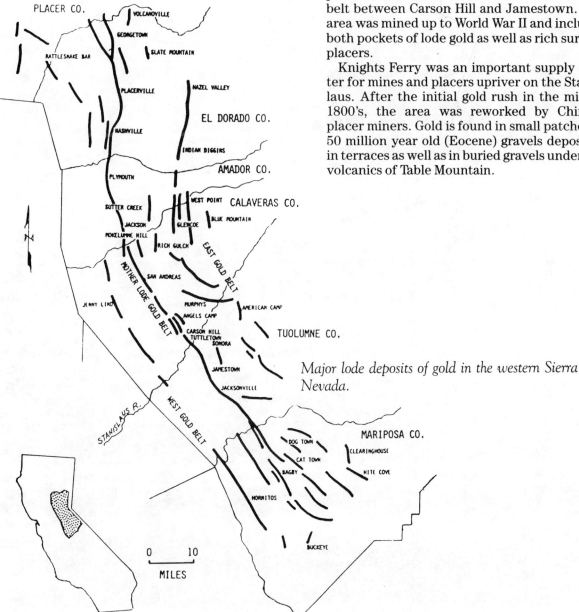

Major lode deposits of gold in the western Sierra Nevada.

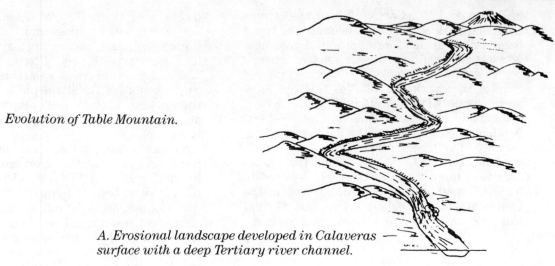

Evolution of Table Mountain.

A. *Erosional landscape developed in Calaveras surface with a deep Tertiary river channel.*

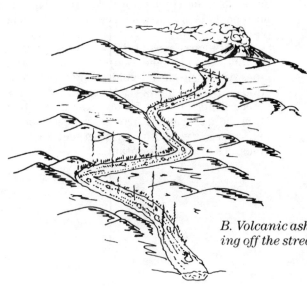

B. *Volcanic ash/lava flows into a channel, choking off the stream and hardening.*

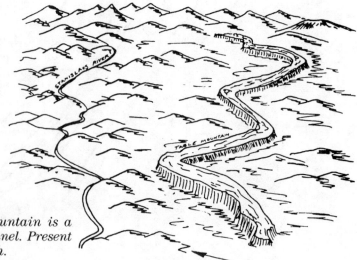

C. *Erosion takes place. Table Mountain is a solid stone cast of the stream channel. Present day Stanislaus River is to the north.*

Geology Along the Route

From the put-in, the first several miles of rocks to be seen are Calaveras schists. Schists can be recognized because they split easily into slabs and have large, shiny, mica crystals. Along this upper stretch, the schists are intruded by hard, darker rocks (ultramafics). These rocks crystallize only under conditions of very high temperature and include minerals rich in iron and magnesium.

During most of the first ten to fifteen miles the canyon walls are very steep and even vertical in places. The lower section of Calaveras appears as a grey limestone and marble. Only three miles downriver from the put-in, the Cataract Limestone Quarry is visible adjacent to the river on the southeast side.

A short distance below Abbeys Ferry, granitic rocks can be seen where they have been intruded into the Calaveras limestone. Nearby, the Columbia Marble Quarry overlooking the river from the southeast was operated in the middle 1800's for tombstones and building construction marble. Today the site is owned by the Merck Company of New Jersey, and the limestone is blasted out for the manufacture of drugs.

North and south of the river for several miles is the distinctive topographic feature, Table Mountain. This mesa lines up parallel to the river and is a classic example of inverted drainage. Heavy erosion of the Calaveras landscape 50 million years ago (early Tertiary period) left deep canyons. Large quantities of gold-bearing gravels from these canyons were deposited as placers eroded from the Mother Lode areas upstream. Deep pits or pockets cut by the streams into the limestone were natural traps for the heavy gold flakes and nuggets.

During this same period, volcanic activity filled many of the canyons with lava and ash, much of it still glowing and incandescent. This hot volcanic material quickly welded itself into a concrete-like mass that hardened in the valley choking off any further river flow or erosion. Below the ash, gold bearing gravels were also buried.

With the passage of time, the rocks adjacent to the old stream eroded more easily than the concrete-like ash filling the river canyons. The latter were left as erosional remnants or "inverted topography." Table Mountain represents the site of an old river valley now standing out in sharp relief.

Gold has been mined from gravels beneath inverted streams by burrowing in from the top or sides, but the monetary returns for this type of mining are marginal. Mining a gold vein may be worthwhile because the vein can be mapped and followed in the subsurface. Buried gravels, unless they are shallow with little covering soil, are often not economic because their exact locations cannot be easily predicted.

Ten miles south of Stanislaus, the river dissects a small "stock" or nob of granite just north of a short section of Calaveras schists at Horseshoe Bend. The end of the run is marked by the town of Melones on the Mother Lode Melones fault. At the take-out, slates of the Mariposa Formation form the lake shore. These smooth, fine-grained, dark rocks are 150 million years old (Jurassic period).

The lower run on the Stanislaus River between Goodwin Irrigation Dam and Knights Ferry is four miles long. On both sides of the river, water diversion ditches run parallel to the river. This stretch is dominated by the complex Gopher Ridge Volcanics which were formed from volcanic debris 150 million years ago. After being consolidated into rock, the unit locally was chopped up by faulting giving it a crumbly appearance in exposures.

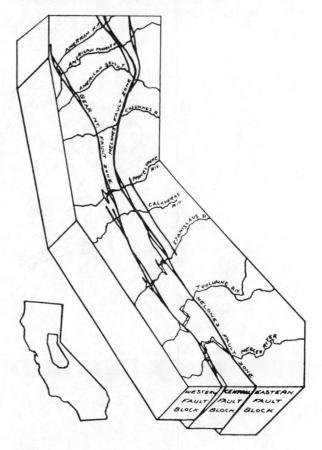

Major fault systems and fault blocks of the western Sierra slope.

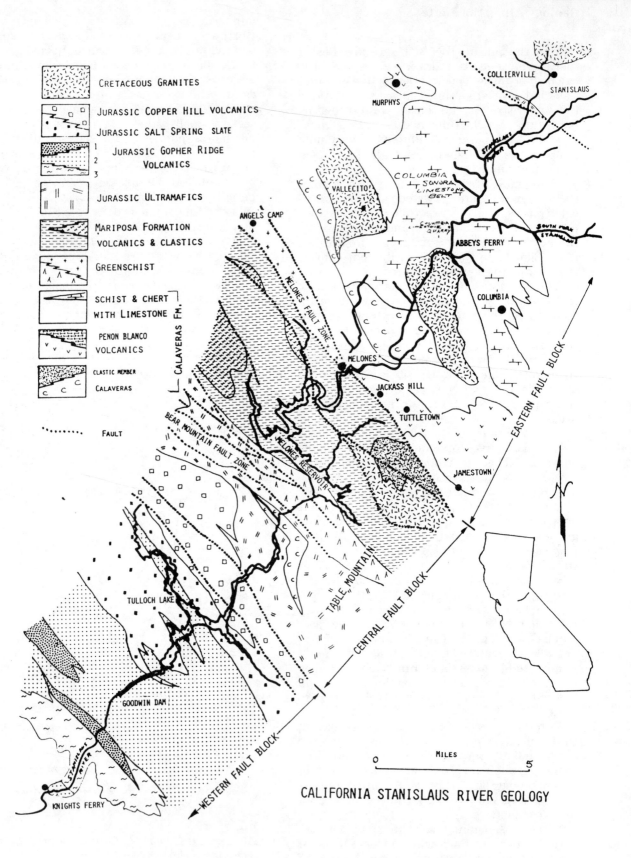

Legend:

- Cretaceous Granites
- Jurassic Copper Hill Volcanics
- Jurassic Salt Spring Slate
- 1 / 2 / 3 Jurassic Gopher Ridge Volcanics
- Jurassic Ultramafics
- Mariposa Formation Volcanics & Clastics
- Greenschist
- Schist & Chert with Limestone Fm. ⎤ Calaveras Fm.
- Penon Blanco Volcanics
- Clastic Member Calaveras
- .... Fault

CALIFORNIA STANISLAUS RIVER GEOLOGY

MILES 0 ─────── 5

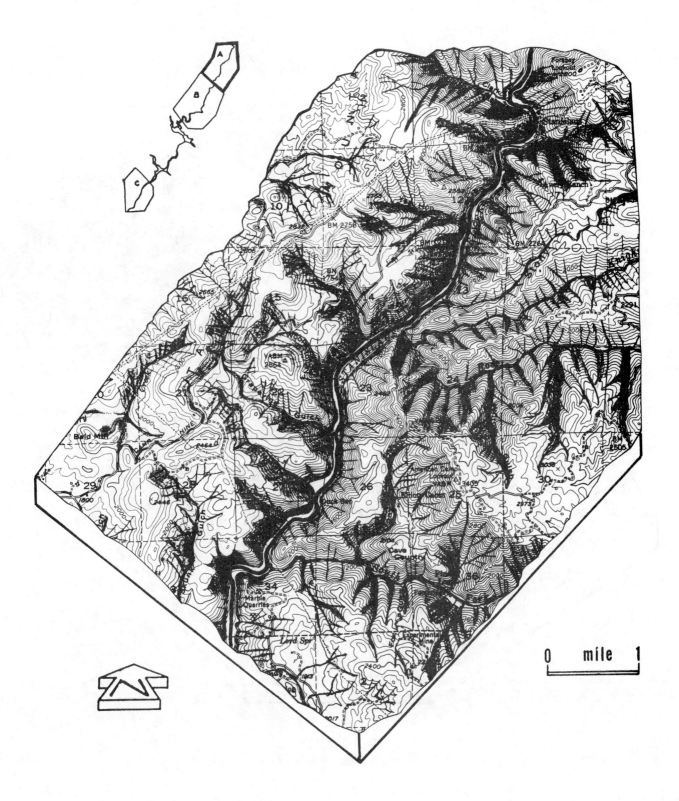

0   mile   1

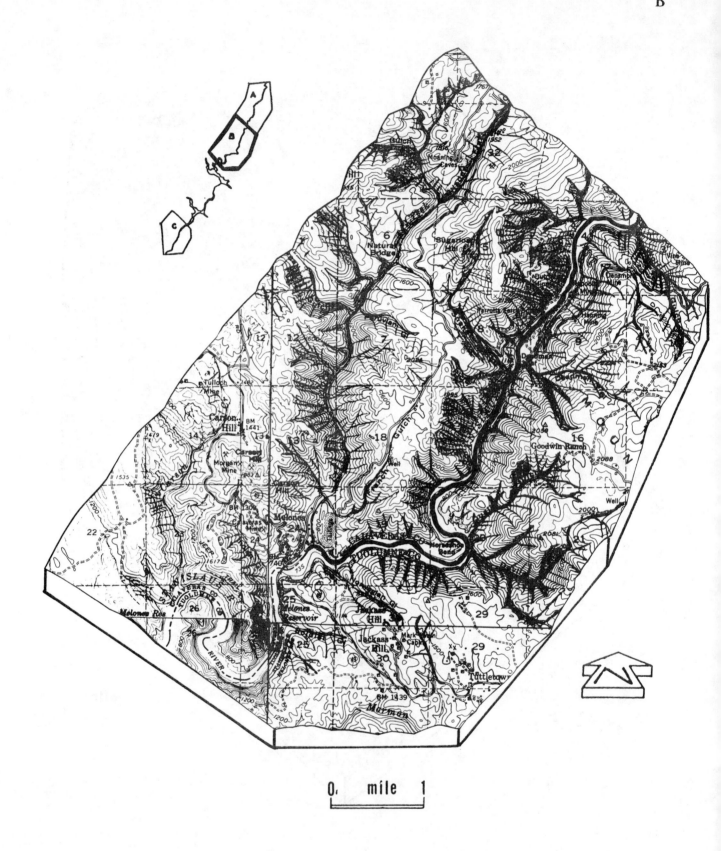

0   mile   1

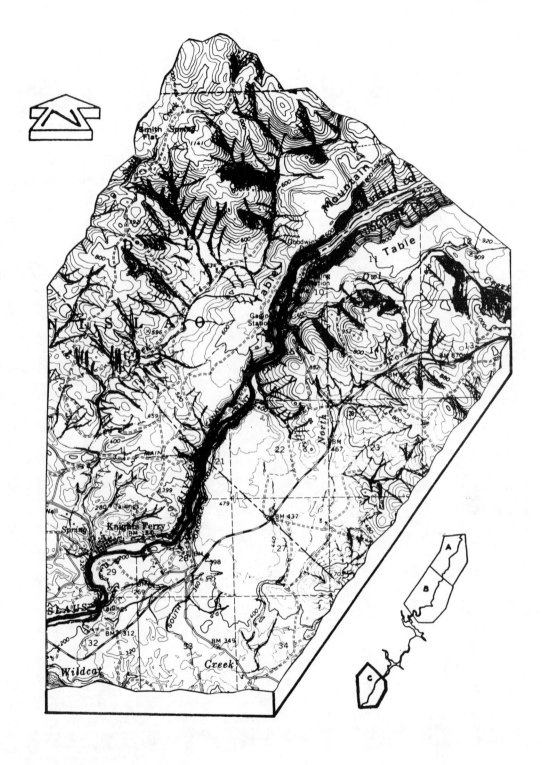

0   mile   1

Tuolumne River

(Lumsden Campground to Wards Ferry Bridge)
Tuolumne County, California
Length of run: 18 miles
Number of days: 1

Originating in Yosemite National Park in the Sierra Nevada Mountains, the Tuolumne River moves westward at great speed through deep canyons of granite and limestone. In its lower reaches it meanders past meadows and pine forests until it meets the plains of the central California valley and the San Joaquin river. Naturalist John Muir described the "glinting and singing exuberance of mountain energy" of the Tuolumne which drops 4,000 feet in four miles through the Grand Canyon of the Tuolumne. Below Lumsden Campground, the river still runs through a wilderness country. Once the forest home of Miwok Indians, the course of the Tuolumne was tied into gold mining history and white settlement after 1849.

Early History of the Tuolumne River

The Spanish and Mexicans didn't extensively penetrate the Tuolumne Valley before 1800 when Miwok Indians there began to harbor refugees from missions as well as steal horses from ranchers. It was almost impossible to track the Indian marauders once they reached the rough terrain of the Sierras. By 1820 the loss of horses reached serious proportions, and a number of expeditions were mounted against the Indians. Sergeant Jose Sanchez chased some thieves to their village on the Mokelumne, and Sebastian Rodriguez in 1828 discovered a large number of stolen, butchered horses in villages along the Stanislaus and Tuolumne rivers.

Although these incursions to trap the Indians were never successful, they opened the area to the Spanish who gave away large land grants in the San Joaquin Valley. The establishment of ranches pushed the Miwok away from the lower hills bordering the valley and into the higher Sierras.

It was during the 1820's that the first fur trappers arrived in the Sierras. Jedediah Smith probably crossed and recrossed more of the West than any other explorer. Unfortunely information on his exact routes was lost with his unexpected death at age 32 by Comanche Indians. From New York, Smith was well-educated and deeply religious. He came west in response to an advertisement seeking fur traders, a pro-

Jedediah Smith was one of the foremost trappers and explorers in the West.

fession he followed for the next ten years.

In a short time, trappers from Hudson's Bay Company in Vancouver appeared. It is not surprising that beaver colonies in the San Joaquin Valley and Sierra rivers were decimated by men operating here through the early 1840's.

Trappers were to be followed shortly by hordes of miners who came with the discovery of gold in Tuolumne County in 1848.

"Frenzy siezed my soul; unbidden my legs performed some entirely new movements of polka steps . . . piles of gold rose before me at every step; castles of marble, dazzling the eye with their rich appliances; thousands of slaves bowing to my beck and call; myriads of fair virgins contending with each other for my love — were among the fancies of my fevered imagination."

Gold fever symptoms described above were responsible for many acts, some humerous and some violent or sad. James Carson, serving in the Army in Monterey, took a furlough to the gold fields. He tells of an 1848 prospector being interred by his fellow miners who were kneeling and listening to a lengthy prayer by a minister. They noticed gold in the freshly dug dirt. At the cry of "gold", the minister opened one eye and yelled, "Gold, by God, and the richest kind of diggin's. The congregation are dismissed." All fell to, later burying their comrade in a less rich spot.

MINING LAWS
OF
SPINGFIELD

Springfield, April 13, 1852.

At an adjourned mass meeting of the miners of Springfield and vicinity; held at Mr. Mc Clure's Hall, according to previous notice,

The following Resolutions and Bounds were adopted, viz. :

BOUNDS.

Commencing at a spring above and near Vaney's steam saw mill, the line running from thence in a westerly direction to the north side of Mr. McKenny's store, from thence to a spring at the head of Dead Man's Gulch, thence following the said gulch to the Stanislaus river, thence down the river to Horse Shoe Bend, thence in an easterly direction to Mormon Creek, intersecting it one hundred feet below the Illinois House, thence up the creek to Sawpit gulch, thence up the Sawpit gulch to the Sullivan's Creek Water Company's reservoir, thence in a straight line to the spring, the place of commencement.

BY-LAWS.

ART. 1. A claim for mining purposes within this district shall not exceed one hundred feet square to each man.

ART. 2. That no man within the bounds of this district shall hold more than one claim.

ART. 3. That each and every man holding a claim within the bounds of this district shall work one day out of every three, or employ a substitute; otherwise such claims shall be forfeited.

ART. 4. That any persons holding claims with dirt thrown up or excavations made on such claims, such claims shall be governed by articles 2 and 3.

ART. 5. That each and every man holding a claim within the bounds of this district, shall designate such claims by erecting good and substantial stakes at each corner of their claims, or dig a ditch around said claim, with a notice, signed by each person or individual of a company holding such claim.

ART. 6. Where two or more claims join together, and are worked by companies, such companies can work on any portion of such claims as they deem expedient, complying with article 3.

ART. 7. That when any dispute arises concerning claims, and either party shall refuse to refer such dispute to a committee of five, two to be chosen by each party and the fifth to be chosen by the four, the party so refusing shall forfeit all right to such claim in dispute.

ART. 8. That all water claims in dispute shall be settled by and according to article 7

ART. 9. That a standing committee of ten be appointed, to whom complaints shall be made in all mining disputes, and it shall be their especial duty to see article 7 enforced, and that said committee be empowered to summon a posse at any time when necessary to assist them in the discharge of their duty.

ART. 10. That all laws passed by this meeting shall be retro-active.

STANDING COMMITTEE.
Chairman.—Mr. J. Swift, Jr.
Mr. James Eddy,   Mr. J. Roberts.
"   Shields,   " E. Smith.
Dr. Hughs,   " W. Crocker,
Mr. Vincent.   " J. Montgomery
W. Sudworth, *Secretary.*

Resolved, That the proceedings of this meeting be published in the Sonora Herald, and that 500 copies be printed in hand bills.

Resolved, That the Standing Committee have the supervision of printing the above.

J. HARRINGTON, President
W. L. Thayer, **Secretary.**

Mining districts drew up their own laws.

Hawkins Bar, below Jacksonville on the Tuolumne, was the first claim to be filed in 1849 by Old Hawkins. Somthing of an eccentric, Hawkins was reputed to fill pickle jars with gold and bury them under his cabin. Cal Gardnier, from Long Island, found miners there friendly but discouraging when he arrived. Everyone was finding some gold, but no one was getting rich. Cal and his friends, digging at some grass hanging over the riverbank, noticed flakes of gold in the roots. Thinking this was normal they discontinued, gathered their gear, and went back to camp. The next morning they discovered that the $20 they had collected from grass roots was unusual. Rushing back from town, they found their claim had already been jumped, since they had failed to leave any tools to mark the place as theirs. This claim proved to be one of the richest on Hawkins Bar. In April, 1849, the population of Hawkins Bar was 15 persons. Seven hundred people had moved there by September, but in 1851 only 150 men remained earning approximately $8 per day.

The same fate befell camps at Don Pedro Bar worked by Don Pedro Sainsevain and those on Wood Creek worked by the Reverend J. Wood of Philadelphia. Don Pedro was once one of the largest towns on the Tuolumne before shrinking to nothing and disappearing completely.

Jacksonville managed to survive the ups and

downs of the gold rush. It was founded in June, 1849, by Colonel Alden M. Jackson, a lawyer, who started a store and settled quarrels among the populace. Mr. W. S. Smart had arrived that Spring with a plow and seeds which he used to produce the famous "Spring Gardens". Smart planted potatoes, carrots, turnips and beets. He received handsome prices for his produce. A pound of potatoes cost $1.00. Vegetable dinners were $2.50. A later fruit crop of peaches, pears, apples, cherries, plums, and assorted berries added to his income. No doubt Mr. Smart unknowingly helped miners to avoid scurvy.

Mail delivery from Jacksonville to the camps was provided by the enterprising Alexander Todd. He collected $1.00 from each miner expecting mail, placed these names on a list of 2,000, and collected mail for persons on this list in San Francisco. At one time Todd was making over $1,000 a day on mail and by selling editions of the New York Herald for $8 each.

Towns were an assembly of scattered tents. One might serve as a store and one as a saloon for gambling and drinking. Whatever the combination, problems developed with the arrival of 1849 miners, "the dregs of the Irish from the eastern cities, the lazy and footloose Mexican, all the rest of the most lawless of classes who . . could . . . . obtain a passage" (Strong, p. 73). Whereas gambling and crime had been minimal in 1848, liberal drinking and the amount of loose gold carried on the person made theft tempting to those who came in 1849. Drifters were commonplace, following rumors of gold nuggets as large as eggs, and looking for gold which could be found without working for it by

Returning miner tells all.

taking advantage of those whose gold "fortune" looked more promising.

Men were shot for minimal reasons, and murder was punishable by hanging, although killing an Indian or Chinese wasn't considered murder. Prejudice in the camps was such by 1851, that it was acceptable to abuse, insult, or take a mining claim from a Chinese without any punishment.

At Sonora Camp, one of the richest placers in Tuolumne County, Chilean miners enticed George Snow to their tent where they stabbed him. Dying he called for help. The Chileans fled only to be caught in a short time, tried, and found guilty. They were taken to the spot where they had killed Snow, hanged, and buried in the grave they had prepared for him.

By contrast, many crimes went unpunished. A debtor who couldn't pay $60 was shot with a "five shooter and putting a ball directly through his chest" (Wooster, p.29). One Dr. Smead was shot in the stomach by the "contents of a French fowling piece," when he questioned one of his partners who called him a thief.

Juries were impromptu, with no qualifications necessary. Jury size varied from camp to camp, some consisting of six people, some twelve or twenty. Juries didn't have the patience for long arguments, so both the trial and judgement didn't take over an hour. An array of punishments included whipping, hanging, branding, ear clipping or banishment. Flogging of 40 to 100 vigorous lashings was standard punishment for lesser crimes not meriting hanging, and banishment from camp followed flogging. Exile could prove serious in wintertime with no supplies or equipment to earn money.

Only one woman, Juanita, was ever hung in the camps, and she was accused of stabbing a drunken miner who had decided to pay her an unwanted visit. Juanita, obligingly, flung herself off a bridge with the rope around her neck.

Arrivals in 1849 included all classes of foreigners, some mixing easily with the majority of Americans, some forming their own social groups. Germans, English, and Irish felt quite at home, whereas, Frenchmen were numerous but clannish. Mexicans and Chileans were respected for their mining knowledge and expertise but treated more shabbily than the European emigrants. Chinese were further down on the social scale, and Indians were at the very bottom.

Arrivals from China numbered 20,000 at the high point of their emigration in 1852. First arrivals were viewed humorously all "bearing long bamboo poles, wearing new cotton

*Chinese battle among themselves in Weaverville,
California, 1854.*

blouses, and baggy breeches . . . and broad bri-
med hats of split bamboo'' (Jackson, p.291).
Soon, their vast numbers, hardworking habits,
and the obvious fact they refused to adopt
American standards set them apart.

Miwok Indians undoubtedly were a source of
unease to the miners since the mines, hence the
gold, was really on Indian lands. Miners were
backed by the U.S. government with the pre-
vailing attitude that the land was conquered
and the Indians ''foreigners''. Indians worked in
the mines until 1849 when sheer numbers of
non-Indians pushed them aside.

Even if miners had been more favorably dis-
posed toward Indians, the very nature of gold
operations would have destroyed their environment. Hostilities developed, naturally enough,
when miners refused to respect native rights,
destroyed their food, and casually murdered
natives everywhere. At Big Oak Flat, south of
the Tuolumne, one '49 Quaker witnessed:

"We had been there only a few days when one
night a band of Mt. Indians made a raid on some
of the miners in the Flat and robbed them of a
horse and other valuables, killing one miner
and wounding another with their arrows. The
miners followed the Indians for 25 miles up into
the mountains, then they found their settlement, and killed old men, squaws, and children, the bucks having fled. I am thankful that I
did not join them as their acts were more foul
than the Indians.''

As the number of foreigners multiplied in the
camps, the California legislature approved a
Foreign Miners Tax in 1851 directed primarily at
Chinese miners. The tax was $20 per person a
month, and tax collectors appeared in the heavily foreign camps near Sonora in May and June.
French, Mexicans and Chileans gathered in an
armed group of 3,000 to 5,000. Americans sent
for help from nearby camps, and 500 armed
men confronted the foreigners by evening of
that day. Nervous merchants, fearing they
might lose money if the foreign population fled
the area, lobbied to have the tax reduced. The
legislature repealed the tax shortly thereafter.

With the arrival of the second wave of gold seekers, easy gold sources had dried up, and many schemes were developed to divert river water in order to expose inaccessible gold on the bottom. Nature, in the form of floods in 1849 and 1850, thwarted these efforts. On January 20, the Jacksonville Damming Company was organized to change the course of the Tuolumne River by digging a canal 2,380 feet in length, 20 feet in width, and 2 feet deep. A dam was constructed over the river at a cost of $16,000, but all was later lost to flooding.

Ferries across the river provided transportation at a price. Wards Ferry, operating between Sonora and Deer Flat was constructed of logs by Joseph Ward who charged .25 cents for persons afoot and .50 cents for horsemen. Unfortunately for Ward, he was murdered for his gold. The same fate was met by James Berger and Sam While, who took over the ferry and added a store to the operation in 1860. Their profits were kept in a baking powder can which was stolen at the time of their murder. One can only wonder at the coincidence that the Mr. Tuttle, later owner of the ferry, came to the same end.

Finally a bridge was built to replace the ferry, presumably to end the murders. However, the toll keeper and a friend were killed here in the 1870's, and the bridge burned by robbers who failed to retrieve any gold. The Tuolumne wasn't fated to have a crossing at this spot.

A toll bridge at Stevens Bar, constructed in 1857, served traffic from Chinese Camp south to Jacksonville. The road came down Woods Creek, which was forded, passed through Jacksonville, and over the Tuolumne which was also forded. At this spot, Stevens built his bridge connecting the road which reached the 2,000 miners working along Moccasin Creek.

Lumsden Bridge, up river, was not built until 1934 at the site of an earlier bridge placed along an old trail by the brothers, David and James Lumsden. The Lumsden's explored and cleared the first trails in the upper Tuolumne as well as maintained a series of ditches and flumes supplying many camps with water from the river. In 1862 the Tuolumne flooded, the rushing waters taking out most of the bridges along the way. Seventy-two to eighty inches of rain fell during November, December, and January.

By 1850 the golden dream was over. The easy gold had been found and taken out. One emigrant wrote to his brother "the stories you hear so frequently in the States are the most extravagant lies imaginable-the mines are humbug." Mining camp refugees were rag-tag. Camps were overcrowded, living costly. Gambling, drinking and murder were rampant. A young storekeeper, protesting he was innocent of los-

ing a gold nugget in his keeping, was beaten to death. A drunken miner shot his wife. "Gambling, drinking and houses of ill fame are the chief amusements of the country," Lucius Fairchild wrote his parents.

Stories of fabulous discoveries replaced those real life events. The Gold Lake story circulated through the camps originating with a man named Stoddard. It was told that Indians living around the lake used gold arrowheads and fish hooks. Gold boulders lay on the lake bottom and sides. Hostile Indians kept miners away. Stoddard led thousands of miners into the Sierras where they found a lake but no gold. There was a motion to hang Stoddard on the spot, but his protests that he was lost saved his life.

There wasn't enough accessible gold left to justify living under these conditions, and by 1851 the majority of miners took their earnings and went home. Some stayed on and took up ranching or farming. Others found that businesses begun during the gold rush were more lucrative than mining and stuck with them.

Indian Life Along the River

There are a number of places along the Tuolumne River near Don Pedro Reservoir where evidence of Indian and early American dwellings can be found. Anthropologists recorded 41 prehistoric and 26 historic sites, although a number have been destroyed by construction and mining operations.

Gambling bones are thrown onto a grass mat. Scores are counted by the position of the bones.

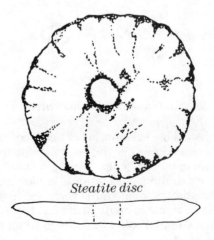

Steatite disc

"Garbage" mounds, or middens, petroglyphs, bedrock mortars along with evidence left behind by miners were typical of these locations. Indian tools included obsidian arrowheads, metates, worked tools, bone tools, and gambling bones about the size of a man's first finger.

Excavations to a depth of three feet at burial places south of Jacksonville showed them to be seasonal summer camps with no permanent structures built.

On Rogers Creek south of the reservoir, Indian material was dated from an early occupation time of 550 A.D. to when the Indians left around 1450 A.D. This camp had to have been used only in winter months. Rogers Creek is intermittent and lack of water in late Summer would hamper acorn leaching and preparation. The obsidian supply for artifacts here was in the Sierras to the east where Washo Indians lived and was no doubt obtained in trade.

A rockshelter with petroglyphs on Moccasin Creek southeast of the Tuolumne would have provided cramped quarters and minimal protection from the rain for two persons at the most and was probably not used for any length of time. An interesting steatite disc three inches in diameter and 1/2 inch thick was found in the cave. We can only guess at its use.

Rock art along the Tuolumne

A village at Wards Ferry near the take-out was occupied most of the year with plentiful water and game available. Mining remains seen here are building foundations, cables marking the crossing point of Wards Ferry, camp stoves, bottles and cans. Many artifacts showed the Indians stopped using the village when the miners arrived.

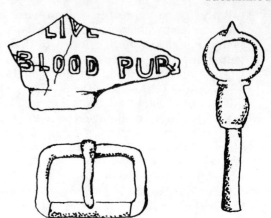

Mining relics along the Tuolumne River (After Moratto, 1971).

Geology of the Tuolumne River

The Tuolumne is a river which has just passed an interesting turning point. Very recent studies, tests, and testimony to evaluate the possibility of damming much of the river led to a decision not to disturb the canyon further at this time. The proposals were to divert much of the remaining water to aqueducts for irrigation downstream. This proposal was based on future estimated growth of the Turlock and Modesto urban areas. Additional power and irrigation needs for this development do not, at this time, exist, and the initial decision to proceed was largely political. Changes proposed for the river took the form of power dams to create reservoirs and diversion tunnels (aqueducts) to "dewater" the upper reaches of the watershed allowing only enough water in the stream valley to maintain fish stocks. At present, the Tuolumne River supplies Los Angeles and San Francisco with portions of their water year-round.

It is interesting to note that the Tuolumne drainage was the scene of one of the earliest Sierra Club struggles led by John Muir himself. Muir's efforts to prevent development of the Hetch Hetchy Valley water system began in 1901. The eventual passage of the Raker Act in 1913 assured the construction project.

The runnable stretch of the Tuolumne extends for eighteen miles from Lumsden campground to the waters of Don Pedro Reservoir at Wards Ferry Bridge. The gradient of the Tuolumne here is only 40 feet per mile, but along much of the river the channel lies between narrow canyon walls of resistant rock making the current very fast and turbulent. Because of this, several of the rapids are difficult even at low water.

The Tuolumne flows from east to west across the west face of the Sierra Nevadas. The west slope of the Sierras is divided into a series of fault terranes referred to as the eastern, central, and western blocks. These massive blocks are oriented northwest by southeast, and the major stream channels flow across the blocks at right angles. As the stream moves from east to west, it cuts progressively younger rocks. The oldest rocks exposed lie just west of the Sierra crest and are over 200 million years old. Further west, younger rocks of 150 million years in age form the western flanks.

Although the Tuolumne cuts across all of the major western Sierra geologic features, the runnable portion of the river is restricted to the Sierra Nevada Batholith and the eastern fault block almost up to the Melones fault zone. The latter is one of the most important geologic features of the western Sierras for its role in the natural occurrence of gold.

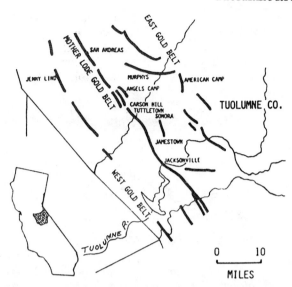

Major lode deposits of gold in the western Sierra Nevada.

Like the famous San Andreas fault, the Melones fault is not a single cohesive structure. Instead it is series of small overlapping faults somewhat like a deck of cards. The fault was active over a considerable period of time 140 million years ago. During and immediately after this period, the fractured rocks of the fault itself became a route for hot, mineral laden, volcanic fluids which rose from deep in the earth, cooled, and precipitated in veins. Gold was one of the minerals deposited in the veins near the surface.

Earthquake activity never involved the entire length of the fault. Instead pressure was relieved along the fault here and there over long periods of time. Tracing the Melones fault along the western Sierras is in effect to map out the Mother Lode of Sierra gold. In addition to gold in the fault itself, there are amounts of gold in the Sierra granites just to the east. Erosion of the Melones fault and the granites during the Tertiary period around 50 million years ago removed the gold and redeposited it downstream to form the placer deposits. Gold is very heavy, almost twenty times heavier than water. When it was eroded and entered the streams it did not go far even in the turbulent, high velocity torrents of that period.

Although the length of the Tuolumne has been prospected, the principal gold districts are at the east end near Soulsbyville and at the western end near Jamestown. Jamestown was initially worked for its placer deposits, but most of the gold it eventually yielded was from sedimentary rocks which had been changed by heat and pressure (metamorphism). Massive quartz veins, some as much as 10 feet thick, contain

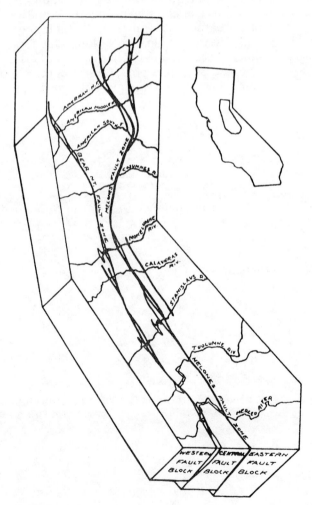

Western, central, and eastern blocks bounded by the Melones and Bear Mountain faults.

The local geology here can be pictured simply as molten rock deep in the ground working its way upward into the rocks of the Calaveras Formation. Much of the local gold can be found in veins of quartz up to 5 feet thick that have forced their way into the Calaveras with the molten granites. In addition to gold, sulfide ores, mainly lead, occur in some abundance.

Beginning in 1849, Chinese Camp in western Tuolumne County saw intensive mining by Chinese nationals. Dominantly placer mined, the district bears extremely rich quartz gravels. Chinese Camp and its buildings are maintained today as an historic center.

Geology Along the Route

The Tuolumne River Valley provides an excellent traverse from granite rocks of the Sierra Nevada Batholith through the Calaveras Formation and up to the Melones fault. At the Lumsden campground put-in, the granitic rocks of the batholith are obvious for their gray color and crystalline texture. Along the first few miles, large boulders of hard, black chert in the riverbed are apparent. These boulders either came downriver or are out of the overlying Ca-

Ancient river channel deposits containing placer gold in the western Sierra Nevada.

quantities of gold. The total output of the Jamestown district was approximately 9 million dollars.

Like Jamestown, the Soulsbyville district was initially placer mined before rich gold ore was found in a lode deposit in 1858. Of the entire east gold belt of the Sierra Nevadas, the Soulsbyville district is the richest having produced nearly 20 million in gold.

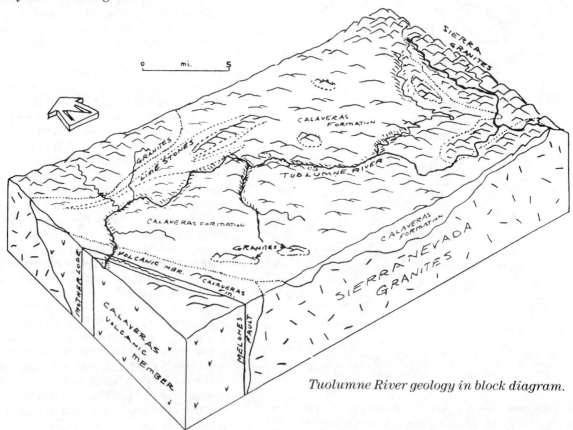

Tuolumne River geology in block diagram.

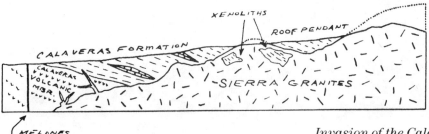

Invasion of the Calaveras Formation by the Sierra granites.

laveras Formation in the canyon above.

Also visible in this initial stretch are veins of quartz and feldspar in the granites. Quartz is one of the last minerals to cool and crystallize in a granite so that it is found in veins which filled up last and hardened. Because gold has roughly the same temperature of crystallization as quartz, these veins are often rich in gold.

In sharp contrast to the grey granites and milky white quartz veins, darker masses of rock are visible in this interval in the canyon walls. These darker rocks are pieces of the original surrounding rock into which the granites forced their way. Often chunks, "xenoliths," of these darker Calaveras Formation rock are visible as small pieces that were frozen by crystallization before they could be completely absorbed by the molten granite.

At the confluence of Clayey River and the Tuolumne, bedded cherts can be seen in the river channel walls. The origin of this exceedingly hard, flinty rock is not well understood. It is either precipitated directly from the ocean as a fine-grained rock, or it is converted to chert from some other parent rock by fluids after the original rock was deposited. Cherts are part of the varied rock types which make up the Calaveras Formation.

Rafting eight to ten miles down river from the put-in, basalt dikes of a dark black color begin to show up in the granites. These dikes are intrusions of lava which occurred after the granites had cooled. Further on, limestones and marble of the Calaveras Formation appear in boudinage structures. Boudinage is a sausage-like "pinched" effect in folded contorted rocks that develops where the rock has been deformed under very hot conditions. In such an event, the layers being folded become semiplastic and are pulled apart to develop a series of pinched folds through its length. When the rock is sliced through, as by a stream, it looks like a string of attached sausages.

After drifting 13 to 15 miles into the canyon, quantities of banded, bluish marble stand out in the valley walls. This marble was originally Calaveras limestone which has been cooked by contact with hot granites. The marble is quarried elsewhere for decorative stone to face buildings. At the Wards Ferry take-out, an outcrop of mica schist can be seen. The latter was originally mudstone or shale that has been heated to such a degree that the mica begins to form large crystals again. On a freshly broken rock face the schist sparkles as it reflects sunlight.

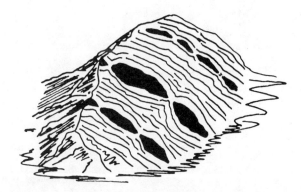

Limestones of the Calaveras Formation have been stretched and pinched off (boudinage structure) in the process of being converted to marble by extreme heat from the underlying granites.

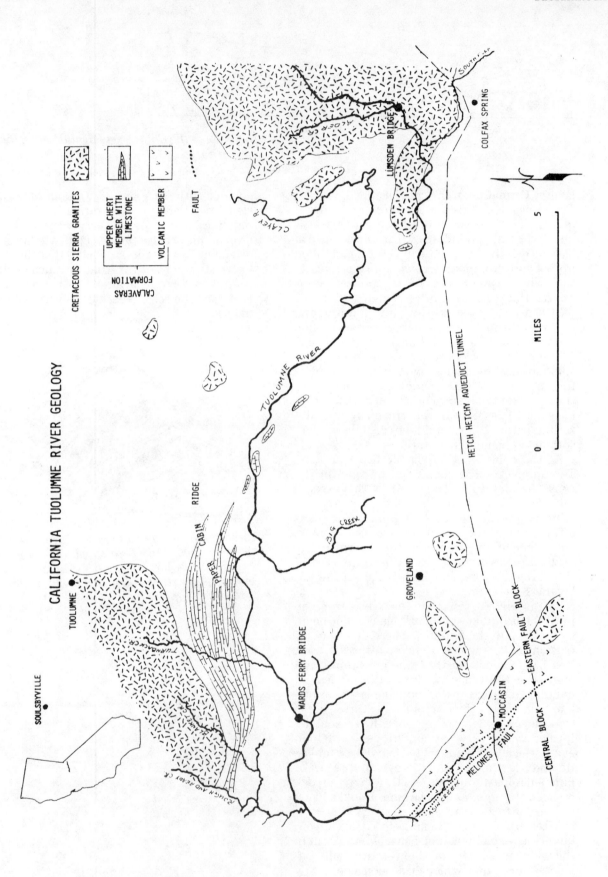

CALIFORNIA TUOLUMNE RIVER GEOLOGY

CRETACEOUS SIERRA GRANITES

CALAVERAS FORMATION

UPPER CHERT MEMBER WITH LIMESTONE

VOLCANIC MEMBER

FAULT

MILES

0   5

312

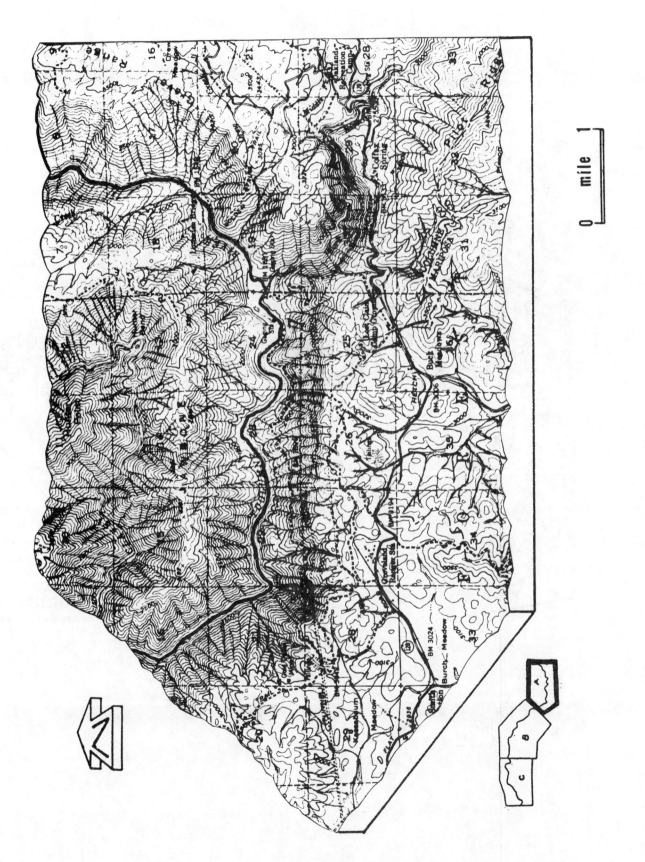

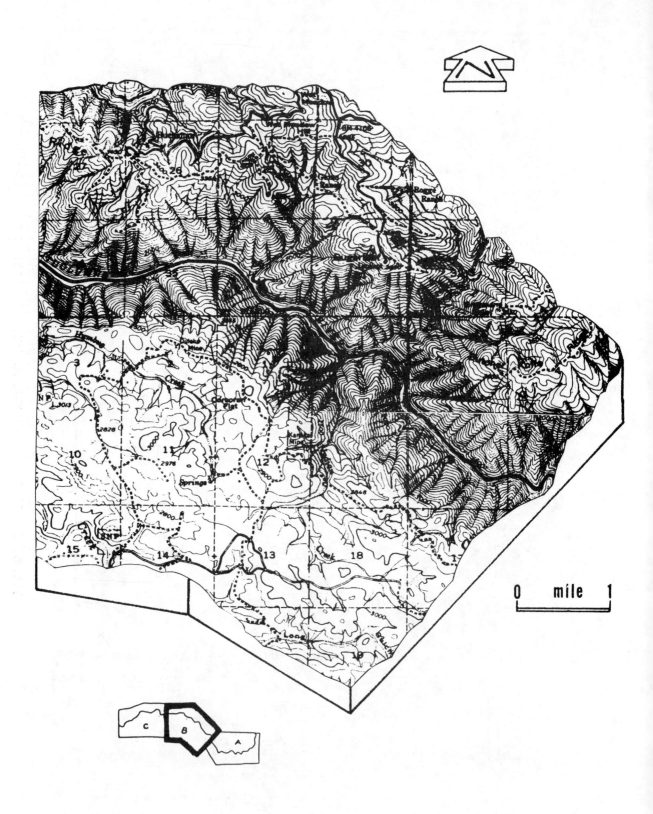

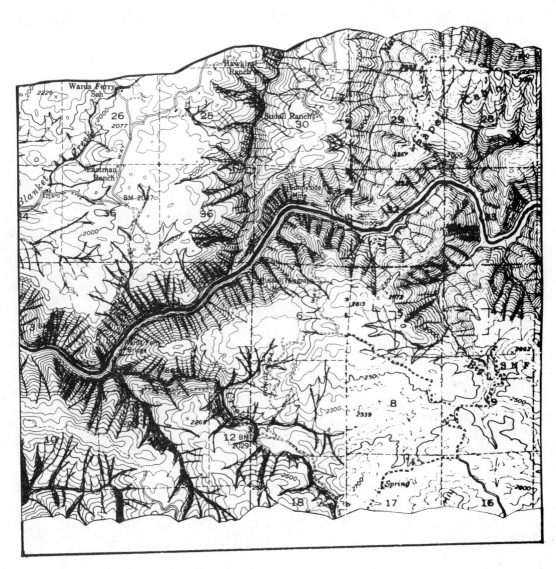

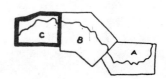

The Water Environment

Water and the River

Doing a river in a raft for the first time is frightening because whitewater is not an environment we normally encounter. After a few days on a river, however, we quickly learn by experience some of the events to anticipate within certain situations. Often at high or turbulent water, even highly experienced guides will receive surprises from the river.

The Erosion Cycle

One of the most elegant ideas to ever come out of the study of rivers was the "cycle of erosion". Very briefly this theory suggests that a landscape undergoes an erosional cycle from youth to maturity and finally old age as it is progressively attacked by erosive forces of rivers and streams.

Rapidly flowing, youthful streams cut sharply into flat landscapes leaving steep-sided, V-shaped canyons and features such as waterfalls and rapids. Mature landscapes are characterized by maximum relief and deep canyons. Streams in the mature stage have thoroughly dissected the flat plateaus and are just beginning to meander back and forth. In the old age stage downcutting streams have given way to lateral erosion as the meandering stream swings back and forth across an ever widening floodplain. At the latest stage of old age, streams may be choked with their own sediment giving them a braided appearance in map view.

Problems arose with this nomenclature of youth, maturity, and old age when it was realized that a single stream must be several ages at one time throughout its length. For example, it could be youthful near its headwaters, mature in its middle, and old age near its mouth. The cycle of erosion scheme is used here because it is easy to understand, and a single phrase suggests a complete picture of the local features of a stream.

Some Characteristics of Moving Water

Understanding a few easy principles of water movement will help to show that riffles, eddies, and meanders are not as complicated

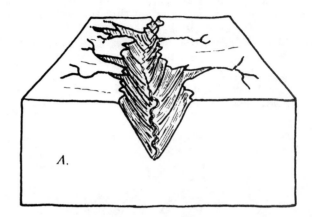

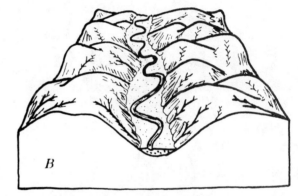

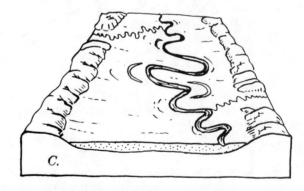

Life cycle of a stream from youth (A) to maturity (B) and old age (C).

as they may appear. Water flows by gravity. Its velocity will be a function of the gradient or number of feet drop per mile. To understand flow, it is well to first look at meanders. The exact origin of these looping patterns is not fully understood even by experts. In one sense, meanders may be a way the river has of temporarily storing the moving water in the watershed to wait its turn at the main channel. Similarly in order to lay out a rope, if you could not lay one coil on top of another, the best solution would be a series of looping curves. The line of maximum water velocity along a meander series crosses the channel after each meander cutting the outside of the meander bends much like a race car on a track trying for the highest speed possible. By following this pathway, experienced guides are able to move a raft through fast moving water in a meander with barely a stroke on the oars.

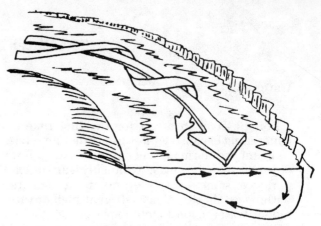

Corkscrew pathway followed by currents through a stream meander. Erosion bank is on the right, deposition bank on the left.

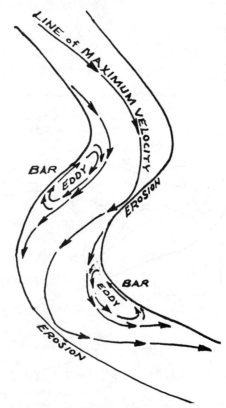

Line of maximum velocity and eddies along stream meanders.

sition of sediment and rock by water are related to water velocity. Increased speed of the water will erode and dislodge even massive boulders. When speed is at a maximum, on the outside of the meander bends, a cutbank or erosion cliff is created. Velocity is at a minimum on the inside of the meander, where sand and gravel are deposited as clean beaches and bars. Movement of sediment downriver is primarily a process of the debris being passed from one meander bend to the next.

This picture of lines of maximum velocity is only half the story, however. Within the water mass itself, the meandering channel sets up a rolling or rotating cycle. The primary effect of this is that as water moves into a meander it rolls over and downward toward the outside bank. For the rafter, this means that the river thrusts the raft into the outside of the meander wall with a strong undertow at the wall. Occasionally the results are disasterous.

Looking at the maximum velocity diagram, we can also understand the erosional/depositional pattern of the river. Erosion and depo-

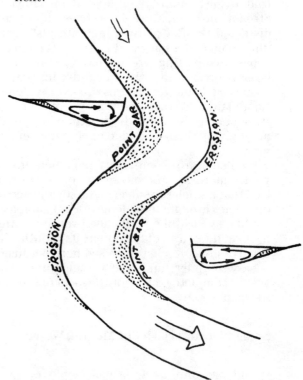

Deposition and erosion sites along stream meanders.

Another intriguing facet of meander development is the "rule of ten channel widths." This observation is simply that a given stream or river will begin a new meander turn within the distance of ten stream widths. A stream with a channel 100 feet wide will proceed in a straight path no more than 1000 feet before turning.

There are two exceptions to this rule. The first is that the river meanders too often, and the second is that the river proceeds straight for a distance beyond 10 widths. These possibilities suggest the stream channel is responding to "geologic structure." This means that something else is controlling the river's pathway such as a fault, fracture, folded rocks, or hard and soft layers.

The stream surface displays turbulence in several ways. Eddies are one of the commonest signs of turbulence. An eddy is a situation where a smaller side current of the stream turns back on itself and flows back upstream. Eddies may be vertical as well as horizontal. Horizontal eddies develop where a

Vortex action from rough streambeds generating standing waves as well as a vacuum effect to lift rocks from the stream bottom.

water mass is moving downstream at a rapid rate when the flow is constricted or blocked. Water in the channel is forced through the gap at an increased speed, and the remaining water moves backward into an eddy. The water is, in effect, "waiting its turn" to move through the main channel. Eddies occur at every meander as well as at areas of constrictions. Vertical eddies are setup as turbulence where large rocks block the stream and impede flow. In this way eddies develop in front and in back of obstructing rocks.

Turbulence is set up by any obstruction in the bottom of the river channel. A vortex or spinning action thus generated by irregularities in the streambed has tremendous erosive power and is able to lift large size particles straight up through the water. At the stream surface the vortex appears only as a small riffle or as a large standing wave.

Rapids form in several ways. Basically a rapid is a local area in the stream where speed increases markedly. A narrow channel or steeper gradient will increase the velocity. Any obstruction in a channel that restricts flow will create rapids. Often just the presence of large rocks in a stream channel will generate rapids around the rocks. Landslides into a stream will produce rapids, but the most common cause of rapids is smaller creeks coming into the main channel. Water in the smaller creek flows faster and may move larger stones. This tributary acts much like a landslide, dumping sediment and water into the main channel. This mass of rock and surplus water will clog the main channel to create rapids.

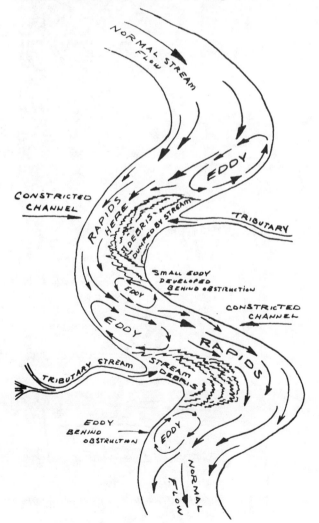

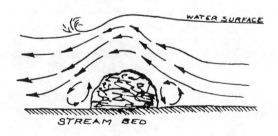

Horizontal eddies developed around a large stone on the streambed.

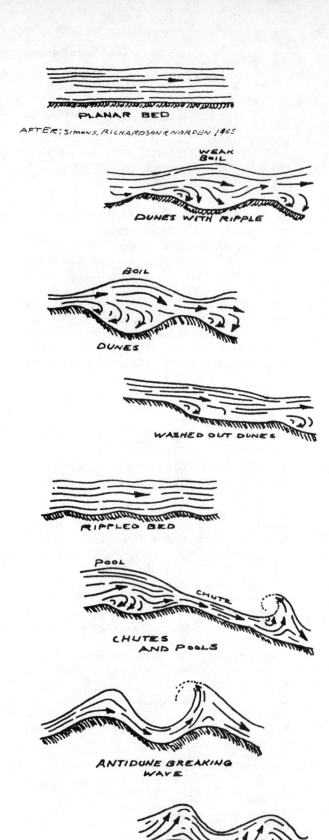

PLANAR BED

AFTER: SIMONS, RICHARDSON & NORDEN 1965

WEAK BOIL

DUNES WITH RIPPLE

BOIL

DUNES

WASHED OUT DUNES

RIPPLED BED

POOL

CHUTE

CHUTES AND POOLS

ANTIDUNE BREAKING WAVE

ANTIDUNE STANDING WAVES

Wave types generated by streambed roughness, gradient and velocity.

Often the local geology will cause rapids. For example, if the stream moves from soft sediment into a hard rock, the channel will narrow down in the hard interval, and water will pick up speed causing a rapid. Erosion patterns of a river will depend on the rock being cut. Harder rocks support steep gradients, whereas softer rocks have more gentle gradients.

Drainage patterns of streams, when viewed on a map, tell a great deal about the gradient and rock type being cut. A dendritic stream pattern — like the branches of a tree — results when the stream is cutting uniform, flat-lying rocks. In such a pattern, the stream is permitted to develop its own tributaries with little influence from the local geology.

Stream drainage patterns

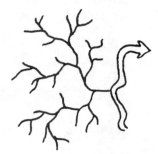

A. Dendritic drainage — homogeneous, flat-lying strata.

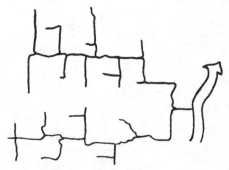

B. Rectangular drainage — jointed rocks broken in long, even fractures.

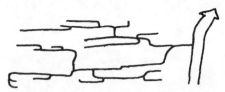

C. Trellis drainage — rocks, folded and eroded to expose hard and soft layers, yielding long valleys and ridges.

Other stream patterns are not as symmetrical as this dendritic pattern. Drainage patterns, developed in rocks that are highly jointed or fractured, form a "squarish" pattern called rectangular drainage. Trellised drainage is produced by a stream as it follows hard and soft layers in between layers of up-turned, folded rocks.

The shape of a river valley is also controlled by the type or rock being eroded. As mentioned above, soft rocks tend to develop wide open valleys with gentle slopes and rolling hills, whereas, very hard rocks have deep channels, steep slopes and cliffs. The weathering pattern has a rather strong effect here. In a humid climate where weathering is thorough, rocks will decompose and erode out into smooth, rounded slopes. Where dry conditions prevail, the same rocks may support steep slopes, cliffs, bluffs, and deep channels.

Erosional patterns of the same rocks in humid (A) and arid (B) regions.

As a stream proceeds to carve and shape a valley, it is both depositing as well as eroding. Loose sand and gravel in the inside of meanders will be in smooth, flat-lying deposits. In addition to side-to-side meandering, the stream is down-cutting. If the down-cutting process gets ahead of side-to-side erosion not all of these old gravel deposits are flushed out leaving some of them high and dry as terraces along the canyon walls. In any given valley there may be small pockets of gravel left as remnant terraces clinging to the valley wall hundreds of feet above the river water. In many old valleys, these pockets of gravel are actively explored and mined out for their gold content. Terraces are easy to see from the river because of their flat tops and abrupt snouts.

Although the gravels and sands deposited in a riverbed and on its bank appear to be chaotic, they are actually deposited according to rigid patterns. Most river deposits show "grading." That is, they blend from coarse, large pebbles at the bottom to small, fine pebbles or sand at the top. This grading may be seen across a single channel or bar. The size of pebbles moved by a stream is dictated by the velocity of the flow.

Another curious effect of river deposition is pebble imbrication. Flat pebbles will tend to be imbricated or shingled, lying one upon the other. Imbricate pebbles invariably tilt upcurrent or upstream. Looking at the pebbles in a stream, most are slightly flattened like a squashed football. Rounded, perfectly spherical pebbles are actually a rarity. Such pebbles wear down faster because they are more easily tumbled along by the current. The flattened pebble tends to stay put in anything but a very strong current thereby slowing its wear. Most pebbles have three separate dimensions.

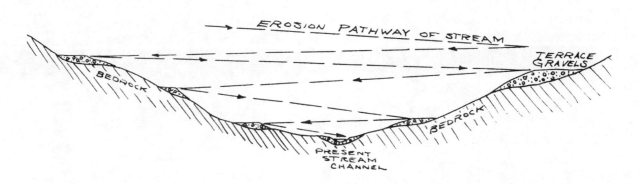

The size of pebbles also tells something of their history. Very small pebbles of one rock type, for example basalt/lava, may mean the rock wears easily or that the source for the pebbles is many miles upstream. Often in mapping a region, geologists will simply follow pebble types upstream to their source. The pebbles in a streambed are, in effect, an inventory of all the varieties of rock in the watershed. Examining streambeds for rock types is often more productive than climbing to the tops of peaks. In addition to being cooler than in the dry canyons, streambeds don't have the annoying soil cover, and the rocks are fresh and unweathered.

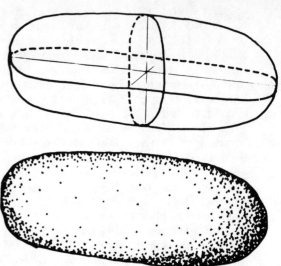

Stream pebble geometry with three axes each of different length.

Shingled arrangement (imbrication) of pebbles in a streambed. Darkened pebbles display imbrication. Arrow shows current direction.

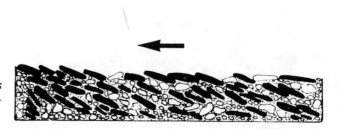

Gold Mining

Gold has unique chemical and physical properties which influence mining and recovery operations. The four fundamental characteristics of gold to consider are the weight or specific gravity, its temperature of crystallization (cooling), the reactivity of the metal, and its workability.

The most striking characteristic of gold is its weight. Almost twenty times heavier than water, gold is twice as heavy as most other metals. More important, however, is that gold is almost ten times heavier than the common mineral quartz. Because of its extreme weight, gold quickly separates in any kind of medium, such as water, sinking to the bottom of streams.

Gold concentrates in channel bottoms in areas where turbulence of the water is low. Most mining techniques exploit this characteristic by seeking the metal in the deepest crevices and cracks where it accumulates. Nuggets in placer deposits will usually be found in association with much larger stones of equal weight. Gold pans, sluices, and long toms all capitalize on the heavy weight of gold and its tendency to be trapped along the very bottom of a moving stream of water.

Large nuggets invariably occur in coarse gravels often out of reach of the small mining operation. Large scale dredging operations are geared to handle gold in the smaller flake and dust sizes but are not set up for the really large nuggets. Massive nuggets several inches across have, on occasion, been discovered cast upon gravel heaps by gold dredges. Although dredges could be adapted to handle larger nuggets, these are so rare that it is economically better to mine for the finer particles in any dredging operation.

Another important characteristic of gold is its low temperature of crystallization. Long after most other minerals have cooled and hardened, gold and quartz are still in a liquid state. When these two minerals finally crystallize they do so in the only available space left — in veins and cracks. If gold were to crystallize at higher temperatures, it would be evenly scattered or diffused through rocks making it much harder to mine.

Most large gold nuggets were originally portions of solid veins of the metal that have been freed by erosion then pounded and shaped by other pebbles in streams. The first clue that lode (vein) gold may be near is the presence of thick quartz veins. Other low temperature minerals that are a tipoff to gold are potassium, feldspar (often a pleasing pink color), and white mica (muscovite). As a granite mass cools, the quartz, gold, potassium feldspar, and mica frequently crystallizes at a very late stage in deposits known as pegmatites. Other minerals that occur in the pegmatite are tourmaline and garnet.

Gold/quartz veins are often confined to the periphery of a granite mass. These veins are forced out into the surrounding rock along fractures and faults. Any cracks or breaks in the rocks around the granite mass may eventually receive these late-stage cooling fluids as an end product to granite crystallization. Faults are a common recipient of these quartz/gold veins. Because most rocks fracture according to a predictable pattern, the mining of vein gold in faults and fractures is conducted according to well- defined structural rules.

Yet another characteristic of gold is its low reactivity with other chemicals. The noble metals, gold and silver, are so called because they are slow to react with even the most caustic chemicals. As rocks decompose by ordinary weathering processes, the high temperature, dark-colored minerals (those rich in iron and magnesium) break down first due to their instability. White mica and quartz are last to decompose releasing gold. Gold reacts only very slowly and thus enters the stream system in a metallic form. There it continues to resist chemical reaction, and the shiny yellow, metallic luster of the metal is often the first clue to its presence.

Finally the ability of gold to resist shattering impacts in the turbulent fluvial (stream) environment prevents it from being pulverized to a fine powder. Gold is remarkably malleable. It can be worked and flattened into extremely fine sheets (leaf). Nuggets and gold flakes invariably have dents and rounded edges from the pounding they absorb from adjacent stream pebbles and cobbles. If it were not extremely malleable, gold would be easily shattered into tiny fragments.

Gold will form an "amalgam" with mercury. The latter metal mixes with and "wets" the gold. Often very fine flour gold is extracted from sand by mixing in mercury. Gold sinks to the bottom of the mercury, and the sand floats to the top where it is skimmed off. Squeezing

Long tom with a rocker and pan (Siskiyou County Historical Society).

the gold/mercury amalgam mixture through a chamois bag will then remove most of the mercury.

The physical/chemical characteristics of gold contribute to many unusual occurrences of the metal, but most can be classified as either lode or placer deposits. Lode gold owes its presence almost entirely to the late stage cooling (crystallization) of molten rock. Hot fluids from a volcanic source will invade the surrounding rock to deposit quartz and gold along any possible fractures.

Mining lode gold is a process of identifying these late stage crystallization areas around the volcanic rocks. Once located, the areas are prospected by seeking out the veins, faults, and fractures that gold laden fluids might have invaded. Gold-bearing veins are mined by driving tunnels into the rock to remove the ore. After graphically projecting a vein into the ground, vertical shafts are then driven into the rock to intersect the vein at some point underground.

Once located, the vein is systematically mined out.

Placer gold has been freed from the rock in which it crystallized to be redeposited in a stream. This gold is located by predicting from its heavy weight where it would be located in a streambed.

Quite naturally, placer gold is usually more dispersed than lode gold, yet incredibly rich, gold-bearing gravels are occasionally discovered. Although placer gold is more scattered than vein gold, it has one primary attraction for miners. The placer gold-bearing rocks have already been broken down by nature. Lode mining ordinarily requires the difficult process of driving tunnels into hard rock. The accompanying hazards of tunnel collapse, the use of explosives and flooding make lode mining a high risk, expensive occupation. While placer mining is not without hazards, these can be minimized.

Placer gold in streams can be expected to accumulate at any point where the current drops off enough to deposit the metal. River bars may have flakes or even nuggets in some of the coarse gravels, but the largest nuggets are to be found along the stream bottoms in any natural trap such as a crack or low spot.

In addition to modern streams, gold is found in older abandoned deposits of prehistoric streams. As a river carves out a valley, it often leaves behind terrace deposits abandoned as the stream eroded to lower levels. Often these ancient stream gravels high up in the canyon walls are rich in gold.

Today terrace placer deposits are still mined out hydraulically by directing a powerful stream of water at them to loosen and wash out the gravel for processing. Many of the richest gravels in California Sierra rivers were covered by a thick layer of volcanic mud or even lava that had filled and choked off the stream. Miners extracted these gold-bearing gravels by either tunneling in from the side or with shafts straight down through the volcanics. Gravels located in this way were brought to the surface to be processed for their gold.

To separate gold from other minerals and rocks, the process is usually a variation on the gold pan strategy. Panning gold is a simple oper-

ation of swirling water and a handful of gravel and sand around in a pan to wash out the lighter rock fragments and minerals. The heavy gold is left to accumulate near the bottom of the pan.

Often black sands of the mineral magnetite occur with gold. As the panning process removes the lighter minerals such as quartz, the black sands are left at the bottom with the flakes of gold.

A sluice for gold works on the same principle as the gold pan, but instead of swirling water in a circle it travels down a wooden channel. Gold-bearing gravels dumped into the sluice at the upper end are washed through the trough to give the heavier gold the chance to separate at the bottom. As the lighter gravels are carried off by the water, the gold is trapped behind wooden riffles built into the bottom of the sluice trough. A sluice may be operated for some period of time before the flow of water is stopped and each riffle carefully cleaned to remove the gold. Often miners took a short cut to riches by helping themselves to their neighbor's sluice. Usually robbing sluice boxes was punishable by hanging.

A cruder variation of the the sluice was the long tom. This was an elongate trough with a

Logan Placer Mine, No. 396, near Waldo, Oregon, in the late 1800's (Ramp, 1979).

Stamp mill (Calif. Div. Mines).

mill-like device that pounded the ore between rotating stones. When gold became more and more rare, eventually the stamp mill was devised to break up and pulverize gravels. The stamp mill is still popular today and consists of a row of heavy trip hammers that are raised to fall by gravity on the ore.

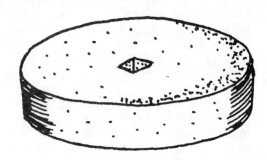

Arrastra — a device to crush gold-bearing rocks.

perforated iron sheet at the base underlain by a riffle box. Gold-bearing gravel dumped in the upper end washed through the iron sheet to trap the gold in the riffles below. Both the sluice and long tom required a steady stream of water to operate.

In areas where water was limited, as it often was, the cradle was used. This consisted of a hopper for the gravel that fed into a rocking box below with riffles. With all the ingenuity which went into these devices, it was the simple gold pan that performed the final separation of the gold.

Like any sedimentary particle, gold comes in various sizes from nuggets to flakes and finally dust or even very fine flour gold. The vast amounts of gold mined in placers were extracted as flakes or dust. Often miners would extract gold-bearing quartz gravels from placer deposits then crush the quartz pebbles to free gold trapped in the quartz. A common device to break up quartz was an arrastra. This was a grist

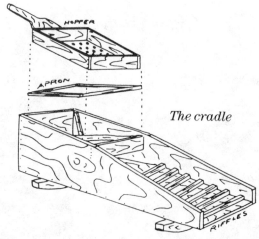

The cradle

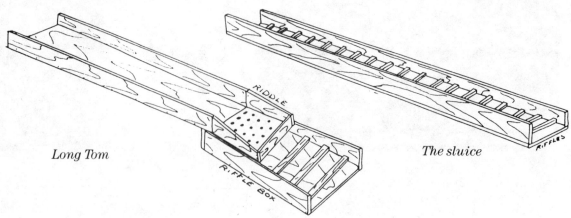

Long Tom

The sluice

Pictographs and Petroglyphs

Rock Art

Rock art is fascinating because it represents the symbols and ideas of one person which have come to us from hundreds of years in the past. There has been much speculation about what the artist meant by the drawings, but unfortunately this may never be known.

". . . what a spiral, a snake, a lizard, or a grid design actually signified in the maker's mind at the time he fashioned it we are not, and probably never will be able to say (Heizer and Baumhoff, p.239).

Rock art designs are made up of individual elements. The meaning of these elements has been interpreted in many ways, and no one specific meaning can be attributed to one element. To add to the problem, the same design element may have a different meaning in a different area. Anthropologists feel "even with the help of Indians" it would be difficult to correctly understand the meaning of these elements. Meanings might differ from tribe to tribe and from person to person.

"It is obvious that all such aboriginal carvings and paintings were simply a medium for expressing some emotion over a spiritual or concrete event. That the primitive artist knew what was intended and his friends or tribal group understood the significance of the pictures cannot be doubted . . . It is equally certain that they were neither applied for the edification of strange peoples nor intended to provide some permanent record for future generations " (Heizer and Clewlow, p.52).

Whether or not you agree with this, ideas have been put foreward to give meaning to the designs. It has been suggested the individual elements are a primitive form of writing, doodling, signatures, or even that they symbolize cosmic mysteries. While any of these may be true, much of this art work seems to have been associated with special magical-religious events. For the ceremony, design elements were drawn on the rock face, and over a period of years changes and additions were made. Very often designs are superimposed over others.

In California and Nevada, Indian art may have been associated with mystical ceremonies involving weather control, hunting, puberty rites, or with shamanistic healing practices. In one locality, for example, women inscribed certain rocks to ensure pregnancy.

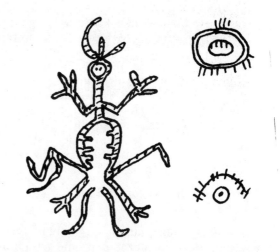

In Oregon and Idaho, rock art localities were records of dreams, traditions, puberty rites, special occasions, battles, visits, and time markings. Some sites were associated with weather conditions, as rain, wind or storms where the rock art itself was supposed to cause these weather disturbances.

The rapid advance of Euro-Americans destroyed Native American culture so completely that the meaning of rock art was lost. The population of Indians had been so reduced by 1870 that in California Native Americans professed ignorance of these hieroglyphic inscriptions and who had created them. One researcher reported in 1876 that "it was a mystery with their fathers who made these inscriptions" (Heizer and Clewlow, p.4). In the Northwest the creation of rock art was attributed, by early recorded accounts of Native Americans, to ancestors, spirits, the Great Spirit, a cultural hero, or to an unknown source.

Rock art is divided into petroglyphs and pictographs. Petroglyphs are patterns or marks incised, grooved, or pecked into rock surfaces. Pictographs are colored pigments applied to surfaces to create a design. Pictographs were painted on rock, wood tablets, tree bark, and animal skins. In a general way, pictograph and petroglyph elements can be lumped into groups. Most of the signs can be grouped into categories resembling human figures, animals, circles, dots, curvilinear lines, angular lines, or geometric patterns.

Damage to pictographs and petroglyphs by humans is a serious problem. Conditions of natural weathering are destructive enough without additional help from human vandals. Spray paint has already been used to destroy drawings. Damage to native rock art also results from persons trying to duplicate a petroglyph by rubbings. Carelessness has left chalks, paints, plaster, and other materials on the rock surface. Photographs are the only approved form for reproducing these delicate antiques.

"We do not know why vandals are so energetic in searching out and destroying petroglyphs and pictographs, but these spots number fewer each year. The American public is said to have developed a deep interest in archaeology, but if this is true, the public's practice has all too often taken the form of do-it-yourself site-wrecking" (Heizer and Clewlow, p.48).

Rock Art in Oregon and Idaho

Because most Indian activity in Idaho and Oregon, including permanent and temporary villages, was near water sources, rock art was located there too. Approximately 44 percent of petroglyphs and pictographs are along the Salmon, Snake, or Columbia rivers. Fifty-one percent are near other streams, rivers or lakes, while the remaining 5 percent are to be found near dry lake beds or seasonal springs.

In the Northwest, rock art doesn't occur randomly, as might be thought, but can be divided into three broad areas. Petroglyphs occur along the coast, the Columbia River, and in southeastern Oregon and southwestern Idaho. Pictographs are found in the interior mountainous regions, and a combination of the two types occurs in northwestern Idaho.

Symbols resembling four-legged animals, stars, flowers and curved lines predominate in Oregon. However, the Klamath-Modoc region in southern Oregon and northern California is an exception. A unique type of art found there contains both pictographs and petroglyphs. These are colored with one pigment outlining another. This style of art seems to have developed without influence from surrounding neighbors.

The heavy vegetation cover obscures most rock art in western Oregon, and few examples occur west of the Cascades. One major pictograph locality in an overhang along the South Santiam River was excavated in 1964. Indians had lived in the cave 8,000 years before the present. Whether the art symbols were made by the occupants at that early date is uncertain. Most rock art in the Pacific Northwest was created within the last 300 years, but the work is difficult to date with certainty. Figures on horseback or carrying guns are one of the best ways to date the drawings.

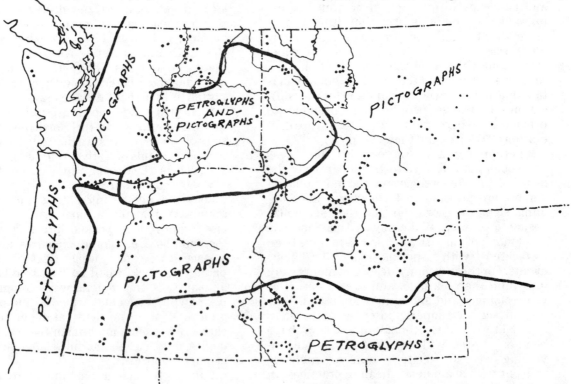

Rock art localities in Idaho and Oregon (After Boreson, 1976).

California and Nevada

Petroglyphs are the most common type of rock art found in California and Nevada. Investigation and mapping of a large number of these sites has shown they were not located near common fishing or pine nut gathering spots but in mountainous areas or in dry gulches. This surprising find led to the conclusion that the petroglyphs were drawn in association with hunting activities and were located along game migration routes. The designs may have had a magical-religious meaning to ensure successful hunting. Hunting blinds or rock walls were erected near the petroglyphs to direct game toward groups of 50 or more hunters hiding behind them.

California can be divided into general areas based on rock art patterns. In northern California and along the eastern part of the state, pecked stone art features angular and curvilinear elements. The northern coastal area has only 8 recorded sites, with 177 elements. Random scratches on soft rock were ceremonial having to do with weather changes. A considerable amount of modification was made to the incisions over a period of time.

Along the Sierra Nevadas petroglyphs can be found which are 3,000 years old, but around 1,500 A.D. the use of petroglyphs gave way to rock painting. Rock painting, or pictographs, lasted until the arrival of Europeans. In the Sierras, 39 rock art locations with 2,549 elements were found.

Pictographs in California and Nevada are not common and are concentrated along the coast from Santa Barbara south, in the southern Sierra Nevadas, and in the northeastern part of

Rock art localities in California (After Heizer and Whipple, 1951).

California. Most of the rock paintings in the northeast use red color only and are of geometric designs. Human, animal, and circle-dots, are rare. These sites were used for puberty rites and fertility ceremonies and not for hunting activities.

Glossary

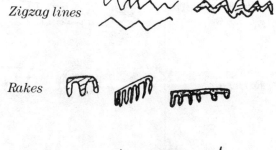

Stars

Oblong shapes

Half circles

Zigzag lines

Rakes

Ladders

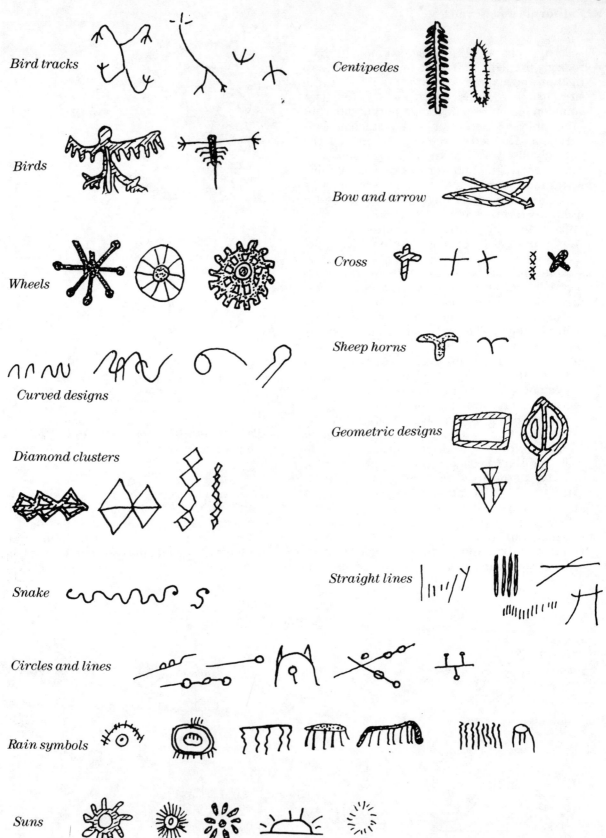

Bird tracks

Centipedes

Birds

Bow and arrow

Wheels

Cross

Curved designs

Sheep horns

Diamond clusters

Geometric designs

Straight lines

Snake

Circles and lines

Rain symbols

Suns

Circles

Hands and feet

Faces

Lizards or frogs

Human figures

Quadrupeds

Grids

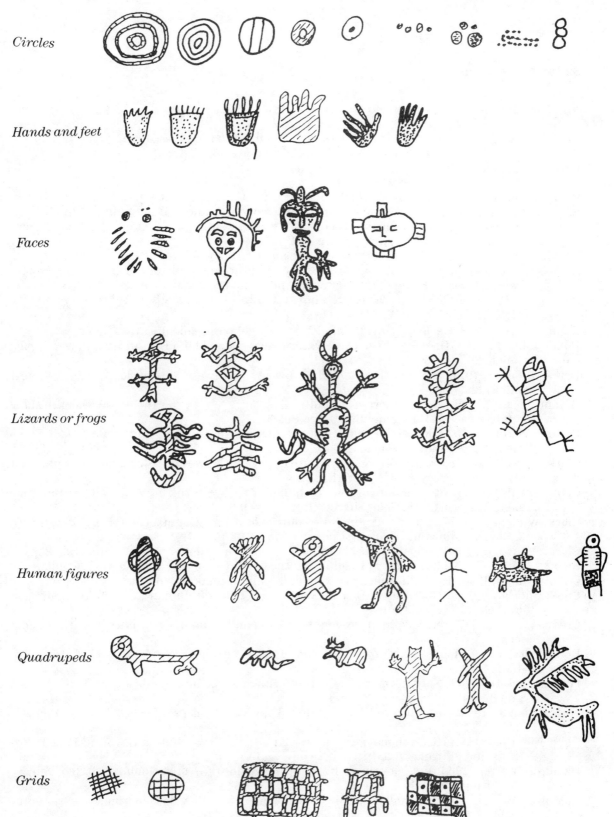

Suggested readings

An Illustrated History of Union and Wallowa Counties, 1902. Portland, Western Historical Publ. Co.

Arnold, R.R., 1932, Indian wars of Idaho. Caldwell Idaho, Caxton Press, 406p.

Baldwin, E.M., 1976. Geology of Oregon. 3rd ed., Dubuque, Iowa, Kendall/Hunt Pub. Co., 170p.

Barrett, S.A., and Gifford, E.W., 1933. Miwok material culture. Public Museum of the City of Milwaukee, Bull. v.2, no.4, p.117-376.

Bailey, Edgar, ed., 1966. Geology of northern California. Calif. Div. Mines and Geol., Bull. 190.

Beal, M.D., 1942. A history of southeastern Idaho. Caldwell, Idaho, Caxton Publ., 443p.

Beaulieu, John D., 1974. Environmental geology of western Linn County, Oregon. Portland, Dept. Geology and Mineral Indus., Bull.84, 117p.

Beckham, S.D., 1977. The Indians of western Oregon. Coos Bay, Arago Books.

Beckham, S.D., Minor, R., and Toepel, K.A., 1981. Prehistory and history of BLM lands in west-central Oregon. Eugene, Univ. Oregon Anthropological Papers, no.25.

Bledsoe, A.J., 1881. History (of) Del Norte County, California with a business directory and traveler's guide. Eureka, Wyman and Co. Publ.

Bond, John G., 1978. Geologic map of Idaho. Moscow, Idaho Bureau Mines and Geology.

Bond, John G., 1963. Geology of the Clearwater embayment. Idaho Bur. Mines and Geol., pamphlet 128.

Boreson, Keo, 1976. Rock art of the Pacific Northwest. Northwest Anthro. Res. Notes, v.1, no.1, p.90-122.

Bryant, Richard, et al., 1978. Report of the cultural resource survey: northeastern Klamath Marsh study area. Vol.1. Eugene, Pro-Lysts, Box 3761.

Cater, F.W., et al., 1973. Mineral resources of the Idaho Primitive area and vicinity, Idaho. U.S. Geol. Survey Bull., 1304, 431p.

Clark, Lorin D., 1964. Stratigraphy and structure of part of the western Sierra metamorphic belt. U.S. Geol. Survey, Prof. Paper 410.

Clark, Lorin D., 1976. Stratigraphy of the north half of the western Sierra Nevada metamorphic belt, California. U.S. Geol. Survey, Prof. Paper 923, 26p.

Clark, William B., 1977. Mines and mineral resources of Alpine County, California. Sacramento, Calif. Div. Mines and Geology, County Report 8, 48p.

Clark, William B., 1970. Gold districts of California. Calif. Div. Mines and Geol., Bull.19, 186p.

Cleland, Robert G., 1922. A history of California: the American period. New York, Macmillan Co.

Colton, Walter, 1948. The California diary. Oakland, Calif., Biobooks, 261p.

Cook, Sherburne F., 1943. The conflict between the California Indian and white civilization. Berkeley, Univ. Calif. Press.

Cressman, L.S., 1937. Petroglyphs of Oregon. Univ. Oregon Monographs, Studies in Anthro., no.2, 78p.

Dodge, Orvil, 1898. Pioneer history of Coos and Curry counties, Or. Salem, Capital Print. Co., 103p.

Forbes, Jack D., 1969. Native Americans of California and Nevada. Healdsburg, Calif., Naturegraph Publ., 197p.

Foster, George, 1944. A summary of Yuki culture. Univ. Calif. Anthro. Records, v.5, no.3, p.155-244.

Gehr, Elliot, et al., 1978. Archaeological investigtions: Wallowa-Whitman Natural Forest. Pro-Lysts, Inc., Eugene, P.O. Box 3761, Univ. Station.

Hamilton, Warren, 1963. Metamorphism in the Riggins region of western Idaho. U.S. Geol. Survey, Prof. Paper 436, 95p.

Hamilton, Warren, 1969. Reconnaissance geologic map of the Riggins Quadrangle, west-central Idaho. U.S. Geol. Survey, Misc. Geol. Inves. Map, I-579.

Harris, Albert D., 1978. Klamath River geology: Curly Jack Camp to Ti Bar, Siskiyou County, California. Calif. Geology, v.31, no.5, p.108-115.

Heizer, R.F., and Baumhoff, M., 1962. Prehistoric rock art of Nevada and eastern California. Berkeley, Univ. Calif.

Heizer, R.F., and Clewlow, C.W., 1973. Prehistoric rock art of California. Ramona, Ballena Press, 149p.

Heizer, R.F., and Elasser, A.B., 1979. The Natural World of the California Indians. Berkeley, Univ. Calif. Press, 217p.

Heizer, R.F., and Whipple, M.A., comp. and ed., 1951. The California Indians; a source book. Berkeley, Univ. Calif. Press, 487p.

Hill, Mary, 1975. Geology of the Sierra Nevada. Berkeley, Calif., Univ. Calif. Press, 232p.

History of Amador County, California, 1881. Oakland, Calif., Thompson and West, 344p.

History of El Dorado County, California, comp. by P. Sioli. 1883, Sioli Publ., 272p.

History of Merced County, California, with illustrations. 1881, San Francisco, Elliott and Moore Publ.

History of Siskiyou County, California. Oakland, Calif., 1881. D.J. Stewart Publ., 240p.

Holt, Catharine, 1946. Shasta ethnography. Univ. Calif. Publ. in Anthroplogy. Anthropological records, v.3, no.4, p.299-349.

Hoopes, Chad L., 1966. Lure of Humboldt Bay region. Dubuque, Iowa, Brown Co., 260p.

Irwin, William P, 1960. Geologic reconnaissance of the northern Coast Ranges and Klamath Mountains, California. Calif. Div. Mines., Bull.179, 80p.

Jackson, Donald D., 1980. Gold dust. Lincoln, Neb., Nebraska Press, 361p.

Josephy, Alvin M., 1971. The Nez Perce Indians and the opening of the Northwest. New Haven, Yale Univ. Press, 667p.

Kittleman, L.R., 1973. Guide to the geology of the Owyhee region of Oregon. Eugene, Univ. Oregon Museum Natl. Hist., Bull.21, 61p.

Kroeber, A.L., 1929. The valley Nisenan. Univ. Calif. Publ. in American Arch. and Ethnology, v.24, no.4, p.253.

Loring, J.M., and Loring, L., 1982. Pictographs and petroglyphs of the Oregon country, part I: Columbia River and northern Oregon. Los Angeles, Univ. Calif. Inst. of Archaeology, monograph XXI, 325p.

Minor, Rick, et al., 1979. Cultural resource overview of the BLM Lakeview district, south-central Oregon. Eugene, Univ. Oregon Anthropological Papers, no.16.

Minor, Rick, et al., 1980. Cultural resource overview of the BLM Salem district, northwestern Oregon. Eugene, Univ. Oregon Anthropological Papers, no.20.

Moore, James G., 1961. Preliminary geologic map of Lyon, Douglas, Ormsby, and part of Washoe Counties, Nevada. U.S. Geol. Survey Mineral Invest. Field Studies Map MF-80.

Moratto, M.J., ed., 1971. A study of prehistory in the Tuolumne River Valley, California. Treganza Anthropological Papers, no.9, 177p.

Morisawa, Marie, 1968. Streams, their dynamics and morphology. New York, McGraw-Hill, 175p.

Pavesic, Max G., 1978. Archaeological overview of the Middle Fork of the Salmon River corridor, Idaho Primitive Area. Boise State Univ., 104p.

Paz, Ireneo, 1937. Life and adventures of the celebrated bandit, Joaquin Murrieta his exploits in the state of California. Trans. from the Spanish by Frances P. Belle. Chicago, Powner Co., 174p.

Peck, Dallas, comp., 1961. Geologic map of Oregon west of the 121st meridian. U.S. Geol. Survey. Misc. Geol. Inv. series, Map I-325.

Peterson, E.R., and Powers, A., 1952. A century of Coos and Curry. Portland, Binfords and Mort Publ., 599p.

Powers, Steven, 1877. Contributions to North American ethnology. U.S. Geographical and Geological Survey of the Rocky Mountain Region, 635p.

Price, John A., 1962. Washo economy. Nevada State Museum, Anthropological papers, no.6.

Purdom, W.B., 1977. Guide to the geology and lore of the wild reach of the Rogue River, Oregon. Eugene, Museum Natural Hist., Bull.22, 67p.

Ramp, Len, 1979. Geology and mineral resources of Josephine County, Oregon. Portland, Dept. Geol. and Mineral Indus., Bull.100, 45p.

Rawls, James, 1984. Indians of California. Norman, Okla., Univ. Oklahoma Press.

Redding Sheet, 1962. Geologic map of California. Calif. Div. Mines and Geology.

Spier, Leslie, 1976. Klamath ethnography. Univ. Calif. Publ. in American Arch. and ethnology, v.30, 338p.

Strong, Philip., 1957. Gold in them hills. New York, Doubleday, 209p.

Swanton, John R., 1971. The Indian tribes of North America. Wash., D.C. Smithsonian, Bull.145, 726p.

Toepel, Kathryn A., Willingham, W.F., and Minor, R., 1980. Cultural resource overview of BLM lands in north-central Oregon. Eugene, Univ. Oregon Anthropological Papers, no. 17.

Ukiah Sheet, 1960. Geologic map of California. Calif. Div. Mines and Geology.

Wagner, D.L., et al., comp., 1981. Geologic map of the Sacramento Quadrangle. Calif. Div. Mines and Geology.

Walker, D.E., 1973. American Indians of Idaho. Vol.1, Aboriginal cultures. Anthropological monog., Univ. Idaho, no.2, 290p.

Walker, G.W., 1977. Geologic map of Oregon east of the 121st meridian. U.S. Geol. Survey, Misc. Inv. Series, Map l- 902.

Walker Lake Sheet, 1963. Geologic map of California. Calif. Div. Mines and Geology.

Weed Sheet, 1964. Geologic map of California. Calif. Div. Mines and Geology.

Weis, Paul L., et al., 1972. Mineral resources of the Salmon River Breaks Primitive Area, Idaho. U.S. Geol. Survey Bull., 1353-C.

Wells, Merle W., 1962? Rush to Idaho. Moscow, Idaho Bur. Mines and Geology, Bull.19, 57p.

Wells, Merle W., 1963. Gold camps and silver cities. Moscow, Idaho Bur. Mines and Geology, Bull.22, 86p.

Williams, Howell, 1949. Geology of the Macdoel quadrangle, Calif. Calif. Div. Mines, Bull.151, 78p.

Wooster, David, 1972. The gold rush; letters of David Wooster from California to the Adrian, Michigan, Expositor, 1850-1855. Mount Pleasant, Mich., Cumming Press, 85p.

Wright, Terry, 1983. Rocks and rapids of the Tuolumne River; a guide to natural and human history. Wilderness Interpretation Publication.

Wright, Terry, 1981. Guide to geology and rapids, South Fork American River. Wilderness Interpretation Publication, 66p.